Control Perspectives on Numerical Algorithms and Matrix Problems

Advances in Design and Control

SIAM's Advances in Design and Control series consists of texts and monographs dealing with all areas of design and control and their applications. Topics of interest include shape optimization, multidisciplinary design, trajectory optimization, feedback, and optimal control. The series focuses on the mathematical and computational aspects of engineering design and control that are usable in a wide variety of scientific and engineering disciplines.

Series Volumes

Bhaya, Amit, and Kaszkurewicz, Eugenius, *Control Perspectives on Numerical Algorithms and Matrix Problems*

Robinett III, Rush D., Wilson, David G., Eisler, G. Richard, and Hurtado, John E., *Applied Dynamic Programming for Optimization of Dynamical Systems*

Huang, J., *Nonlinear Output Regulation: Theory and Applications*

Haslinger, J. and Mäkinen, R. A. E., *Introduction to Shape Optimization: Theory, Approximation, and Computation*

Antoulas, Athanasios C., *Approximation of Large-Scale Dynamical Systems*

Gunzburger, Max D., *Perspectives in Flow Control and Optimization*

Delfour, M. C. and Zolésio, J.-P., *Shapes and Geometries: Analysis, Differential Calculus, and Optimization*

Betts, John T., *Practical Methods for Optimal Control Using Nonlinear Programming*

El Ghaoui, Laurent and Niculescu, Silviu-Iulian, eds., *Advances in Linear Matrix Inequality Methods in Control*

Helton, J. William and James, Matthew R., *Extending H^∞ Control to Nonlinear Systems: Control of Nonlinear Systems to Achieve Performance Objectives*

Control Perspectives on Numerical Algorithms and Matrix Problems

Amit Bhaya
Eugenius Kaszkurewicz

Federal University of Rio de Janeiro
Rio de Janeiro, Brazil

siam

Society for Industrial and Applied Mathematics
Philadelphia

10 9 8 7 6 5 4 3 2 1

Library of Congress Cataloging-in-Publication Data

Bhaya, Amit.
Control perspectives on numerical algorithms and matrix problems / Amit Bhaya, Eugenius Kaszkurewicz.
p. cm. – (Advances in design and control)
Includes bibliographical references and index.
ISBN 0-89871-602-0 (pbk.)
1. Control theory. 2. Numerical analysis. 3. Algorithms. 4. Mathematical optimization. 5. Matrices. I. Kaszkurewicz, Eugenius. II. Title. III. Series.

QA402.3.B49 2006
515'.642–dc22 2005057551

Dedication

The authors dedicate this book to all those who have given them the incentive to work, including teachers, students, colleagues, and family. We cannot name one without naming them all, so they shall remain unnamed but sincerely appreciated just the same.

Contents

List of Figures

List of Tables

Preface

Control theory is a powerful body of knowledge. It can model complex objects and produce adaptive solutions that change automatically as circumstances change. Control has an impressive track record of successful applications and increasingly it lends its basic ideas to other disciplines.

—J. R. Leigh [Lei92]

General philosophy of the book

In the spirit of the quote above, this book makes a case for the application of control ideas in the design of numerical algorithms. This book argues that some simple ideas from control theory can be used to systematize a class of approaches to algorithm analysis and design. In short, it is about building bridges between control theory and numerical analysis.

Although some of the topics in the book have been published in the literature on numerical analysis, the authors feel, however, that the lack of a unified control perspective has meant that these contributions have been isolated and the important role of control and system theory has, to some extent, not been sufficiently recognized. This, of course, also means that control and system theory ideas have been underexploited in this context.

The book is a control perspective on problems mainly in numerical analysis, optimization, and matrix theory in which systems and control ideas are shown to play an important role. In general terms, the objective of the book is to showcase these lesser known applications of control theory not only because they fascinate the authors, but also to disseminate control theory ideas among numerical analysts as well as in the reverse direction, in the hope that this will lead to a richer interaction between control and numerical analysis and the discovery of more nontraditional contexts such as the ones discussed in this book. It should also be emphasized that this book represents a departure from (or even the "inverse" of) a strong drive in recent years to give a numerical analyst's perspective of various computational problems and algorithms in control. The word "Perspectives" in the title of the book is intended to alert the reader that we offer new perspectives, often of a preliminary nature: We are aware that much more needs to be done. Thus this is a book that stresses perspectives and control formulations of numerical problems, rather than one that develops numerical methods. It is natural to expect that new perspectives will lead to new methods, and indeed, some of these are proposed in this book.

As a general statement, one may say that designing a numerical algorithm for a given problem starts by finding an iterative method that, in theory, converges to the solution of the problem. In practice, further, often difficult, analyses and modifications of the method

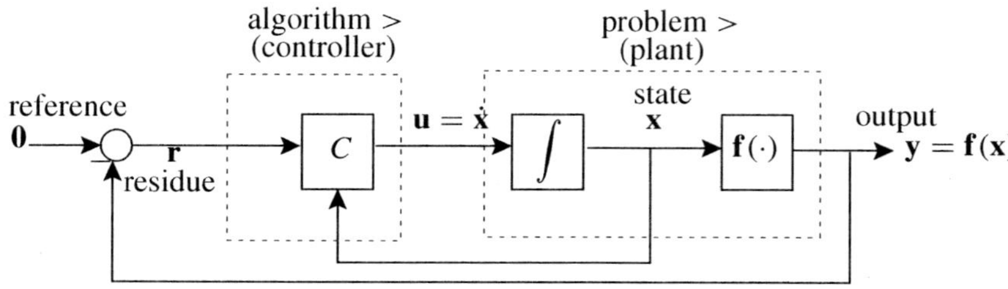

Figure 1. *A continuous realization of a general iterative method to solve the equation* $\mathbf{f}(\mathbf{x}) = \mathbf{0}$ *represented as a feedback control system. The plant, the object of the control, represents the problem to be solved, while the controller C, a function of* $\mathbf{x}$ *and* $\mathbf{r}$, *is a representation of the algorithm designed to solve it. Thus the choice of an algorithm corresponds to the choice of a controller.*

are required to show that, under the usual circumstances of implementation such as finite precision, discretization errors, etc., the proposed iterative method still works, which means that it displays a property called robustness. These modifications, if they can be implemented, inspire confidence in the usefulness of the method and, if they cannot be found, usually condemn it. However, apart from many general principles that are known today, there does not seem to be a systematic way to analyze the behavior of algorithms or to propose modifications to eliminate known undesirable behavior. As a result of this, more algorithms are being proposed and (ab)used than are being analyzed rigorously in order to be classified as reliable or unreliable. This phenomenon is widespread in many areas of scientific and technological research and, as massive amounts of desktop computing power are now routine, is likely to become even more prevalent.

The above paragraph can be rewritten using the terms *discrete dynamical system* instead of *iterative method* and the term *perturbation* for all the different types of errors to which the algorithm is subject. If this is done, then the problem of designing a usable or good iterative method can briefly be described as that of influencing a given dynamical system such that its convergence properties remain unchanged in the presence of perturbations, thus achieving robustness. The area of control theory deals with the problem exactly like this, where the term that influences the dynamical system generally enters in a specific way (linearly, affinely, nonlinearly, etc.) and is referred to as a control. Figure 1 summarizes this concept pictorially in the form of a block diagram, familiar to engineers.

Description of contents

Chapter 2 shows how standard iterative methods for solving nonlinear equations can be approached from the point of view of control. In other words, Figure 1 is given a full explanation and it is shown how, in the standard iterative methods for finding zeros of linear and nonlinear functions, a suitable dynamical system is defined by the problem data and, furthermore, how the algorithm may be interpreted as a controller for this dynamical system. This leads to a unified treatment of these methods: The conjugate gradient method is seen to be a manifestation of the well-known and popular proportional-derivative controller, and furthermore a very natural taxonomy of iterative methods for linear systems results. The proposed approach also leads to new zero finding methods that have properties different from the classical methods in ways that are significant in certain situations, which are presented together with illustrative examples.

Another point worth mentioning is that, following the usual practice in control, both the discrete-time iterative methods as well as their continuous-time counterparts are considered. The latter are quite natural in a control context, and in some specific ways, their analysis is easier. It should also be recalled that, historically speaking, continuous methods, first called analog methods, formed the basis of computing in science and engineering. From this perspective, this book revisits "analog computing," not only in order to reach a better understanding of discrete computing, but also to be considered as a serious alternative to the latter in some cases, such as in neural networks. Today, of course, continuous-time algorithms suggested by the theory would usually be implemented on a digital computer or, in some cases, through an efficient VLSI circuit implementation.

Chapter 3 examines the closely related ideas of optimal and variable structure control in the design and analysis of iterative methods for finding zeros of nonlinear functions, establishing connections with the methods examined in Chapter 2. Methods for finding minima of unconstrained functions are also examined from the viewpoint of optimal and variable structure control. The highlights are the unified view of these somewhat disparate topics and methods, as well as new analyses of methods that have been proposed in the literature, such as one for finding zeros of polynomials with complex coefficients.

In recent years, there has been an explosion of interest in different aspects of so-called artificial neural networks. Their ability to perform large-scale numerical computations as well as some optimization tasks efficiently is one of the features that has attracted attention in many fields. Chapter 4 examines neural networks as well as other gradient dynamical systems that can be used to solve linear and quadratic programming problems. The approach taken is an exact penalty function approach which leads to gradient dynamical systems (GDSs) with discontinuous right-hand sides, already met with in the previous chapter. These systems are analyzed using a Persidskii-type control Liapunov function (CLF) approach introduced by the authors. The highlights of this analysis are simple GDS solvers for several problems currently very popular in the neural network and pattern recognition community, such as support vector machines and k-winners-take-all networks. Once again, the emphasis is on the unified and simple control approach that is being advocated, rather than on these specific applications.

Chapter 5 discusses control aspects in the numerical solution of initial value problems for ordinary differential equations and matrix problems. One of the success stories of control which is not widely known is the stepsize control of numerical integration methods for solving ordinary differential equations, originated by Gustafsson, Söderlind, and coworkers, who showed that the classical proportional-integral (PI) controller paradigm can be used to design very effective adaptive stepsize control algorithms.

Shooting methods for boundary value problems can also be recast as feedback control. The consequences of this are examined and a connection to the iterative learning control (ILC) paradigm of control theory is also discussed.

The chapter closes with an exploration of some control and system-theoretic ideas that are intimately related to and throw light on some problems in matrix theory.

A preconditioner is a matrix $\mathbf{P}$ with a simple structure that pre- or postmultiplies a given matrix $\mathbf{A}$. Preconditioners play an important role in making some classes of iterative methods more efficient by increasing rates of convergence. In order for preconditioning to be efficient, it is standard to impose restrictions on **P**. For instance, a question that has attracted much attention among numerical analysts is that of determining an optimal *diagonal* preconditioner **P** (i.e., **P** is restricted, for computational simplicity, to be a diagonal matrix). In section 5.3, this problem is shown to be equivalent to the problem of clustering the eigenvalues of an appropriately chosen dynamical system using what is known as decentralized feedback. This problem has been much studied under the name of *decentralized control* and this theory is used to show how far it is possible to go with diagonal preconditioning, as well as how to formulate, in control terms, the problem of computing an optimal diagonal preconditioner.

Section 5.4 discusses the problem of D-stability. This problem, which arose in the context of price stability in economics, is one of determining conditions on a Hurwitz matrix **A** (i.e., one that has all its eigenvalues in the left half of the complex plane) such that it remains Hurwitz when pre- or postmultiplied by any diagonal matrix with positive entries. A control approach, similar to that taken for the preconditioning problem in some aspects, is used to characterize D-stability, presently limited to the case of low dimensions.

Section 5.5 shows how the concept of realization of a dynamical system that is controllable but not observable for a given parameter value leads to the resultant of a system of two polynomial equations in two unknowns. This, in turn, leads to an algorithm for finding zeros of such a polynomial system.

Previous work in the field and disclaimers

This book makes a case for more (and more systematic) use of control theory in numerical methods. This is done basically by presenting several successful examples. The authors hasten to add that they are not experts in many of the example areas or in numerical analysis and also make no pretense of developing a "metatechnique" that will guide future applications of control to numerical methods. We do, however, hope that, despite these shortcomings, this book will motivate more practitioners of each field to learn the other and eventually lead to an evolution of the perspective proposed here.

Some further disclaimers are in order. We should point out that we are certainly not the first to propose such a perspective.

Tsypkin, in two seminal and wide-ranging books [Tsy71, Tsy73], approached the problem of adaptation and learning from a control perspective, emphasizing the central role of gradient dynamical systems, both deterministic and stochastic, and pointing out that the gradient systems "cover many iterative formulas of numerical analysis." He also clearly identified the importance of studying both continuous and discrete algorithms on an equal footing and was one of the first to consider the question of optimality of algorithms. Thus, even though the subject and scope are quite different, Tsypkin's books should be considered the precursors of this one, certainly in regard to the contents of Chapters 2–4. Another source of inspiration for Chapters 2 and 3 was the important but little known book of Krasnosel'skii, Lifshits, and Sobolev [KLS89], which mentions the possibility of approaching iterative methods from a control viewpoint and gives the spurt method (see section 3.2.2) as an example of variable structure control applied to a Krylov-type method.

Stuart and Humphries, in many papers culminating in their book [SH98], propose a dynamical systems approach to numerical analysis. This is close to what we are proposing, the main difference being that they do not consider control inputs and therefore do not consider the question of modifying the dynamics of a given iterative method, but rather the different question of when iterations that are discrete approximations of a continuous dynamical system have the same qualitative behavior. This theme is also important in Chapter 2, but since it has already been treated exhaustively, we refer the reader, where appropriate, to these more advanced works.

Helmke and Moore in several papers, also consolidated in a book [HM94], showed how to construct continuous-time dynamical systems that solve various problems in linear algebra and control. Chu, Brockett, and others [Chu88, Bro89, Bro91] have proposed, in a similar vein, continuous-time realizations of many of the common iterative methods used in linear algebra. In both cases, the continuous-time dynamical systems approach provides many new insights into old problems. Finally, Pronzato, Wynn, and Zhigljavsky [PWZ00] carry out a sophisticated study of applications of dynamical systems in search and optimization, using a unifying renormalization idea. In terms of suggesting that the perspective of the present book should be developed, an essay by Campbell [Cam01] recently came to our attention. It is entitled "Numerical Analysis and System Theory" and in section 3, entitled "System theory and its impact on numerical analysis," Campbell writes that "the application of control ideas in numerical analysis" is a direction of interaction between the two areas that is "less well developed." He cites the application of PI control to stepsize control and optimal control codes as examples of successful instances of interaction in the direction of control to numerical analysis, and then goes on to say: "Generally, numerical analysts are not familiar with control theory. It is natural to ask whether one can use control theory to design better stepsize strategies. One can ask for even more. Can one design control strategies that can be applied across a family of numerical methods?" This book can be regarded as a step in the direction of an affirmative answer to Campbell's question.

Prerequisites

Although there are many points of contact between the research cited above and the point of view taken in this book, we have tried to avoid overlap with the books mentioned above by emphasizing dynamical systems with an explicit control input and a feedback loop, chosen so as to improve the performance of the "controlled" iterative method in some prespecified way. Another difference is that we aim at an audience that is not necessarily well versed, in for example, the language of manifolds and differential geometry, and we demand few mathematical prerequisites of the interested reader: just linear algebra and matrix theory, as well as basic differential and difference equation theory. Of course, this means that this book is elementary and concentrates much more on the perspective that is being developed, rather than technical details. It is appropriate to mention here that we have kept the mathematical level as simple as possible and alerted the reader, wherever necessary, to the existence of more rigorous or mathematically sophisticated presentations.

Chapter 1 provides a quick overview of the control and system theory prerequisites for this book. It is intended to help potential readers who are not familiar with control to acquaint themselves with the basic control and systems theory terminology, as well as to provide a list of references in which the readers can find all the details that we omit.

For readers who have a control background and wish to review numerical linear algebra and numerical analysis, comprehensive references for numerical linear algebra are

[Dat95, Dem97], and a recent text on numerical analysis is [SM03]. For those wishing to review optimization theory, there are several excellent textbooks on optimization: Some of our favorites are [BV04, NW99, NS96, Lue69, Lue84].

Notation and acronyms

Standard notation is used consistently, as far as possible. Uppercase boldface letters, Greek and Roman, represent matrices, while lowercase boldface letters, Greek and Roman, represent vectors. For typographical reasons, column vectors are written "lying down" in parentheses with commas separating the components, e.g., $\mathbf{x} = (x_1, \ldots, x_n) \in \mathbb{R}^n$. Lowercase lightface letters, Greek and Roman, represent scalars. Calligraphic letters and uppercase lightface letters, Roman and Greek, usually denote sets. The reader who wishes to find the meaning of an acronym, a symbol, or notation should consult the index.

Acknowledgments

Both of the authors would like to acknowledge various Brazilian funding agencies that have supported our work over the years: CNPq, the Brazilian National Council for Scientific and Technological Development; CAPES, the Quality Improvement Program for Universities of the Ministry of Education and Culture; FINEP, the Federal Scientific and Technological Research Funding Agency; FAPERJ, the Scientific and Technological Research Funding Agency of the state of Rio de Janeiro; and finally, COPPE/UFRJ, the Graduate School of Engineering of the Federal University of Rio de Janeiro, our home institution. We make special mention of the electronic Portal Periódicos of CAPES, supported by the Ministries of Education and Culture and Science and Technology: Electronic access to the databases of major publishers and citation indexing companies was of fundamental importance in doing the extensive bibliographical research for this book.

This book was written using Aleksander Simonić's magnificent WinEdt 5.4 software running Christian Schenk's MiKTeX 2.4 implementation of LaTeX 2E. The figures were prepared using Samy Zafrany's TkPaint 1.6 freeware. The PostScript® and PDF files were prepared, using Ghostscript® and GhostView, by Russell Lang. To all these individuals, we express our sincere appreciation and heartfelt thanks.

During the time that the manuscript for this book was being written, several people gave us critical commentary, which was much appreciated and, as far as our limitations permitted, incorporated into the book. We would especially like to thank Prof. José Mario Martinez of Imecc/Unicamp, the Institute of Mathematics, Statistics and Scientific Computation of the State University of Campinas, São Paulo, and Prof. Daniel B. Szyld of the Department of Mathematics of Temple University, Philadelphia. Some of our doctoral students, past and present, also participated, through their theses and computational implementations, in the development of some of the topics in this book, and we would like to acknowledge Christian Schaerer, Leonardo Ferreira, Oumar Diene, and Fernando Pazos. Of course, the usual disclaimer applies: Infelicities and outright blunders are ours alone.

We would like to thank our acquisitions editor at SIAM, Elizabeth Greenspan, for patiently bearing with our receding horizon approach to deadlines that went, in Douglas Adams's phrasing, "whooshing by."

Our families have to be thanked for suspending disbelief whenever the topic of finishing the book came up, which was practically every day: "*A Chean, Lucia, Barbara, Asmi, e Felipe, nosso muito obrigado, por entenderem nossos dilemas e nossas obsessões com este livro.*"

In closing, we cannot do better than to repeat, with one small change, the following words from the epilogue to Professor Tsypkin's inspiring book [Tsy73], mentioned earlier in this preface, which were written about learning theory more than three decades ago, but could just as well have been written about the contents of this book:

> *All new problems considered in this book contain the elements of old classical problems: the problems of convergence and stability, the problems of optimality. Thus, in this respect we have not followed the fashion of moving away from the reliable classical results: "Extreme following of fashion is always a sign of bad taste." It seems to us that even now the theory of [numerical algorithms] greatly needs further development and generalization of these classical results. To date they have provided a great service to the ordinary systems, but now they also have to serve [numerical algorithms].*[1]

Amit Bhaya and Eugenius Kaszkurewicz
Rio de Janeiro, Brazil
October 10, 2005

[1] In the original, instead of the bracketed words *numerical algorithms*, the words *learning systems* appear.

Chapter 1

Brief Review of Control and Stability Theory

All stable processes we shall predict. All unstable processes we shall control.

—John von Neumann

This chapter gives an introduction to the bare minimum of the control terminology and concepts required in this book. We expect readers who are well versed in control to only glance at the contents of this chapter, while readers from other fields can use it to get an idea of the material touched upon, but must expect to look at the references provided for details.

1.1 Control Theory Basics

This section establishes the basic terminology from control theory that will be used in what follows. A brief mention is made of selected elementary definitions and results in linear system theory as well as of the internal model principle and proportional, integral, and derivative (PID) control.

1.1.1 Feedback control terminology

> *The term feedback is used to refer to a situation in which two (or more) dynamical systems are connected together such that each system influences the other and their dynamics are thus strongly coupled. Simple causal reasoning about such a system is difficult because the first system influences the second and the second system influences the first, leading to a circular argument. This makes reasoning based on cause and effect tricky and it is necessary to analyze the system as a whole. A consequence of this is that the behavior of a feedback system is often counterintuitive and therefore it is necessary to resort to formal methods to understand them.*
>
> —R. Murray and K. J. Åström [MÅ06]

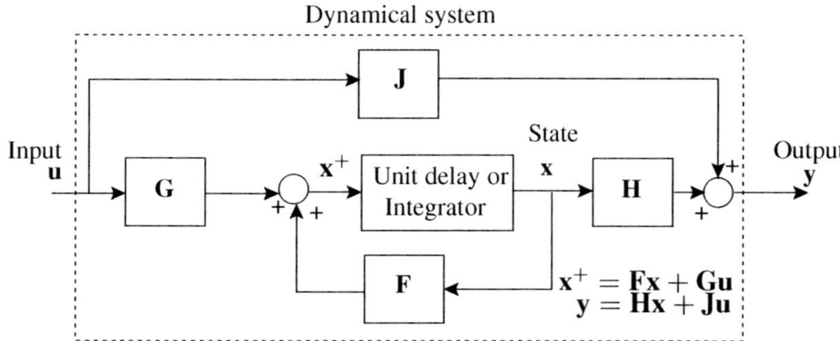

Figure 1.1. *State space representation of a dynamical system, thought of as a transformation between the input* $\mathbf{u}$ *and the output* $\mathbf{y}$*. The vector* $\mathbf{x}$ *is called the state. The quadruple* $\{\mathbf{F}, \mathbf{G}, \mathbf{H}, \mathbf{J}\}$ *will denote this dynamical system and serves as the building block for the standard feedback control system depicted in Figure* 1.2*. Note that* $\mathbf{x}^+$ *can represent either* $d\mathbf{x}/dt$ *or* $\mathbf{x}(k+1)$.

One branch of control theory, called *state space control*, is concerned with the study of dynamical systems of the form (see Figure 1.1)

$$\begin{aligned} \mathbf{x}^+ &= \mathbf{F}\mathbf{x} + \mathbf{G}\mathbf{u}, \\ \mathbf{y} &= \mathbf{H}\mathbf{x} + \mathbf{J}\mathbf{u}, \end{aligned} \tag{1.1}$$

and their feedback interconnections in order to achieve some basic properties, such as stability of the interconnected system, often referred to as a *closed-loop system*, especially when in the form of Figure 1.2. Note that $\mathbf{x}^+$ can represent either $d\mathbf{x}/dt$ or $\mathbf{x}(k+1)$. In the former case, the variable t is thought of as time, (1.1) is said to be a *continuous-time system*, and the central block in Figure 1.1 represents an integrator; in the latter case, k is a discrete-time variable, the system is said to be a *discrete-time system* and the central block represents a delay of one unit.

If the dynamical system is *linear*, the transformations $\{\mathbf{F}, \mathbf{G}, \mathbf{H}, \mathbf{J}\}$ are all linear and representable by matrices: $\mathbf{F}$ is called the *system matrix*, $\mathbf{G}$ the *input matrix*, $\mathbf{H}$ the *output matrix*, and $\mathbf{J}$ the *feedforward matrix*. If one or more of the mappings $\mathbf{F}, \mathbf{G}, \mathbf{H}, \mathbf{J}$ is nonlinear, the dynamical system is nonlinear, and the nonlinear transformations will be denoted by the corresponding lowercase boldface letters. If the transformations vary as a function of time, this is denoted either by the appropriate subscript t or k, or within parentheses in the usual functional notation.

If the matrices $\mathbf{F}$, $\mathbf{G}$, $\mathbf{H}$, and $\mathbf{J}$ are constant, the system is called *time invariant* (or *autonomous* or *stationary*); otherwise the system is called *time varying* (or *nonautonomous* or *nonstationary*). Finally, if the matrices $\mathbf{F}$ and $\mathbf{G}$ are zero, the system is said to be *static* or *memoryless*; otherwise, it is called *dynamic*. Thus the quadruple $\{\mathbf{F}, \mathbf{G}, \mathbf{H}, \mathbf{J}\}$ is used as a convenient shorthand for (1.1). The word *decoupled* is used to indicate that a certain matrix is diagonal. For instance, a controller described by the quadruple $\{\mathbf{0}, \mathbf{0}, \mathbf{0}, \mathbf{I}\}$ would be called static and decoupled. The term *multivariable* is sometimes used to denote the fact that some matrix is not diagonal. Finally, if the state feedback loop is present (Figure 1.2),

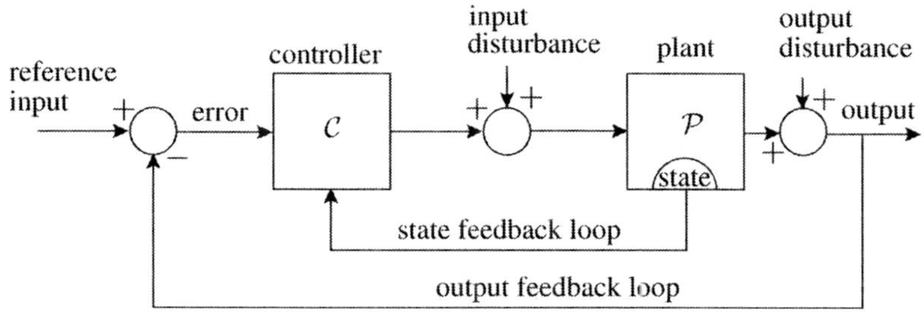

Figure 1.2. *A standard feedback control system, denoted as $S(\mathcal{P}, \mathcal{C})$. The plant, object of the control, will represent the problem to be solved, while the controller is a representation of the algorithm designed to solve it. Note that the plant and controller will, in general, be dynamical systems of the type* (1.1)*, represented in Figure* 1.1*, i.e.,* $\mathcal{P} = \{\mathbf{F}_p, \mathbf{G}_p, \mathbf{H}_p, \mathbf{J}_p\}$, $\mathcal{C} = \{\mathbf{F}_c, \mathbf{G}_c, \mathbf{H}_c, \mathbf{J}_c\}$. *The semicircle labeled* state *in the plant box indicates that the plant state vector is available for feedback.*

and we wish to emphasize, for example, the dependence of the controller on the state vector, the controller quadruple will then be written as $\mathcal{C} = \{\mathbf{F}(\mathbf{x}), \mathbf{G}(\mathbf{x}), \mathbf{H}(\mathbf{x}), \mathbf{J}(\mathbf{x})\}$. If the outputs are not of interest, a dynamical system will sometimes be referred to as just the pair $\{\mathbf{F}, \mathbf{G}\}$. Similarly, if $\mathbf{J}(\mathbf{x}) = \mathbf{0}$, we will refer just to the triple $\{\mathbf{F}(\mathbf{x}), \mathbf{G}(\mathbf{x}), \mathbf{H}(\mathbf{x})\}$.

A standard feedback control system consists of the interconnection of two systems of the type (1.1) in the configuration shown in Figure 1.2. Note that there are two feedback loops, corresponding to state and output feedback, respectively. In a given feedback system, one or both of the feedback loops may be present.

In closing, we mention, extremely briefly, some of the main problems that control theory deals with in the context of the system $S(\mathcal{P}, \mathcal{C})$ of Figure 1.2. The problem of *regulation* is that of designing a controller $\mathcal{C}$ such that the output of the system always returns to the value of the reference input, usually considered constant, in the face of some classes of input and output disturbances. The closely related problem of *asymptotic tracking* is that of choosing a controller that makes the output follow or track a class of time-varying inputs asymptotically. The problem of *stabilization* is that of choosing a controller so that a possibly unstable plant $\mathcal{P}$ (i.e., one for which, in the absence of the feedback loops in Figure 1.2 (i.e., *in open loop*), a bounded input can lead to an unbounded output) leads to a stable configuration $S(\mathcal{P}, \mathcal{C})$. Another type of stabilization problem has to do with the notion of Liapunov stability discussed in section 1.4: one basic problem is to choose the plant input as a linear function of the plant state so that the resulting system with this *state feedback* is stable. Succinctly, given $\mathbf{x}^+ = \mathbf{A}\mathbf{x} + \mathbf{B}\mathbf{u}$, we set $\mathbf{u} = \mathbf{K}\mathbf{x}$, so that the resulting system is $\mathbf{x}^+ = (\mathbf{A} + \mathbf{B}\mathbf{K})\mathbf{x}$, and the question then is, can $\mathbf{K}$ be chosen so that the eigenvalues of $\mathbf{A} + \mathbf{B}\mathbf{K}$ can be "placed" within regions in the complex plane that correspond to stable behavior of the closed-loop system? The answer is yes, provided that the matrices $\mathbf{A}, \mathbf{B}$ satisfy a certain rank condition, which, surprisingly, is equivalent to the property of being able to choose a control that takes the system (1.1) from an arbitrary initial state to an arbitrary final state. The latter property is called *controllability*.

Some additional details on the concepts mentioned above are given in the sections in which they are used below, but the interested reader without a control background is referred to [Son98] for a mathematically sophisticated introduction or to [Kai80, Del88, CD91, Ter99, Che99] for more accessible approaches.

The internal model principle

As stated in the previous paragraph, input signals (functions of time), denoted u, to a given dynamical system $\mathcal{S}$ may be thought of as disturbances to be rejected or signals to be tracked, depending on the application. Suppose that it is known that a certain quantity $y(t)$ associated to the system, called a regulated variable, has the property that $y(t) \rightarrow 0$ as $t \rightarrow \infty$ whenever the system is subject to an input signal from the class U. In control terminology, one says that $\mathcal{S}$ regulates against all external input signals u which belong to the given class U of time-functions.

Roughly speaking, the internal model principle states that the system $\mathcal{S}$ must necessarily contain a subsystem $\mathcal{S}_{\text{IM}}$ which can itself generate all disturbances in the class U; i.e., $\mathcal{S}_{\text{IM}}$ is thought of as a "model" of the system that is capable of generating the external signals, and this also explains the name of the principle. A specific example of the principle, which is the one that occurs throughout this book, is as follows.

Let $y(t) \rightarrow 0$ as $t \rightarrow \infty$ whenever the system is subject to any external constant signal (i.e., the class U consists of all constant functions); then the system $\mathcal{S}$ must contain a subsystem $\mathcal{S}_{\text{IM}}$ which generates all constant signals (typically an integrator, since constant signals are generated by the differential equation $\dot{u} = 0$). The choice of $y = 0$ as the regulation or reference value is just a matter of convention since a change of variables reduces a given regulation objective $y(t) \rightarrow y^*$, where y^* is some predetermined value, to the special case $y^* = 0$.

Since there are few accessible descriptions of the internal model principle available in control textbooks, a brief derivation of a simple form of the internal model principle for a single input, single output discrete-time linear system is given below. Consider the system

$$\mathbf{x}_{k+1} = \mathbf{F}\mathbf{x}_k + \mathbf{g}u_k, \tag{1.2}$$

$$y_k = \mathbf{h}^T\mathbf{x}_k + ju_k. \tag{1.3}$$

The theorem below is a simple statement of the internal model principle. It shows that if any constant input $u_k = u$ (for all k and for some $u \in \mathbb{R}$) is to lead to a zero output $y_k = 0$ for all k sufficiently large, then the system must contain "integral action"; namely, there must exist $z_k = \mathbf{v}^T\mathbf{x}_k$ such that $z_{k+1} = z_k + y_k$ (which is a discrete integrator of the sequence y_k, because $z_n = z_0 + \sum_{k=0}^{n-1} y_k$). The precise statement of the theorem is as follows.

Theorem 1.1. *Consider the system* (1.2), (1.3) *in which* $\mathbf{F}$ *is nonsingular,* $\mathbf{g} \neq \mathbf{0}$, *and* $[\mathbf{h}^T \;\; j] \neq \mathbf{0}$. *If* $y_k = 0$ *for all constant inputs* $u_k = u$ *for all* k *and for some* $u \in \mathbb{R}$, *then there exists* $\mathbf{v}$ *such that* $z_k = \mathbf{v}^T\mathbf{x}_k$ *satisfies the difference equation*

$$z_{k+1} = z_k + y_k. \tag{1.4}$$

Proof. When the dynamical system (1.2), (1.3) reaches a steady state $\mathbf{x}^*$, under the influence of a constant input $u_k = u$, the following equation must hold:

$$\mathbf{x}^* = \mathbf{F}\mathbf{x}^* + \mathbf{g}u,$$

$$y^* = \mathbf{h}^T\mathbf{x}^* + ju.$$

This means that $y^* = 0$ for all $u \in \mathbb{R}$ if the following equation is satisfied:

$$\begin{bmatrix} \mathbf{F}-\mathbf{I} & \mathbf{g} \\ \mathbf{h}^T & j \end{bmatrix} \begin{bmatrix} \mathbf{x}^* \\ u \end{bmatrix} = \begin{bmatrix} \mathbf{0} \\ 0 \end{bmatrix}. \tag{1.5}$$

From this equation and the hypotheses on $\mathbf{F}$, $\mathbf{g}$, and the pair $[\mathbf{h}^T \quad j]$, it follows that

$$y^* = (j - \mathbf{h}^T(\mathbf{F}-\mathbf{I})^{-1}\mathbf{g})u.$$

Thus $y^* = 0$ for all constant $u \in \mathbb{R}$ implies that we must have

$$(j - \mathbf{h}^T(\mathbf{F}-\mathbf{I})^{-1}\mathbf{g}) = 0,$$

which is equivalent to

$$\det\begin{bmatrix} \mathbf{F}-\mathbf{I} & \mathbf{g} \\ \mathbf{h}^T & j \end{bmatrix} = 0,$$

which in turn implies that there exists $\mathbf{v}$ such that

$$[\mathbf{v}^T \quad -1]\begin{bmatrix} \mathbf{F}-\mathbf{I} & \mathbf{g} \\ \mathbf{h}^T & j \end{bmatrix} = \mathbf{0}.$$

Thus

$$y_k = [\mathbf{h}^T \quad j]\begin{bmatrix} \mathbf{x}_k \\ u_k \end{bmatrix} = \mathbf{v}^T[\mathbf{F}-\mathbf{I} \quad \mathbf{g}]\begin{bmatrix} \mathbf{x}_k \\ u_k \end{bmatrix} = \mathbf{v}^T(\mathbf{x}_{k+1} - \mathbf{x}_k). \tag{1.6}$$

Defining z_k as $\mathbf{v}^T\mathbf{x}_k$, (1.6) becomes

$$y_k = z_{k+1} - z_k$$

or, equivalently,

$$z_{k+1} = z_k + y_k$$

as claimed. □

This section is based on [Son03]. Theorem 1.1 is a discrete-time version of the corresponding continuous-time result stated in [YHSD00]. The first statement and proof of the internal model principle is due to Francis and Wonham [FW75].

Integral control

In control theory, *integral action* is a term used to refer to a linear dynamical system in which the output $y(t)$ is a constant (gain) times the integral of the input $u(t)$, i.e.,

$$y(t) = k\int_0^t u(\tau)\,\mathrm{d}\tau + y(0), \tag{1.7}$$

which can also be written as

$$\frac{\mathrm{d}y}{\mathrm{d}t} = ku(t). \tag{1.8}$$

If the output $y(t)$ is a constant, then (1.8) implies that $u(t) = 0$ for all $t > 0$. More generally, if the output becomes constant over some time interval, the input must be identically zero over this interval. This simple observation is at the heart of integral controllers, which are ubiquitous in control theory and practice. Suppose that the controller in Figure 1.2 is an integrator of type (1.7) and, furthermore, that an output disturbance introduces an error

$$e(t) = r - y(t).$$

Suppose, furthermore, that the closed-loop system is asymptotically stable so that, if reference and disturbance signals are constants, all signals eventually reach constant steady state values. In particular, the integrator output reaches a constant value and, from the preceding discussion, this means that the integrator input, which is exactly the tracking (or regulation) error, must become zero. Note that this argument is extremely simple and general and does not depend on any special assumptions: The plant may be linear or nonlinear, and neither initial conditions nor the values of the constant reference and disturbance affect the validity of the argument. Finally, observe that this argument is a special case of the internal model principle discussed above.

1.1.2 PID control for discrete-time systems

This section gives a brief account of the simplest case of the so-called proportional-integral-derivative (PID) control for a single-input, single-output discrete-time plant which is also chosen to have a particularly simple structure. The development is tailor-made for use in Chapter 5.Referring to Figure 1.3, let the plant be given as follows:

$$y_{n+1} = ku_n + d_n. \tag{1.9}$$

An integral controller is described by the recursion:

$$u_{n+1} = u_n + k_I(r - y_{n+1}). \tag{1.10}$$

The reason for the name *integral control* becomes clear if the recursion (1.10) is solved to yield

$$u_{n+1} = u_n + k_I \sum_{m=1}^{n} (r - y_m). \tag{1.11}$$

Clearly the second term in (1.11) is the discrete equivalent of the integral of the error ($e_m := r - y_m$). Actually, it is the sum of all control errors up to instant m, scaled by the gain k_I. The closed-loop dynamics, consisting of the plant together with this integral controller, is found by substituting the plant dynamics (1.9) in (1.10):

$$u_{n+1} = u_n + k_I(r - (ku_n + d_n)), \tag{1.12}$$

which can be written as the (closed-loop) recursion

$$u_{n+1} = (1 - kk_I)u_n + k_I(r - d_n). \tag{1.13}$$

The solution of this recursion is given by

$$u_n = (1 - kk_I)^n u_0 + k \sum_{m=1}^{n} (1 - kk_I)^{n-m}(r - d_{m-1}). \tag{1.14}$$

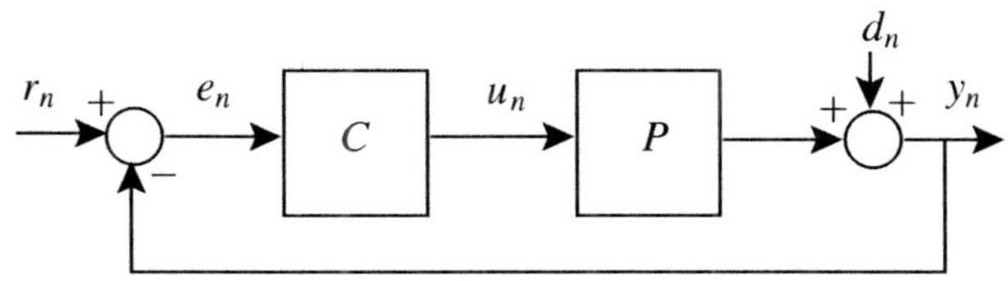

Figure 1.3. *Plant P and controller C in standard unity feedback configuration.*

The first term on the right-hand side of (1.14) is the contribution of the initial condition, whereas the second term represents the contribution of the disturbance input and is a discrete-time convolution of the error sequence and the impulse response.

Stability of this recursion requires

$$|1 - kk_I| < 1 \Rightarrow kk_I \in (0, 2). \tag{1.15}$$

The term *damping* is used in reference to the experiment of using a step input and observing the output of the system. When $kk_I \in (1, 2)$, the control is *underdamped* and this leads to a fast rise and oscillations around the steady state until equilibrium is reached. If $kk_I \in (0, 1)$, then the control is said to be *overdamped*, and the response is slower and nonoscillatory. The choice $kk_I = 1$ is called *deadbeat control* and corresponds to a so-called critically damped system. In this case, the output rises monotonically to a constant steady state value corresponding to the value of the step, in a time that is in between the times taken in the over- and underdamped cases.

For the same plant P of Figure 1.3, assume that the controller is of the proportional-integral (PI) type, which is written as

$$u_n = u_0 + k_I \sum_{m=1}^{n} (r - y_m) + k_P(r - y_n), \tag{1.16}$$

where, on the right-hand side, the second term is the integral term, as before, while the third term is the one proportional to the error. This controller can be written in recursive or state space form:

$$u_{n+1} = u_n + k_I(r - y_{n+1}) + k_P(y_n - y_{n+1}). \tag{1.17}$$

The closed-loop dynamics, found by substituting the plant dynamics (1.9) in (1.17), is

$$u_{n+1} = (1 - kk_I - kk_P)u_n + kk_P u_{n-1} - (k_I + k_P)d_n - k_P d_{n-1} + k_I r, \tag{1.18}$$

so that the characteristic equation is given by

$$q^2 - (1 - kk_I - kk_P)q - kk_P = 0. \tag{1.19}$$

The introduction of the proportional term has led to second-order dynamics, and the additional parameter k_P influences the coefficients of the characteristic equation, the roots of which determine the error dynamics. Thus, positioning or placing the roots of the characteristic equation by appropriate choice of the parameters k_I and k_P is the task that must be carried out in order to design a PI controller for the simple plant (1.9). From (1.19), it is clear that if the coefficient of the term in q is chosen as α and the constant coefficient as β, then PI controller design in this case amounts to solving the simultaneous equations below for the unknowns k_I and k_P:

$$\begin{aligned} 1 - kk_I - kk_P &= \alpha, \\ kk_P &= \beta. \end{aligned} \tag{1.20}$$

This is always possible, yielding $k_P = \beta/k$ and $k_I = [(1-\alpha) - \beta]/k$.

1.2 Optimal Control Theory

Control design problems, like most other design problems, must deal with the issue of "best" designs. This is done, as usual, with respect to an objective function, usually referred to as a *performance index* in the context of control. The controller that meets the other design objectives as well as minimizes the performance index is referred to as an *optimal controller*.

The elements of the optimal control problem are described for the continuous-time linear dynamical system

$$\dot{\mathbf{x}}(t) = \mathbf{F}\mathbf{x}(t) + \mathbf{G}\mathbf{u}(t), \qquad \mathbf{x}(t_0) = \mathbf{x}_0, \tag{1.21}$$

$$\mathbf{y}(t) = \mathbf{H}\mathbf{x}(t). \tag{1.22}$$

Suppose that the time interval of interest is defined as $[t_0, t_f] \subset \mathbb{R}$. A quadratic performance index is defined on this time interval as

$$J := \frac{1}{2}\int_{t_0}^{t_f} (\mathbf{x}^T(t)\mathbf{Q}\mathbf{x}(t) + \mathbf{u}^T(t)\mathbf{R}\mathbf{u}(t))\mathrm{d}t + \frac{1}{2}\mathbf{x}^T(t_f)\mathbf{S}\mathbf{x}(t_f), \tag{1.23}$$

where $\mathbf{R}$ is a symmetric positive definite matrix and $\mathbf{Q}$ and $\mathbf{S}$ are symmetric positive semidefinite matrices. The matrices $\mathbf{Q}$, $\mathbf{R}$, $\mathbf{S}$ are referred to as state, input, and final state weighting matrices. This problem is known as a *fixed final time, free final state problem*, since there is no constraint on the final state. Observe also that there are no explicit constraints on the input.

The *linear quadratic* (LQ) control problem is that of minimizing J defined in (1.23), subject to the dynamics (1.21) and (1.22), i.e., minimizing J over all possible choices of the control $\mathbf{u}(t)$, $t \in [t_0, t_f]$. Of course, each choice of $\mathbf{u}$ results in a state trajectory $\mathbf{x}(t)$, computable by integration of (1.21) starting from the given initial condition, and thus determines the associated value of the performance index.

There are many ways of solving the LQ problem and one of them will be given below, since it will be used in Chapters 3 and 5. The *Hamiltonian function*, also sometimes called the Pontryagin H-function, for the LQ problem is defined as

$$H := \frac{1}{2}\mathbf{x}^T\mathbf{Q}\mathbf{x} + \frac{1}{2}\mathbf{u}^T\mathbf{R}\mathbf{u} + \boldsymbol{\lambda}^T(\mathbf{F}\mathbf{x} + \mathbf{G}\mathbf{u}), \tag{1.24}$$

where $\boldsymbol{\lambda}$ is the *costate vector*, also referred to as the vector of *Lagrange multipliers*. The celebrated *Pontryagin minimum principle (PMP)* [PBGM62] states, roughly speaking, that the optimal controller minimizes the Hamiltonian function. More formally, since the control vector is unconstrained, the necessary condition for the optimality of the control $\mathbf{u}$ is

$$\frac{\partial H}{\partial \mathbf{u}} = \mathbf{0}, \tag{1.25}$$

from which it follows that the optimal control can be expressed in feedback form as

$$\mathbf{u} = -\mathbf{R}^{-1}\mathbf{G}^T\boldsymbol{\lambda}. \tag{1.26}$$

Substituting this value of $\mathbf{u}$ in (1.21) yields

$$\dot{\mathbf{x}} = \mathbf{F}\mathbf{x} - \mathbf{G}\mathbf{R}^{-1}\mathbf{G}^T\boldsymbol{\lambda}, \quad \mathbf{x}(t_0) = \mathbf{x}_0. \tag{1.27}$$

The *costate equation* must also be satisfied:

$$\dot{\boldsymbol{\lambda}} = \frac{\partial H}{\partial \mathbf{x}} = -\mathbf{Q}\mathbf{x} - \mathbf{F}^T\boldsymbol{\lambda}. \tag{1.28}$$

Finally, the state vector at time t_f must satisfy

$$\boldsymbol{\lambda}(t_f) = \frac{\partial}{\partial \mathbf{x}}\left(\frac{1}{2}\mathbf{x}^T(t)\mathbf{S}\mathbf{x}(t)\right)\bigg|_{t=t_f} = \mathbf{S}\mathbf{x}(t_f). \tag{1.29}$$

Note that (1.27), (1.28), and (1.29) define a two-point boundary value problem (TPBVP) that can be written as

$$\begin{bmatrix} \dot{\mathbf{x}} \\ \dot{\boldsymbol{\lambda}} \end{bmatrix} = \begin{bmatrix} \mathbf{F} & -\mathbf{G}\mathbf{R}^{-1}\mathbf{G}^T \\ -\mathbf{Q} & -\mathbf{F}^T \end{bmatrix} \begin{bmatrix} \mathbf{x} \\ \boldsymbol{\lambda} \end{bmatrix}, \quad \boldsymbol{\lambda}(t_f) = \mathbf{S}\mathbf{x}(t_f). \tag{1.30}$$

It can be shown, from (1.30), that the costate must satisfy the equation

$$\boldsymbol{\lambda}(t) = \mathbf{P}(t)\mathbf{x}(t), \tag{1.31}$$

where $\mathbf{P}(t)$ satisfies the matrix Riccati differential equation

$$\dot{\mathbf{P}} = \mathbf{P}\mathbf{G}\mathbf{R}^{-1}\mathbf{G}^T\mathbf{P} - \mathbf{F}^T\mathbf{P} - \mathbf{P}\mathbf{F} - \mathbf{Q}, \quad \mathbf{P}(t_f) = \mathbf{S}. \tag{1.32}$$

This means that the optimal state feedback controller is given by

$$\mathbf{u}(t) = -\mathbf{R}^{-1}\mathbf{G}^T\mathbf{P}(t)\mathbf{x}(t). \tag{1.33}$$

Finally, for ease of reference, the use of the PMP is summarized in two general cases as follows. Consider the plant defined by a sufficiently smooth dynamical system with control:

$$\dot{\mathbf{x}}(t) = \mathbf{f}(\mathbf{x}(t), \mathbf{u}(t), t), \quad \mathbf{x}(t_0) = \mathbf{x}_0 \in \mathbb{R}^n, \tag{1.34}$$

and the performance index defined as

$$J = S(\mathbf{x}(t_f), t_f) + \int_{t_0}^{t_f} V(\mathbf{x}(t), \mathbf{u}(t), t)\mathrm{d}t, \tag{1.35}$$

with appropriate boundary conditions specified.

Then the optimal control $\mathbf{u}^*$ that minimizes J is found from the following five steps:

1. The Hamiltonian function

$$H(\mathbf{x}(t), \mathbf{u}(t), \boldsymbol{\lambda}) := V(\mathbf{x}(t), \mathbf{u}(t)) + \boldsymbol{\lambda}^T\mathbf{f}(\mathbf{x}(t), \mathbf{u}(t), t)$$

is formed.

Table 1.1. *Boundary conditions for the fixed and free final time and state problems.*

Type of problem	Boundary conditions
Fixed final time t_f and state $\mathbf{x}_f$	$\mathbf{x}(t_0) = \mathbf{x}_0,\ \mathbf{x}(t_f) = \mathbf{x}_f$
Fixed final time t_f, free final state	$\mathbf{x}(t_0) = \mathbf{x}_0,\ \boldsymbol{\lambda}^*(t_f) = \left.\frac{\partial H^*}{\partial \mathbf{x}^*}\right\vert_{t_f}$

2. The Hamiltonian function is minimized with respect to $\mathbf{u}(t)$, i.e., by solving $\partial H/\partial \mathbf{u} = \mathbf{0}$ to obtain the optimal control $\mathbf{u}^*(t)$ as a function of the optimal state $\mathbf{x}^*$ and costate $\boldsymbol{\lambda}^*$.

3. Using the results of the previous two steps, the optimal Hamiltonian function H^* is found as a function of the optimal state $\mathbf{x}^*$ and costate $\boldsymbol{\lambda}^*$.

4. The set of $2n$ differential equations

$$\dot{\mathbf{x}}^* = \frac{\partial H^*}{\partial \boldsymbol{\lambda}^*}, \tag{1.36}$$

$$\dot{\boldsymbol{\lambda}}^* = -\frac{\partial H^*}{\partial \mathbf{x}^*} \tag{1.37}$$

   is solved, subject to an appropriate set of boundary conditions (two cases of interest are given in Table 1.1).

5. The optimal state and costate trajectories, $\mathbf{x}^*(t)$ and $\boldsymbol{\lambda}^*(t)$ are substituted in the expression for the optimal control $\mathbf{u}^*(t)$ found in step 2.

Table 1.1 gives the boundary conditions for two cases of interest in what follows.

Further details on this and many other optimal control problems can be found in [PBGM62, Żak03, Nai03].

1.3 Linear Systems, Transfer Functions, Realization Theory

Consider a single-input, single-output, continuous-time, linear dynamical system $\{\mathbf{F}, \mathbf{g}, \mathbf{h}\}$ with $\mathbb{R}^n$ as the state space:

$$\dot{\mathbf{x}}(t) = \mathbf{F}\mathbf{x}(t) + \mathbf{g}u(t), \tag{1.38}$$

$$y(t) = \mathbf{h}^T\mathbf{x}(t). \tag{1.39}$$

The initial condition of the dynamical system $\mathbf{x}(0)$ is assumed to be $\mathbf{0}$, unless otherwise specified.

In the so-called input–output approach in system theory, it is assumed that the state vector $\mathbf{x}$ of the dynamical system is inaccessible and that only the input u and output y are accessible (measurable). This means that the state must be inferred or estimated from the measurements of the input and the output. An easy way of calculating the output given the input is by the introduction of the *Laplace transform*, denoted $\mathcal{L}$, which is a mapping from the space of time functions to the space of functions of the complex variable s. Let a time function $f(t)$ be given. Then $\mathcal{L} : f(t) \to \hat{f}(s)$, where $\hat{f}(s)$ is defined as

$$\hat{f}(s) := \int_0^\infty e^{-s\tau} f(\tau) \mathrm{d}\tau. \tag{1.40}$$

From this definition, it is easy to show that

$$\mathcal{L}\frac{\mathrm{d}f}{\mathrm{d}t} = s\hat{f}(s) - f(0). \tag{1.41}$$

Using (1.41) and taking the Laplace transforms of (1.38) and (1.39) yields

$$s\hat{\mathbf{x}}(s) - \mathbf{x}(0) = \mathbf{F}\hat{\mathbf{x}}(s) + \mathbf{g}\hat{u}(s), \tag{1.42}$$

$$\hat{y}(s) = \mathbf{h}^T\hat{\mathbf{x}}(s), \tag{1.43}$$

which, on solving (1.42) for $\hat{\mathbf{x}}$ and substituting in (1.43), leads to the *transfer function*

$$\hat{w}(s) = \frac{\hat{y}(s)}{\hat{u}(s)} = \mathbf{h}^T(s\mathbf{I} - \mathbf{F})^{-1}\mathbf{g}. \tag{1.44}$$

The transfer function $\hat{w}(s)$ can be expanded in the series

$$\hat{w}(s) = \sum_{k=0}^{\infty} (\mathbf{h}^T\mathbf{F}^k\mathbf{g})/s^{k+1}. \tag{1.45}$$

The series representation is convergent for $|s|$ large enough, and the coefficients $\{\mathbf{h}^T\mathbf{F}^k\mathbf{g}\}_{k=0}^{\infty}$ are called the *Markov parameters* or *impulse response* of the system. The triple $\{\mathbf{F}, \mathbf{g}, \mathbf{h}\}$ is called a *realization* of the transfer function $\hat{w}(s)$. Note that (1.44) implies that, for zero initial conditions, the transfer function can be calculated from complete knowledge of the Laplace transforms of an input and the corresponding output.

The realization problem is that of determining all realizations $\{\mathbf{F}, \mathbf{g}, \mathbf{h}\}$ that yield a given rational function or, equivalently, a given set of Markov parameters.

From the well-known formula for the inverse of a matrix, (1.44) can be written, for all s not in the spectrum of $\mathbf{F}$, as

$$\hat{w}(s) = \mathbf{h}^T\mathrm{adj}[s\mathbf{I} - \mathbf{F}]\mathbf{g}/\det(s\mathbf{I} - \mathbf{F}) = \mathbf{h}^T\mathrm{adj}[s\mathbf{I} - \mathbf{F}]\mathbf{g}/p_{\mathbf{F}}(s) =: n(s)/d(s), \tag{1.46}$$

where $\mathrm{adj}(\mathbf{M})$ stands for the classical *adjoint matrix* (sometimes called the *adjugate matrix*) made up of cofactors of dimension $(n-1)$ of the matrix $\mathbf{M}$ and $p_{\mathbf{F}}(s)$ is the characteristic polynomial of the matrix $\mathbf{F}$. Observe also that $n(s)$ and $d(s)$ are polynomials in s, called, for obvious reasons, the numerator and denominator polynomials, respectively.

The singularities of the function $\hat{w}(\cdot)$ (of the complex variable s) are called *poles of the transfer function*, while the zeros of the numerator are called *zeros of the transfer function*. Clearly, if no cancellation of common factors between the numerator and denominator polynomials occurs, then the poles of the transfer function are the roots of the characteristic polynomial $p_{\mathbf{F}}$, i.e., the eigenvalues of the matrix $\mathbf{F}$. Otherwise, when cancellation of common factors occurs, not every eigenvalue of $\mathbf{F}$ is a pole of $\hat{w}$. Such a cancellation is called a *pole-zero cancellation*. Thus it is possible to have a triple $\{\mathbf{F}_1, \mathbf{g}_1, \mathbf{h}_1\}$ with $\mathbf{F}_1 \in \mathbb{R}^{m\times m}$ and $m < n$, giving rise to the same transfer function $\hat{w}(s)$ that is obtained with the triple $\{\mathbf{F}, \mathbf{g}, \mathbf{h}\}$. This leads to the following definition.

Definition 1.2. *A* minimal realization $\{\mathbf{F}, \mathbf{g}, \mathbf{h}\}$ *of* $\hat{w}(s)$ *is one in which* $\mathbf{F}$ *has minimal dimension or order.*

This minimal order is called the *McMillan degree* of both the transfer function $\hat{w}$ and the Markov parameter sequence $\mathbf{h}^T\mathbf{F}^i\mathbf{g}$. Determination of minimal realizations is one of the important tasks of system theory and one way to achieve it is to associate Hankel matrices (and quadratic forms in Hankel matrices) to a sequence of Markov parameters. The Hankel matrices are defined as

$$\mathbf{H}_l^{(k)} := \left[\mathbf{h}^T\mathbf{F}^{k+i+j-2}\mathbf{g}\right] \quad (i, j = 1, \dots, l) \tag{1.47}$$

$$= \begin{bmatrix} \mu_k & \mu_{k+1} & \mu_{k+2} & \cdot \\ \mu_{k+1} & \mu_{k+2} & \cdot & \cdot \\ \mu_{k+2} & \cdot & \cdot & \cdot \\ \cdot & \cdot & \cdot & \cdot \end{bmatrix} \quad (i, j = 1, \dots, l), \tag{1.48}$$

$$\mathbf{H}^{(k)} = \mathbf{H}_\infty^{(k)}. \tag{1.49}$$

The defining property of Hankel matrices is that the (i, j) entry depends only on the value of $i + j$. It can be shown [CD91] that the rank of $\mathbf{H}^{(0)}$ is the McMillan degree of the transfer function $\hat{w}$.

Hankel matrices, Krylov matrices, and invariant subspaces

In order to discuss factorization of Hankel matrices, it is necessary to introduce the so-called Krylov matrices and two special Krylov matrices known as the controllability and observability matrices, respectively.

Definition 1.3. *Given a triple* $\{\mathbf{F}, \mathbf{g}, \mathbf{h}\} \in (\mathbb{R}^{n\times n}, \mathbb{R}^n, \mathbb{R}^p)$, *the* Krylov matrix $\mathbf{K}_m(\mathbf{g}, \mathbf{F})$ *is defined as*

$$\mathbf{K}_m(\mathbf{g}, \mathbf{F}) := [\mathbf{g}\ \mathbf{F}\mathbf{g}\ \cdots\ \mathbf{F}^{m-1}\mathbf{g}] \in \mathbb{C}^{n\times m}. \tag{1.50}$$

The controllability matrix $\mathbf{K}(\mathbf{g}, \mathbf{F})$ *is defined as* $\mathbf{K}_\infty(\mathbf{g}, \mathbf{F})$. *The dual matrices are defined in terms of* (1.50) *as*

$$\mathbf{K}_m(\mathbf{h}^T, \mathbf{F}) := \mathbf{K}_m(\mathbf{h}, \mathbf{F}^T)^T, \tag{1.51}$$

as well as the observability matrix

$$\mathbf{K}(\mathbf{h}^T, \mathbf{F}) = \mathbf{K}_\infty(\mathbf{h}^T, \mathbf{F}). \tag{1.52}$$

A fundamental fact relating Hankel matrices to the Krylov matrices is as follows.

Lemma 1.4. *For any* $j \in \mathbb{N}$,

$$\mathbf{H}_m^{(j)} = \mathbf{K}_m(\mathbf{h}^T, \mathbf{F})\mathbf{F}^j\mathbf{K}_m(\mathbf{g}, \mathbf{F}), \tag{1.53}$$

$$\mathbf{H}^{(j)} = \mathbf{K}(\mathbf{h}^T, \mathbf{F})\mathbf{F}^j\mathbf{K}(\mathbf{g}, \mathbf{F}). \tag{1.54}$$

Proof. The (i, k) entry on each side is $(\mathbf{h}^T\mathbf{F}^{i-1})\mathbf{F}^j(\mathbf{F}^{k-1}\mathbf{g})$. □

Observing that $\mathbf{H}^{(j)}$ is a submatrix of $\mathbf{H}^{(0)}$, a straightforward corollary of the above lemma is as follows.

Corollary 1.5.

$$\text{rank}\,[\mathbf{H}^{(j)}] \le \text{rank}\,[\mathbf{H}^{(0)}] \le \min\{\text{rank}\,[\mathbf{K}(\mathbf{h}^T, \mathbf{F})], \text{rank}\,[\mathbf{K}(\mathbf{g}, \mathbf{F})]\} \le n. \tag{1.55}$$

Finally, as preparation for the Kalman–Gilbert canonical form, the following lemma on smallest invariant subspaces is needed.

Lemma 1.6. *The smallest* $\mathbf{F}$*-invariant subspace of* $\mathbb{C}^n$ *that contains* $\mathbf{g}$ *is* range$[\mathbf{K}(\mathbf{g}, \mathbf{F})]$. *The smallest* $\mathbf{F}$*-invariant subspace of* $\mathbb{C}^n_*$ *that contains* $\mathbf{h}^T$ *is* range$[\mathbf{K}(\mathbf{h}^T, \mathbf{F})]$.

Proof. If $\mathcal{S}$ is $\mathbf{F}$-invariant and contains $\mathbf{g}$, then it must contain $\mathbf{Fg}$, $\mathbf{F}(\mathbf{Fg})$, etc. In other words, range$[\mathbf{K}(\mathbf{g}, \mathbf{F})] \subset \mathcal{S}$. Moreover,

$$\mathbf{F}\,\text{range}[\mathbf{K}(\mathbf{g}, \mathbf{F})] = \text{range}[\mathbf{Fg}, \mathbf{F}(\mathbf{Fg}), \ldots] \subset \text{range}[\mathbf{K}(\mathbf{g}, \mathbf{F})],$$

showing that $[\mathbf{K}(\mathbf{g}, \mathbf{F})]$ is $\mathbf{F}$-invariant. The dual argument proves the assertion for range$[\mathbf{K}(\mathbf{h}^T, \mathbf{F})]$. □

Kalman–Gilbert canonical form and minimal realization

Four subspaces of $\mathbb{R}^n$ are associated to the dynamical system (1.38).

Definition 1.7.

(i) *The* controllable subspace *is defined as*

$$\mathcal{K}(\mathbf{g}, \mathbf{F}) := \text{range of } [\mathbf{K}(\mathbf{g}, \mathbf{F})] = \mathcal{S}_c.$$

It is the smallest $\mathbf{F}$*-invariant subspace containing* $\mathbf{g}$.

(ii) *The* unobservable subspace *is defined as*

$$\mathcal{N}(\mathbf{h}^T, \mathbf{F}) := \text{null space of } \mathbf{K}(\mathbf{h}^T, \mathbf{F}) = \mathcal{S}_{\bar{o}}.$$

It is the largest $\mathbf{F}$*-invariant subspace annihilated by* $\mathbf{h}^T$.

The complements in $\mathbb{R}^n$ *of these two spaces can be chosen as the unobservable and controllable subspaces of the dual system* $\{\mathbf{F}^T, \mathbf{h}, \mathbf{g}\}$ *as follows.*

(iii) *The* unobservable subspace for the dual system $\{\mathbf{F}^T, \mathbf{h}, \mathbf{g}\}$ *is defined as*

$$\mathcal{N}(\mathbf{g}^T, \mathbf{F}^T) := \text{null space of } \mathbf{K}(\mathbf{g}^T, \mathbf{F}^T) = \mathcal{S}_{\bar{c}}$$

and is the largest $\mathbf{F}^T$*-invariant subspace annihilated by* $\mathbf{g}^T$.

(iv) *The* controllable subspace for the dual system *is defined as*

$$\mathcal{K}(\mathbf{h}, \mathbf{F}^T) := \text{range of } [\mathbf{K}(\mathbf{h}, \mathbf{F}^T)] = \mathcal{S}_o$$

and is the smallest $\mathbf{F}^T$*-invariant subspace containing* $\mathbf{h}$.

The motivation for the names of these subspaces is that the controllable subspace consists of all states $\mathbf{x}$ for which a suitable control $u(t)$ exists that transfers the zero state to $\mathbf{x}$ at some time t. The unobservable space is so named because it consists of initial states $\mathbf{x}(0)$ such that $y(t) \equiv 0$ when no input is applied, meaning that measurement of the output alone does not allow the calculation of the initial state (which is thus an unobservable state). Note that the subscripts $\bar{c}$ and $\bar{o}$ signify uncontrollable and unobservable, respectively.

By taking the intersection of these four invariant subspaces in an appropriate order, the *Kalman–Gilbert canonical form* is obtained. The starting point is to define four intersections of these subspaces:

$$\begin{aligned}
\mathcal{S}_{c\bar{o}} &:= \mathcal{N}(\mathbf{h}^T, \mathbf{F}) \cap \mathcal{N}(\mathbf{h}^T, \mathbf{F}) = \mathcal{S}_c \cap \mathcal{S}_{\bar{o}}, \\
\mathcal{S}_{co} &:= \mathcal{K}(\mathbf{g}, \mathbf{F}) \cap \mathcal{K}(\mathbf{h}, \mathbf{F}^T) = \mathcal{S}_c \cap \mathcal{S}_o, \\
\mathcal{S}_{\bar{c}\bar{o}} &:= \mathcal{N}(\mathbf{h}^T, \mathbf{F}) \cap \mathcal{N}(\mathbf{g}^T, \mathbf{F}^T) = \mathcal{S}_{\bar{c}} \cap \mathcal{S}_{\bar{o}}, \\
\mathcal{S}_{\bar{c}o} &:= \mathcal{N}(\mathbf{g}^T, \mathbf{F}^T) \cap \mathcal{K}(\mathbf{h}, \mathbf{F}^T) = \mathcal{S}_{\bar{c}} \cap \mathcal{S}_o.
\end{aligned} \tag{1.56}$$

Clearly,

$$\mathbb{R}^n = \mathcal{S}_{c\bar{o}} \oplus \mathcal{S}_{co} \oplus \mathcal{S}_{\bar{c}\bar{o}} \oplus \mathcal{S}_{\bar{c}o}.$$

The following facts are also immediate:

$$\begin{aligned}
&\mathcal{S}_{c\bar{o}} \text{ and } \mathcal{S}_c = \mathcal{S}_{c\bar{o}} \oplus \mathcal{S}_{co} \text{ are } \mathbf{F}\text{-invariant}, \\
&\mathcal{S}_{\bar{c}o} \text{ and } \mathcal{S}_o = \mathcal{S}_{co} \oplus \mathcal{S}_{\bar{c}o} \text{ are } \mathbf{F}^T\text{-invariant}.
\end{aligned}$$

Note that the controllable, observable subspace $\mathcal{S}_{co}$ is not invariant under either $\mathbf{F}$ or $\mathbf{F}^T$. Finally, by taking any bases for the four subspaces, *in the order listed above*, a new representation $\{\mathbf{F}_{\text{can}}, \mathbf{g}_{\text{can}}, \mathbf{h}_{\text{can}}\}$—the Kalman–Gilbert canonical form—is obtained:

$$\mathbf{F}_{\text{can}} = \begin{bmatrix} \bar{\mathbf{F}}_{11} & \bar{\mathbf{F}}_{12} & \bar{\mathbf{F}}_{13} & \bar{\mathbf{F}}_{14} \\ \mathbf{0} & \bar{\mathbf{F}}_{22} & \mathbf{0} & \bar{\mathbf{F}}_{24} \\ \mathbf{0} & \mathbf{0} & \bar{\mathbf{F}}_{33} & \bar{\mathbf{F}}_{34} \\ \mathbf{0} & \mathbf{0} & \mathbf{0} & \bar{\mathbf{F}}_{44} \end{bmatrix} = \mathbf{T}^{-1}\mathbf{F}\mathbf{T} \tag{1.57}$$

and

$$\mathbf{g}_{\text{can}} = \begin{bmatrix} \bar{\mathbf{g}}^1 \\ \bar{\mathbf{g}}^2 \\ \mathbf{0} \\ \mathbf{0} \end{bmatrix} = \mathbf{T}^{-1}\mathbf{g}, \qquad \mathbf{h}_{\text{can}} = \begin{bmatrix} \mathbf{0} \\ \bar{\mathbf{h}}^2 \\ \mathbf{0} \\ \bar{\mathbf{h}}^4 \end{bmatrix} = \mathbf{T}^T\mathbf{h}. \tag{1.58}$$

Following Parlett [Par92], observe that the usual order of the subspaces $\mathcal{S}_{\bar{c}\bar{o}}$ and $\mathcal{S}_{\bar{c}o}$ used in the system theory literature has been inverted. The order chosen above has the advantage of making $\mathbf{F}_{\text{can}}$ block triangular, while the usual order chosen in system theory simply puts the uncontrollable, unobservable subspace in final position.

Calculation of the transfer function $\mathbf{h}_{\text{can}}^T(s\mathbf{I} - \mathbf{F}_{\text{can}})^{-1}\mathbf{g}_{\text{can}}$ shows that the canonical form also reveals the minimal realization:

$$\hat{w}(s) = \mathbf{h}_{\text{can}}^T(s\mathbf{I} - \mathbf{F}_{\text{can}})^{-1}\mathbf{g}_{\text{can}} = (\bar{\mathbf{h}}^2)^T(s\mathbf{I} - \bar{\mathbf{F}}_{22})^{-1}\bar{\mathbf{g}}^2. \tag{1.59}$$

The system $\{\bar{\mathbf{F}}_{22}, \bar{\mathbf{g}}^2, (\bar{\mathbf{h}}^2)^T\}$ is a minimal realization of the transfer function $\hat{w}$. Moreover, the following equality between Hankel matrices holds:

$$\mathbf{H}^{(0)}(\mathbf{g}_{\text{can}}, \mathbf{h}_{\text{can}}^T, \mathbf{F}_{\text{can}}) = \mathbf{H}^{(0)}(\bar{\mathbf{g}}^2, (\bar{\mathbf{h}}^2)^T, \bar{\mathbf{F}}_{22}). \tag{1.60}$$

Finally,

$$\text{rank}\,[\mathbf{H}^{(0)}] = \dim \mathcal{S}_{co}. \tag{1.61}$$

1.4 Basics of Stability of Dynamical Systems

Consider the vector difference equation (iterative method)

$$\mathbf{x}_{k+1} = \mathbf{f}(\mathbf{x}_k), \quad k \in \mathbb{N}, \ \mathbf{x}_0 \text{ given}, \tag{1.62}$$

where $\mathbf{x}_k \in \mathbb{R}^n$ and $\mathbf{f} : \mathbb{R}^n \to \mathbb{R}^n$ and k indicates the iteration number. It is assumed that $\mathbf{f}(\mathbf{x})$ is continuous in $\mathbf{x}$. Equation (1.62) is said to be *autonomous* or *time invariant*, since the variable k does not appear explicitly in the right-hand side of the equation. A vector $\mathbf{x}^* \in \mathbb{R}^n$ is called an *equilibrium point* or *fixed point* of (1.62) if $\mathbf{f}(\mathbf{x}^*) = \mathbf{x}^*$. It is usually assumed that a convenient change of coordinates allows $\mathbf{x}^*$ to be taken as the origin (zero vector); this equilibrium is called the *zero solution*.

The notation $\mathbf{x}_k$ is used to denote a sequence of vectors (alternatively written $\{\mathbf{x}_k\}$) that starts from the initial condition ($\mathbf{x}_0$) and satisfies (1.62). Such a sequence is called a *solution* of (1.62).

Convergence and stability of discrete-time systems: Definitions

The convergence and stability definitions used in this book are collected below.

First, (1.62) is regarded as an iterative method in order to define the notions of local and global convergence.

Definition 1.8. *The iterative method is called* locally convergent (LC) *to* $\mathbf{x}^*$ *if there is a* $\delta > 0$ *such that whenever* $\|\mathbf{x}_0 - \mathbf{x}^*\| < \delta$ *(in some norm* $\|\cdot\|$*), the solution* $\mathbf{x}_k$ *exists and* $\lim_{k\to\infty} \mathbf{x}_k = \mathbf{x}^*$. *It is* globally convergent (GC) *if* $\lim_{k\to\infty} \mathbf{x}_k = \mathbf{x}^*$ *for any* $\mathbf{x}_0$.

If (1.62) is considered to be a discrete dynamical system (difference equation), then $\mathbf{x}^*$ is called an equilibrium point and stability definitions are given as follows.

Definition 1.9. *The equilibrium point of* (1.62) *is said to be*

(i) stable *or sometimes* stable in the sense of Liapunov *if, given* $\varepsilon > 0$, *there exists* $\delta > 0$ *such that* $\|\mathbf{x}_0 - \mathbf{x}^*\| < \delta$ *implies* $\|\mathbf{x}_k - \mathbf{x}^*\| < \varepsilon$ *for all* $k \geq 0$; *and* unstable *if it is not stable.*

(ii) attractive (A) *if there exists* $\delta > 0$ *such that* $\|\mathbf{x}_0 - \mathbf{x}^*\| < \delta$ *implies*

$$\lim_{k\to\infty} \mathbf{x}_k = \mathbf{x}^*.$$

If $\delta = \infty$, *then* $\mathbf{x}^*$ *is* globally attractive (GA).

(iii) asymptotically stable (AS) *if it is stable and attractive;* globally asymptotically stable (GAS) *if it is stable and globally attractive.*

(iv) exponentially stable *if there exist* $\delta > 0$, $\mu > 0$, *and* $\eta \in (0, 1)$ *such that* $\|\mathbf{x}_k - \mathbf{x}^*\| \le \mu\eta^k$ *whenever* $\|\mathbf{x}_0 - \mathbf{x}^*\| < \delta$; *globally exponentially stable if* $\delta = \infty$.

In general, exponential stability implies all the other types. Note that attractivity does not imply stability. Attractivity and convergence are equivalent concepts. The following equivalences are clear [Ort73]:

$$\text{GAS} \Rightarrow \text{GA} \Leftrightarrow \text{GC} \Rightarrow \text{LC} \Leftrightarrow \text{A} \Leftarrow \text{AS}.$$

Since Liapunov methods are used throughout the book, it will usually be the case that stability results are obtained, rather than convergence results. The reader will notice, however, that the term *convergence* will sometimes be used rather loosely as a synonym for *asymptotic* or *exponential stability*; the appropriate term should be clear from the context.

Liapunov stability theorems for discrete-time systems

The major tool in the stability analysis of nonlinear difference and differential equations was introduced by Liapunov in his famous memoir published in 1892 [Lia49].

Consider the time-invariant equation

$$\mathbf{x}_{k+1} = \mathbf{f}(\mathbf{x}_k), \tag{1.63}$$

where $\mathbf{f} : G \to \mathbb{R}^n$, $G \subset \mathbb{R}^n$, is continuous. Assume that $\mathbf{x}^*$ is an equilibrium point of the difference equation, i.e., $\mathbf{f}(\mathbf{x}^*) = \mathbf{x}^*$.

Let $V : \mathbb{R}^n \to \mathbb{R}$ be defined as a real-valued function. The *decrement* or *variation* of V relative to (1.63) is defined as

$$\Delta V(\mathbf{x}_k) := V(\mathbf{x}_{k+1}) - V(\mathbf{x}_k) = V(\mathbf{f}(\mathbf{x}_k)) - V(\mathbf{x}_k).$$

Note that if $\Delta V(\mathbf{x}) \le 0$, then V is nonincreasing along solutions of (1.63).

Definition 1.10. *The function* V *is said to be a* Liapunov function *on a subset* H *of* $\mathbb{R}^n$ *if* (i) V *is continuous on* H *and* (ii) *the decrement* $\Delta V \le 0$, *whenever* $\mathbf{x}$ *and* $\mathbf{f}(\mathbf{x})$ *are in* H.

Let $B(\mathbf{x}, \rho)$ denote the open ball of radius ρ and center $\mathbf{x}$ defined by $B(\mathbf{x}, \rho) := \{\mathbf{y} \in \mathbb{R}^n : \|\mathbf{y} - \mathbf{x}\| < \rho\}$.

Definition 1.11. *The real-valued function* V *is said to be* positive definite at $\mathbf{x}^*$ *if* (i) $V(\mathbf{x}^*) = 0$ *and* (ii) $V(\mathbf{x}) > 0$ *for all* $\mathbf{x} \in B(\mathbf{x}, \rho)$, *for some* $\rho > 0$.

The first Liapunov stability theorem is now stated.

Theorem 1.12. *If V is a Liapunov function for* (1.63) *on a neighborhood H of the equilibrium point $\mathbf{x}^*$, and V is positive definite with respect to $\mathbf{x}^*$, then $\mathbf{x}^*$ is stable. If, in addition, $\Delta V(\mathbf{x}) < 0$, whenever $\mathbf{x}$ and $\mathbf{f}(\mathbf{x})$ are in H and $\mathbf{x} \neq \mathbf{x}^*$, then $\mathbf{x}^*$ is asymptotically stable. Moreover, if $G = H = \mathbb{R}^n$ and*

$$V(\mathbf{x}) \to \infty \quad as \quad \|\mathbf{x}\| \to \infty,$$

then $\mathbf{x}^*$ is globally asymptotically stable.

For a linear time-invariant system

$$\mathbf{x}_{k+1} = \mathbf{F}\mathbf{x}_k, \tag{1.64}$$

the basic stability result is that the origin is stable if and only if the spectral radius of $\mathbf{F}$ satisfies $\rho(\mathbf{F}) \leq 1$ and any eigenvalues of modulus unity are semisimple (i.e., correspond to Jordan blocks of dimension 1). The remaining parts of Definitions 1.8 and 1.9 are all equivalent if $\rho(\mathbf{F}) < 1$, in which case the matrix $\mathbf{F}$ is often called *Schur stable*, or sometimes just a *Schur matrix*.

Application of the basic Liapunov stability theorem, using a quadratic Liapunov function $V(\mathbf{x}) := \mathbf{x}^T\mathbf{P}\mathbf{x}$, to a linear time-invariant system (1.64) leads to the following basic Liapunov theorem.

Theorem 1.13. *The origin or zero solution of the linear time-invariant system* (1.64) *is exponentially stable if and only if there exists a positive definite matrix $\mathbf{P}$ such that the equation*

$$\mathbf{F}^T\mathbf{P}\mathbf{F} - \mathbf{P} = -\mathbf{Q} \tag{1.65}$$

is satisfied for some positive definite matrix $\mathbf{Q}$.

Equation (1.65) is known as *Stein's equation* or the *discrete-time Liapunov equation*: If it is satisfied, clearly $\rho(\mathbf{F}) < 1$ and $\mathbf{F}$ is a Schur stable matrix. Finally, if (1.65) is satisfied for a positive diagonal matrix $\mathbf{P}$, then $\mathbf{F}$ is referred to as a *(Schur) diagonally stable matrix*.

More discrete-time stability theory: Invariant sets and LaSalle's theorem

Consider the nonautonomous (time-varying) discrete dynamical system

$$\mathbf{x}_{k+1} = \mathbf{f}_k(\mathbf{x}_k), \tag{1.66}$$

where $\mathbf{f}_k : \mathbb{R}^n \to \mathbb{R}^n$ for each k. A function $\mathbf{x}_k(k_0, \mathbf{x}_0)$ is called a solution of the difference equation (1.66) if it satisfies the following three conditions:

(i) $\mathbf{x}_k(k_0, \mathbf{x}_0)$ is defined for $k_0 \leq k \leq k_0 + K$ for some integer $K > 0$;

(ii) $\mathbf{x}_{k_0}(k_0, \mathbf{x}_0) = \mathbf{x}_0$, the initial condition;

(iii) $\mathbf{x}_{k+1}(k_0, \mathbf{x}_0) = \mathbf{f}_k(\mathbf{x}_k(k_0, \mathbf{x}_0))$ for $k_0 \leq k \leq k_o + K - 1$.

A solution to (1.66) can be assumed to exist and be unique for all $k \geq k_0$, and moreover, this solution is continuous in the initial vector $\mathbf{x}_0$. This means that if $\{\mathbf{x}_j\}$ is a sequence of vectors with $\mathbf{x}_j \to \mathbf{x}_0$ as $j \to \infty$, then the solutions through $\mathbf{x}_j$ converge to the solution through $\mathbf{x}_0$:

$$\mathbf{x}_k(k_0, \mathbf{x}_j) \to \mathbf{x}_k(k_0, \mathbf{x}_0) \quad \text{as } j \to \infty.$$

Given a norm $\|\cdot\|$ in $\mathbb{R}^n$ and a nonempty subset A of $\mathbb{R}^n$, let the distance from $\mathbf{x} \in \mathbb{R}^n$ to A be denoted $d(\mathbf{x}, A)$ and defined as

$$d(\mathbf{x}, A) := \inf\{\|\mathbf{x} - \mathbf{y}\| : \mathbf{y} \in A\}. \tag{1.67}$$

Let $\mathbb{R}^* := \mathbb{R} \cup \{\infty\}$ and let $d(\mathbf{x}, \infty) := 1/\|\mathbf{x}\|$. Define $A^* := A \cup \{\infty\}$ and $d(\mathbf{x}, A^*) := \min\{d(\mathbf{x}, A), d(\mathbf{x}, \infty)\}$.

A point $\mathbf{p} \in \mathbb{R}^n$ is called a *positive limit point* of $\mathbf{x}_k$ if there exists a sequence $k_{n+1} > k_n \to \infty$, and $\mathbf{x}_{k_n} \to \mathbf{p}$, as $n \to \infty$. The union of all the positive limit points of $\mathbf{x}_k$ is called the *positive limit set* of $\mathbf{x}_k$.

Definition 1.14. *Let $V_k(\mathbf{x})$ and $W(\mathbf{x})$ be real-valued functions, continuous in $\mathbf{x}$, defined for all $k \geq k_0$ and all $\mathbf{x} \in G$, where G is a (possibly unbounded) subset of $\mathbb{R}^n$. If $V_k(\mathbf{x})$ is bounded below and, for all $k \geq k_0$ and all $\mathbf{x} \in G$,*

$$\Delta V_k(\mathbf{x}) := V_{k+1}(\mathbf{f}_k(\mathbf{x})) - V_k(\mathbf{x}) \leq -W(\mathbf{x}) \leq 0, \tag{1.68}$$

then V is called a Liapunov function *for* (1.66) *on G.*

Let $\overline{G}$ be the closure of G, including ∞ if G is unbounded, and let the set A be defined as

$$A := \{\mathbf{x} \in \overline{G} : W(\mathbf{x}) = 0\}. \tag{1.69}$$

The discrete-time version of LaSalle's theorem and its simple proof are then as follows.

Theorem 1.15 [Hur67]. *If there exists a Liapunov function V for* (1.66) *on G, then each solution of* (1.66) *which remains in G for all $k \geq k_0$ approaches the set $A^* = A \cup \{\infty\}$ as $k \to \infty$.*

Proof. Let $\mathbf{x}(k)$ be a solution to (1.66) which remains in G for all $k \geq k_0$. Then, since V is a Liapunov function, from (1.68), it follows that it is a monotone nonincreasing function, which is, by assumption, bounded from below. Thus $V_k(\mathbf{x}_k)$ must approach a limit as $k \to \infty$, and $W(k)$ must approach zero as $k \to \infty$. From the definition of A^* and the continuity of $W(\mathbf{x})$, it follows that $d(\mathbf{x}_k, A^*) \to 0$ as $k \to \infty$. If G is bounded or if $W(\mathbf{x})$ is bounded away from zero for all sufficiently large $\mathbf{x}$, then all solutions that remain in G are bounded and approach a closed, bounded set contained in A as $k \to \infty$. If G is unbounded and there exists a sequence $\{\mathbf{x}_n\}$ such that $\mathbf{x}_n \in G$ as $\|\mathbf{x}_n\| \to \infty$ and $W(\mathbf{x}_n) \to 0$ as $n \to \infty$, then it is possible to have an unbounded solution under the conditions of the theorem. ☐

This theorem actually contains all the usual Liapunov stability theorems; for example, if G is the entire space $\mathbb{R}^n$ and $W(\mathbf{x})$ is positive definite, then $A = \{\mathbf{0}\}$ and all solutions approach the origin as $k \to \infty$.

If the function $\mathbf{f}_k(\mathbf{x})$ in (1.66) is independent of k, then the discrete dynamical system is said to be *autonomous* and is written as (1.63), studied above. Solutions to (1.63) are essentially independent of k_0, so $k_0 = 0$ is the usual choice and the solution is written as $\mathbf{x}_k(\mathbf{x}_0)$, or as just $\mathbf{x}_k$, if the initial condition does not need to be specified explicitly.

A set B is called an *invariant set* of (1.63) if $\mathbf{x}_0 \in B$ implies that there is a solution of $\mathbf{x}_k^*$ of (1.63) such that $\mathbf{x}_k^* \in B$ for all k and $\mathbf{x}_0^* = \mathbf{x}_0$. A basic lemma about limit sets and invariance is as follows.

Lemma 1.16. *The positive limit set B of any bounded solution of* (1.63) *is a nonempty, compact, invariant set of* (1.63).

For an autonomous or time-invariant difference equation, Theorem 1.15 (also called the Krasovskii–LaSalle theorem) can be strengthened as follows.

Theorem 1.17. *If there exists a Liapunov function $V(\mathbf{x})$ for* (1.63) *on some set G, then each solution $\mathbf{x}_k$ which remains in G is either unbounded or approaches some invariant set contained in A as $k \to \infty$.*

Proof. From Theorem 1.15, $\mathbf{x}_k \to A \cup \{\infty\}$ as $k \to \infty$. If $\mathbf{x}_k$ is unbounded, then Lemma 1.16 does not hold. If $\mathbf{x}_k$ is bounded, then its positive limit set is an invariant set. $\square$

Let M_i be an invariant set of (1.63) contained in A and let the set M be defined as

$$M := \cup_i M_i. \tag{1.70}$$

Then $\mathbf{x}_k \to M$ as $k \to \infty$ whenever $\mathbf{x}_k$ remains in G and is bounded. Note that the set M may be much smaller than the set A.

A common use of Theorem 1.17 is to conclude stability of the origin in the case when $M = \{\mathbf{0}\}$.

A useful corollary of Theorem 1.17 is as follows.

Corollary 1.18 [Hur67]. *If, in Theorem* (1.17), *the set G is of the form*

$$G = G(\eta) = \{\mathbf{x} : V(\mathbf{x}) < \eta\}$$

for some $\eta > 0$, then all solutions that start in G remain in G and approach M as $k \to \infty$.

This corollary can be used to obtain regions of convergence for various iterative methods that can be described by an autonomous difference equation. A *region of convergence* is a set $G \subset \mathbb{R}^n$ such that, if $\mathbf{x}_0 \in G$, then $\mathbf{x}_k \in G$ for all $k \geq 0$ and $\mathbf{x}_k$ converges to the desired vector as $k \to \infty$. The *largest region of convergence* is then defined as the union of all regions of convergence.

Practical stability

Another useful notion is that of *practical stability*, which arises by considering a solution of a discrete dynamical system to be stable if it enters and remains in a sufficiently small set. This notion is particularly appropriate when the discrete dynamical system represents an iterative method subject to disturbances such as roundoff errors: The solution may no longer approach the desired solution, but the method is still considered satisfactory if all solutions get and remain sufficiently close to the desired solution. Practical stability is the subject of the next theorem.

Theorem 1.19 [Hur67]. *Consider the discrete dynamical system* (1.66) *and let a set* $G \subset \mathbb{R}^n$*, possibly unbounded, be given. Let* $V(\mathbf{x})$ *and* $W(\mathbf{x})$ *be continuous, real-valued functions defined on* G *and such that, for all* k *and all* $\mathbf{x}$ *in* G,

$$\begin{aligned} &V(\mathbf{x}) \geq 0, \\ &\Delta V_k(\mathbf{x}) := V(\mathbf{f}_k(\mathbf{x})) - V(\mathbf{x}) \leq W(\mathbf{x}) \leq a \end{aligned} \tag{1.71}$$

for some constant $a \geq 0$*. The sets* S *and* A *are defined as*

$$\begin{aligned} S &:= \{\mathbf{x} \in \overline{G} : W(\mathbf{x}) \geq 0\}, \\ A &:= \{\mathbf{x} \in \overline{G} : V(\mathbf{x}) \leq b + a\}. \end{aligned} \tag{1.72}$$

Then, any solution $\mathbf{x}_k$ *which remains in* G *and enters* A *when* $k = k_1$ *remains in* A *for all* $k \geq k_1$.

The proof of the theorem is by induction: If $\mathbf{x}_k \in A$, then the properties of S, A, and $V(\mathbf{x})$ can be used to show that $\mathbf{x}_{k+1} \in A$.

A corollary of Theorem 1.19 is useful in the problem of studying the effect of roundoff errors.

Corollary 1.20 [Hur67]. *Let* $\delta := \sup\{-W(\mathbf{x}) : \mathbf{x} \in G\backslash A\} > 0$*. Then each solution* $\mathbf{x}_k$ *of* (1.66) *which remains in* G *enters* A *in a finite number of steps. Furthermore, if* $G = G(\eta) := \{\mathbf{x} : V(\mathbf{x}) < \eta\}$*, then all solutions that start in* G *remain in* G *and enter* A *in a finite number of steps.*

Liapunov stability theorems for continuous-time systems

Continuous-time analogs of these theorems are as follows. Consider the dynamical system (ordinary differential equation, or ODE):

$$\dot{\mathbf{x}}(t) = \mathbf{f}(\mathbf{x}(t), t), \quad \mathbf{x}(t_0) = \mathbf{x}_0, \tag{1.73}$$

where $t_0 \geq 0$, $\mathbf{x}(t) \in \mathbb{R}^n$, and $\mathbf{f} : \mathbb{R}^n \times \mathbb{R}_+ \to \mathbb{R}^n$ is continuous. Equation (1.73) is called *time invariant* or *autonomous* when the right-hand side does not depend on t:

$$\dot{\mathbf{x}} = \mathbf{f}(\mathbf{x}). \tag{1.74}$$

It is further assumed below that (1.73), (1.74) have unique solutions corresponding to each initial condition $\mathbf{x}_0$. In the time-invariant case, this happens, for example, if f satisfies a Lipschitz condition

$$\forall \mathbf{x}, \mathbf{y} \in B \subset \mathbb{R}^n, \|\mathbf{f}(\mathbf{x}) - \mathbf{f}(\mathbf{y})\| \leq \ell \|\mathbf{x} - \mathbf{y}\|.$$

The constant ℓ is known as a *Lipschitz constant* for f, and f is sometimes referred to as *Lipschitz continuous*. The terms *locally Lipschitz* and *globally Lipschitz* are used in the obvious manner to refer to the domain over which the Lipschitz condition holds. The Lipschitz property is stronger than continuity (which implies uniform continuity) but weaker than continuous differentiability. A simple global existence and uniqueness theorem in the time-invariant case is as follows.

Theorem 1.21. *Let* $\mathbf{f}(\mathbf{x})$ *be locally Lipschitz on a domain* $D \subset \mathbb{R}^n$, *and let* W *be a compact subset of* D. *Let the initial condition* $\mathbf{x}_0 \in W$ *and suppose it is known that every solution of* (1.74) *lies entirely in* W. *Then there is a unique solution that is defined for all* $t \geq 0$.

Consider the system (1.73), where f is locally Lipschitz, and suppose that $\mathbf{x}^* \in \mathbb{R}^n$ is an equilibrium point of (1.73); that is,

$$\mathbf{f}(\mathbf{x}^*, t) \equiv 0.$$

Since a change of variables can shift the equilibrium point to the origin, all definitions and theorems below are stated for this case. It is also common to speak of the properties of the *zero solution*, i.e., $\mathbf{x}(t) \equiv 0$ for all t.

Definition 1.22. *The equilibrium point* $\mathbf{x} = 0$ *(equivalently, the zero solution) of* (1.73) *is said to be*

(i) stable, *if for arbitrary* t_0 *and each* $\epsilon > 0$, *there is a* $\delta = \delta(\epsilon, t_0) > 0$ *such that* $\|\mathbf{x}(t_0)\| < \delta$ *implies* $\|\mathbf{x}(t)\| < \epsilon$ *for all* $t \geq t_0$.

The idea is that the entire trajectory stays close to zero if the initial condition is close enough to zero.

(ii) unstable, *if not stable.*

(iii) asymptotically stable, *if it is stable and if a convergence condition holds: For arbitrary* t_0, *there exists* $\delta_1(t_0)$ *such that* $\|\mathbf{x}(0)\| < \delta_1(t_0)$ *implies*

$$\lim_{t\to\infty} \mathbf{x}(t) = 0.$$

(iv) uniformly stable *and* uniformly asymptotically stable *if* δ *in* (i) *and* δ_1 *in* (ii) *can be chosen independently of* t_0.

(v) globally asymptotically stable *when* δ_1 *in* (iii) *can be taken arbitrarily large.*

(vi) exponentially stable *when, in addition to stability,* $\|\mathbf{x}(t_0)\| < \delta_1(t_0)$ *implies*

$$\|\mathbf{x}(t)\| \leq K\|\mathbf{x}(t_0)\| \exp[-\alpha(t - t_0)]$$

for some positive α *and* K.

Liapunov theorems for time-invariant systems

Let $V(\mathbf{x})$ be a real scalar function of $\mathbf{x} \in \mathbb{R}^n$ and let D be a closed bounded region in $\mathbb{R}^n$ containing the origin.

Definition 1.23. $V(\mathbf{x})$ *is* positive definite *(*semidefinite*) in D, denoted* $V > 0$ $(V \geq 0)$*, if* $V(0) = 0$*,* $V(\mathbf{x}) > 0$ $(V(\mathbf{x}) \geq 0)$ *for all* $\mathbf{x} \neq 0$ *in D.* $W(\mathbf{x})$ *is negative definite (negative semidefinite) if and only if* $-W(\mathbf{x})$ *is positive definite (positive semidefinite).*

Theorem 1.24. *Consider* (1.74) *and let* $V(\mathbf{x})$ *be a positive definite real-valued function defined on D, a closed bounded region containing the origin of* $\mathbb{R}^n$*. The zero solution of* (1.74) *is*

(i) stable *if* $\dot{V} = \nabla V^T \mathbf{f}(\mathbf{x}) \leq 0$ *(the derivative of V along the trajectories of* (1.74)).

(ii) asymptotically stable *if* $\dot{V}$ *(see item* (i)*) is negative definite or, alternatively,* $\dot{V}(\mathbf{x}) \leq 0$*, but* $\dot{V}$ *is not identically zero along any trajectory except* $\mathbf{x} \equiv 0$.

(iii) globally asymptotically stable *if in item* (ii)*,* $D = \mathbb{R}^n$ *and* $V(\mathbf{x}) \to \infty$ *as* $\|\mathbf{x}\| \to \infty$.

(iv) exponentially stable *if in item* (ii) *there holds* $\alpha_1 \|\mathbf{x}\|^2 \leq V(\mathbf{x}) \leq \alpha_2 \|\mathbf{x}\|^2$ *and* $-\alpha_3 \|\mathbf{x}\|^2 \leq \dot{V}(\mathbf{x}) \leq -\alpha_4 \|\mathbf{x}\|^2$ *for some positive* α_i.

A function $V(\mathbf{x})$ which allows a proof of a stability result using one of the items of this theorem is called a *Liapunov function.*

For the time-varying case (1.73), some modifications are needed. Consider real scalar functions $V(\mathbf{x}, t)$ of the vector $\mathbf{x} \in \mathbb{R}^n$ and time $t \in \mathbb{R}_+$, defined on a closed bounded region D containing the origin.

Definition 1.25. $V(\mathbf{x}, t)$ *is* positive definite *in D, denoted* $V > 0$*, if* $V(0, t) = 0$ *and there exists* $W(\mathbf{x})$ *with* $V(\mathbf{x}, t) \geq W(\mathbf{x})$ *for all* $\mathbf{x}, t$*, and* $W > 0$*.* $V(\mathbf{x}, t)$ *is* nonnegative definite *in D if* $V(0, t) = 0$ *and* $V(\mathbf{x}, t) \geq 0$ *for all* $\mathbf{x}, t$.

Observe that the derivative of V along the trajectories of (1.73) is given by

$$\dot{V} = \nabla V^T \mathbf{f}(\mathbf{x}, t) + (\partial V / \partial t).$$

With these changes, item (i) and the first part of item (ii) of Theorem 1.24 hold. If $V(\mathbf{x}, t) \leq W_1(\mathbf{x})$ for all t and some positive definite W_1, then uniformity holds in both cases. In item (iii), if $W(\mathbf{x}) \leq V(\mathbf{x}, t) \leq W_1(\mathbf{x})$ with $W(\mathbf{x}) \to \infty$ as $\|\mathbf{x}\| \to \infty$, then uniform global asymptotic stability holds. Item (iv) is valid as stated, without change.

LaSalle's invariance principle, also referred to as *LaSalle's theorem*, says that if one can find a function V such that $\dot{V}$ is negative definite and, in addition, it can be shown that no system trajectory stays forever at points where $\dot{V} = 0$, then the origin is asymptotically stable. To formalize this, the following definition is needed.

Definition 1.26. *A set M is said to be an* invariant set *with respect to* (1.74) *if* $\mathbf{x}(0)$ *in* M *implies* $\mathbf{x}(t)$ *in* M *for all* t *in* $\mathbb{R}$.

LaSalle's theorem is now stated.

Theorem 1.27. *Let D be a closed and bounded (compact) set with the property that every solution of* (1.74) *that starts in D remains for all future time in D. Let $V : D \to \mathbb{R}$ be a continuously differentiable function such that $\dot{V}(\mathbf{x}) \leq 0$ in D. Let E be the set of all points in D, where $\dot{V}(\mathbf{x}) = 0$. Let M be the largest invariant set in E. Then every solution starting in D approaches M as $t \to \infty$.* □

It is clear that the second part of Theorem 1.24 (ii) is a special case of LaSalle's theorem—in fact, a very important special case known as the Barbashin–Krasovskii theorem. LaSalle's theorem extends Liapunov's theorem in at least three important ways: First, the negative definite requirement of the latter is relaxed to negative semidefiniteness; second, it can be used when the system has an equilibrium set rather than an isolated equilibrium point; third, the function $V(\mathbf{x})$ does not have to be positive definite.

For a linear time-invariant system,

$$\dot{\mathbf{x}} = \mathbf{F}\mathbf{x}, \tag{1.75}$$

the basic stability result is that the origin or zero solution is stable if and only if each eigenvalue of the matrix $\mathbf{F}$ satisfies $\mathrm{Re}(\lambda_i(\mathbf{F})) \leq 0$ (where Re denotes the real part of a complex number) and eigenvalues that have real part equal to zero are semisimple. Global exponential stability of the zero solution holds if each eigenvalue of $\mathbf{F}$ has real part strictly negative, and the matrix $\mathbf{F}$ is then referred to as *Hurwitz stable*, or sometimes just as a *Hurwitz matrix*.

Application of the basic Liapunov stability theorem, using a quadratic Liapunov function $V(\mathbf{x}) := \mathbf{x}^T\mathbf{P}\mathbf{x}$, to the linear time-invariant system (1.75) leads to the following basic Liapunov theorem.

Theorem 1.28. *The origin or zero solution of the linear time-invariant system* (1.64) *is exponentially stable if and only if there exists a positive definite matrix* $\mathbf{P}$ *such that the equation*

$$\mathbf{F}^T\mathbf{P} + \mathbf{P}\mathbf{F} = -\mathbf{Q} \tag{1.76}$$

is satisfied for some positive definite matrix $\mathbf{Q}$.

Equation (1.65) is known as the *Liapunov equation* or the *continuous-time Liapunov equation*; if it is satisfied, then $\mathrm{Re}(\lambda_i(\mathbf{F})) \leq 0$ and $\mathbf{F}$ is a Hurwitz stable matrix. Finally, if (1.76) is satisfied for a positive diagonal matrix $\mathbf{P}$, then $\mathbf{F}$ is referred to as a *(Hurwitz) diagonally stable matrix*.

Control Liapunov functions and Liapunov optimizing control

These forty years now, I've been speaking in prose without knowing it!

—M. Jourdain in Molière's *The Bourgeois Gentleman*

The quote above summarizes what might be said by most practitioners of control when confronted with the terms *control Liapunov function* and *Liapunov optimizing control*. These concepts have been in use essentially since the earliest days [KB60]; however, we believe that the advantage of using these terms is that the power of Liapunov theory in control design problems is made apparent, since Liapunov theory is well known as a stability analysis tool, but relatively less well known as a tool to be used in the design of systems.

The concept of a control Liapunov function (CLF) is useful in order to provide a framework for many developments that occur in other chapters of this book. It is defined below for the discrete-time case only, since the continuous-time case is completely analogous.

Definition 1.29. *Consider the dynamical system*

$$\mathbf{x}_{k+1} = \Phi(\mathbf{x}_k, \mathbf{u}_k), \tag{1.77}$$

where $\mathbf{x} \in \mathbb{R}^n$*, the control input* $\mathbf{u}$ *is a vector in* $\mathbb{R}^{n_i}$*, and the function* $\Phi : \mathbb{R}^n \times \mathbb{R}^{n_i} \to \mathbb{R}^n$ *is smooth in both arguments with* $\Phi(\mathbf{0}, \mathbf{0}) = \mathbf{0}$*. Consider also a* C^1 *proper function* $V : \mathbb{R}^n - \{\mathbf{0}\} \to \mathbb{R}^+$*, with* $V(\mathbf{0}) = 0$*, which, for all* $\mathbf{x}_k \in \mathbb{R}^n - \{\mathbf{0}\}$*, satisfies*

$$\Delta V := V(\mathbf{x}_{k+1}) - V(\mathbf{x}_k) = V(\Phi(\mathbf{x}_k, \mathbf{u}(\mathbf{x}_k))) - V(\mathbf{x}_k) < 0 \tag{1.78}$$

for suitable values of the control input $\mathbf{u}(\mathbf{x}_k) \in \mathbb{R}^{n_i}$*. Such a function* $V(\cdot)$ *is called a* control Liapunov function *for system* (1.77).

In order to have the stabilizing control given in terms of state feedback, it is also desirable to compute, if possible, a smooth function $\mathbf{G}(\mathbf{x}) : \mathbb{R}^n - \{\mathbf{0}\} \to \mathbb{R}^{n_i}$ (with $\mathbf{G}(\mathbf{0}) = \mathbf{0}$) such that

$$\mathbf{u}_k = -\mathbf{G}(\mathbf{x}_k) \tag{1.79}$$

globally asymptotically stabilizes the zero solution of (1.77), with a specified rate of convergence. In other words, the control Liapunov function is used as a tool to find the appropriate stabilizing state feedback. For more on control Liapunov functions, see [Son89, AMNC97, Son98].

The concept of *Liapunov optimizing control* is to use the first term on the right-hand side of (1.78), in which the control $\mathbf{u}(\cdot)$ occurs, to make the decrement ΔV as negative as possible. The intuitive justification for this is that the more negative the decrement is, the faster the system will stabilize (i.e., reach the equilibrium). This idea is very old, dating back at least to the seminal pair of papers by Kalman and Bertram [KB60], but seems to have been given the descriptive and, in our view, empowering, name much more recently [VG97].

Sector nonlinearities, Lur'e systems, Persidskii systems, and absolute stability

An important class of nonlinear systems, first defined and intensively studied by Lur'e, Aizerman, and coworkers is a class of feedback systems in which, in the scalar case, the only nonlinearity is located in the feedback loop (see Figure 1.4) and has the property that its graph lies in the sector enclosed by straight lines of positive slope $k_2 > k_1 > 0$. If this property holds, the nonlinearity is said to be a sector nonlinearity, belonging to the

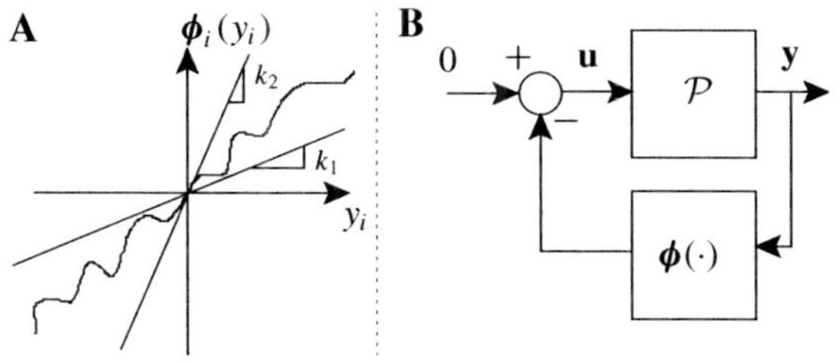

Figure 1.4. *A linear system* $\mathcal{P} = \{\mathbf{F}, \mathbf{G}, \mathbf{H}, \mathbf{J}\}$ *with a sector nonlinearity* $\boldsymbol{\phi}(\cdot)$ (fig. A) *in the feedback loop is often called a Lur'e system* (fig. B).

sector $[k_1, k_2]$; if $k_1 = 0$ and $k_2 = \infty$, then the nonlinearity is said to be a first-quadrant–third-quadrant (or, sometimes, infinite sector) nonlinearity. It is the latter type that will be considered in this book.

The *Lur'e* or *absolute stability problem* can be stated as follows. Given the system $\mathcal{P} = \{\mathbf{F}, \mathbf{G}, \mathbf{H}, \mathbf{J}\}$, with the pair $(\mathbf{F}, \mathbf{G})$ controllable and the pair $(\mathbf{H}, \mathbf{F})$ observable, with a sector nonlinearity $\boldsymbol{\phi}(\cdot)$ in the sector $[k_1, k_2]$ in the feedback loop (Figure 1.4A), determine conditions on the transfer function matrix $\mathbf{M}(s) := \mathbf{H}(s\mathbf{I} - \mathbf{F})^{-1}\mathbf{G} + \mathbf{J}$ and the numbers k_1, k_2, such that the origin is globally asymptotically stable for all functions $\boldsymbol{\phi}(\cdot)$ belonging to the sector $[k_1, k_2]$.

The solution of this problem involves the concepts of positive real functions, passivity, and the celebrated Popov–Yakubovich–Kalman lemma, for which we refer the reader to [Vid93, Kha02]. We note that positive real functions are briefly discussed in Chapter 5.

Persidskii systems and diagonal-type functions were first discussed in [Per69], where absolute stability of such systems was proved using diagonal-type Liapunov functions. More general classes of such systems are treated in detail in the monograph [KB00].

Definition 1.30. *A function* $\mathbf{f} : \mathbb{R}^n \to \mathbb{R}^n$ *is said to be* diagonal *or, alternatively, of* diagonal type *if* f_i *is a function only of the argument* x_i, *i.e.,*

$$\mathbf{f}(\mathbf{x}) = \begin{bmatrix} f_1(x_1) \\ \vdots \\ f_n(x_n) \end{bmatrix}.$$

Definition 1.31. *A system of ODEs is said to be of Persidskii type if it is of the form*

$$\dot{x}_i = \sum_{j=1}^{n} b_{ij} f_j(x_j(t)) \qquad (i = 1, \ldots, n), \tag{1.80}$$

where, for each i, j, $f_j : \mathbb{R} \to \mathbb{R}$ *belongs to the class of infinite sector nonlinearities* $S = \{f_j(\cdot) : \psi f_j(\psi) > 0;\ f_j(0) = 0;\ \int_0^x f_j(\tau)d\tau \to \infty$ *when* $|x_j| \to \infty\}$.

System (1.80) can be written in vector notation as follows:

$$\dot{\mathbf{x}} = \mathbf{B}\mathbf{f}(\mathbf{x}), \tag{1.81}$$

where $\mathbf{B} \in \mathbb{R}^{n\times n}$, $\mathbf{x} \in \mathbb{R}^n$, and $\mathbf{f}(\cdot)$ is a diagonal function that belongs to the class $S^n = S \times S \times \cdots \times S$.

The basic stability result for a Persidskii system (1.81) is as follows.

Theorem 1.32. *The zero solution of* (1.81) *is globally asymptotically stable for all* $\mathbf{f}(\cdot) \in S^n$ *if the matrix* $\mathbf{B}$ *is Hurwitz diagonally stable.*

Observe that the theorem is an absolute stability type of result (also called a robust stability result), since it establishes the stability of an entire class of systems. The Persidskii diagonal-type Liapunov function that is used in the proof of Theorem 1.32 is defined as follows. Let $\mathbf{p}$ be a vector with all components positive and $\mathbf{P}$ be a diagonal matrix with components of the diagonal equal to that of the vector $\mathbf{p}$. Define

$$\boldsymbol{\psi}_S(\mathbf{x}) := \left(\int_0^{x_1} \psi_1(\tau)\mathrm{d}\tau, \ldots, \int_0^{x_n} \psi_n(\tau)\mathrm{d}\tau \right).$$

The Persidskii diagonal-type Liapunov function is then defined as

$$V(\mathbf{x}) = 2\mathbf{p}^T \boldsymbol{\psi}_S(\mathbf{x}), \tag{1.82}$$

and $\dot{V}$ is negative definite, because $\mathbf{B}$ satisfies (1.76) with positive diagonal matrix $\mathbf{P}$, i.e., is diagonally stable.

Three first-quadrant–third-quadrant or infinite sector nonlinearities, the signum (denoted sgn), the half signum (denoted hsgn), and the upper half signum (denoted uhsgn) that occur frequently in this book are described below.

For $x \in \mathbb{R}$, the signum function is defined as

$$\mathrm{sgn}(x) = \begin{cases} 1 & \text{if } x > 0, \\ -1 & \text{if } x < 0. \end{cases} \tag{1.83}$$

Note that this function is not defined at $x = 0$. By abuse of notation, however, we will also use the notation sgn to denote the following relation (no longer a function, since it is not uniquely defined for $x = 0$):

$$\mathrm{sgn}(x) = \begin{cases} 1 & \text{if } x > 0, \\ [-1,\ 1] & \text{if } x = 0, \\ -1 & \text{if } x < 0. \end{cases} \tag{1.84}$$

Note that the signum function defined in (1.84) can be thought of as the *subdifferential* (see Definition 1.33) of the absolute value function $|x|$, which has derivative $+1$ for $x > 0$, derivative -1 for $x < 0$ and, for $x = 0$, all its *subgradients* lie in the interval $[-1, 1]$. In general, no confusion is caused by using the same notation for the function and the relation, although some authors use Sgn for the latter. In a similar fashion, the half signum function/relation is defined as

$$\mathrm{hsgn}(x) = \begin{cases} 0 & \text{if } x > 0, \\ [-1,\ 0] & \text{if } x = 0, \\ -1 & \text{if } x < 0. \end{cases} \tag{1.85}$$

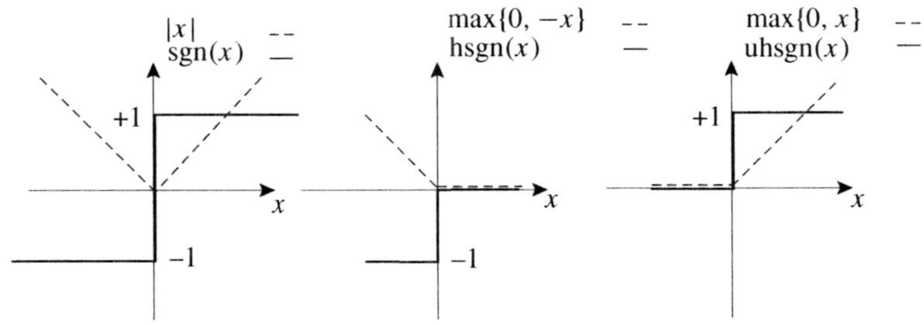

Figure 1.5. *The signum (sgn(x)), half signum (hsgn(x)) and upper half signum (uhsgn(x)) relations (solid lines) as subdifferentials of, respectively, the functions* $|x|$, $\max\{0, -x\} = -\min\{0, x\}$, $\max\{0, x\}$ *(dashed lines).*

Notice that the relation hsgn is the subdifferential of the function $-\min\{0, x\}$. Finally, the upper half signum function/relation uhsgn is defined as

$$\text{uhsgn}(x) = \begin{cases} 1 & \text{if } x > 0, \\ [0,\ 1] & \text{if } x = 0, \\ 0 & \text{if } x < 0 \end{cases} \tag{1.86}$$

and is the subdifferential of the function $\max\{0, x\}$. These three nonlinearities, depicted in Figure 1.5, can all be thought of as belonging to the first and third quadrants in the limit.

Finally, if the argument of any one of these three functions is a vector, then the function is to be understood as applying to each of the components of the vector argument. For example, for $\mathbf{x} \in \mathbb{R}^n$, $\text{sgn}(\mathbf{x}) := (\text{sgn}(x_1), \dots, \text{sgn}(x_n))$.

1.5 Variable Structure Control Systems

A variable structure dynamical system is one whose structure (the right-hand side of an ODE) changes in accordance with the current value of its state. Thus, loosely speaking, a variable structure system can be thought of as a set of dynamical systems together with a state-dependent switching logic between each of them. The idea is to choose the logic in such a way that the overall system with switching combines the desirable properties of the individual systems of which it is composed. A remarkable feature of a variable structure system is that a property, such as stability, may emerge in it even though it is not a property of any of the systems involved in its composition.

To make this description concrete, we give two classic examples from Utkin [Utk78].

Examples of variable structure systems

In the first example, consider a second-order harmonic oscillator, i.e.,

$$\ddot{x} = -\Psi x, \quad x \in \mathbb{R}, \tag{1.87}$$

where Ψ takes one of two values according to the switching logic

$$\Psi = \begin{cases} \omega_1^2 & \text{if } x\dot{x} > 0, \\ \omega_2^2 & \text{if } x\dot{x} < 0, \end{cases} \tag{1.88}$$

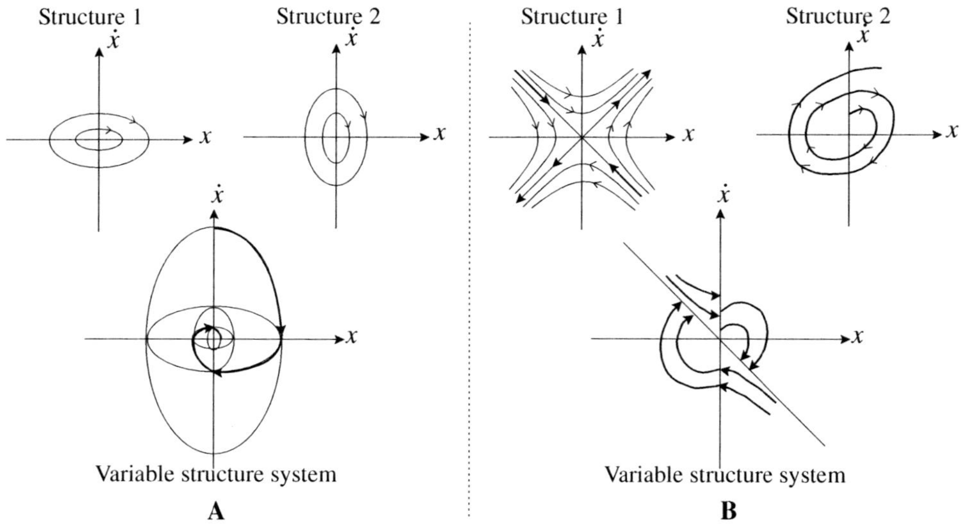

Figure 1.6. *Example of* (i) *an asymptotically stable variable structure system that results from switching between two stable structures (systems)* (Fig. A); (ii) *an asymptotically stable variable structure system that results from switching between two unstable systems* [Utk78] (Fig. B).

and $\omega_1^2 > \omega_2^2$. Observe that each value of Ψ defines a subsystem (1.87) that is stable, but not asymptotically stable. Figure 1.6A shows that, with the switching logic (1.88), the resulting variable structure system is asymptotically stable. In other words, switching between two linear systems, each of which has an equilibrium that is a center, can produce a system that has a globally asymptotically stable equilibrium.

The second example concerns the system

$$\ddot{x} - \xi\dot{x} + \Psi x = 0, \qquad \xi > 0, \tag{1.89}$$

where

$$\Psi = \begin{cases} \alpha & \text{if } xs > 0, \\ -\alpha & \text{if } xs < 0, \end{cases} \quad \text{where } s = cx + \dot{x}, \text{ and} \tag{1.90}$$

$$c = -\frac{\xi}{2} \pm \sqrt{\frac{\xi^2}{4} + \alpha}.$$

For this example, switching takes place across a line through the origin in the phase plane. This defines a linear system that has an equilibrium at the origin which is an unstable focus on one side of the switching line, and a linear system that has an equilibrium which is a saddle point on the other side of this line. Figure 1.6B shows that, with the switching logic (1.90), the resulting variable structure system is asymptotically stable.

A notable feature of the two examples above is that the switching is defined with respect to some manifold on which it is discontinuous. This means that such variable structure systems are dynamical systems with discontinuous right-hand sides and that care must be taken with respect to existence and uniqueness of solutions of such systems.

Another important aspect of variable structure systems is brought out by assuming, for simplicity, that there is only one switching surface. If this is the case, then adequate design of the switching surface can lead to the new dynamical behavior alluded to above as follows. If the switching surface is locally attractive, then all nearby trajectories converge to it. Thenceforth, if all trajectories that attain the switching surface are shown to remain on it, then, constrained to the surface, the dynamical system has reduced order, and this reduced-order system can have new properties vis-à-vis the component systems. The motion of the system during its confinement to the switching surface is referred to as the *sliding phase*. Alternatively, it is said that the system is in *sliding mode*.

Reaching phase, sliding phase, and sliding mode

If the vector fields in the neighborhood of the switching surface are directed towards it, then it becomes locally attractive, and moreover, once a trajectory attains or intersects the switching surface, then it remains on it thereafter. In the literature on variable structure control, the terms *reaching phase* and *sliding phase* are used to describe, respectively, the intervals in which the trajectories attain the switching surface and subsequently the interval in which trajectories remain on the switching surface, which is a lower dimensional submanifold of the original state space.

Let $s(\mathbf{x}) = 0$ describe the switching surface. For the reaching phase to occur, a simple sufficient condition is to ensure that the switching surface is attractive, which can be done using the Liapunov function $V(s) = (1/2)s^2$. Then $\dot{V}(s) = s\dot{s}$ and thus the reaching phase occurs, followed by a sliding phase, if, in the neighborhood of the switching surface $s(\mathbf{x}) = 0$,

$$s\dot{s} < 0. \tag{1.91}$$

Desiderata for Filippov solutions

Motivated by applicability to a broad class of ODEs with discontinuous right-hand sides that arise in practice, Filippov [Fil88] proposed some reasonable mathematical desiderata for a new solution concept.

1. If the right-hand side is continuous, then the Filippov solution should reduce to the usual solution.

2. For the equation $\dot{\mathbf{x}} = \mathbf{f}(t)$, the Filippov solution should be of the form

$$\mathbf{x} = \int \mathbf{f}(t)dt + \text{constant}.$$

3. For any initial condition $\mathbf{x}(t_0) = \mathbf{x}_0$, the Filippov solution must exist at least for $t > t_0$ and should admit continuation.

4. The limit of a uniformly convergent sequence of Filippov solutions should be a Filippov solution.

5. Changes of variable should not affect the property of being a Filippov solution.

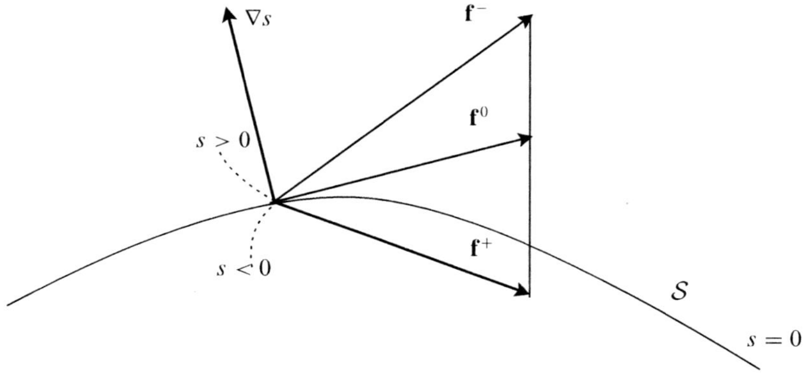

Figure 1.7. *Pictorial representation of the construction of a Filippov solution in* $\mathbb{R}^2$.

Description of Filippov solution

For simplicity, consider a single input dynamical system $\dot{\mathbf{x}} = \mathbf{f}(\mathbf{x}, u)$ and a switching surface $s(\mathbf{x}) = 0$. Assume that the input u is defined as follows

$$u = \begin{cases} u^- & \text{if } s < 0, \\ u^+ & \text{if } s > 0. \end{cases} \tag{1.92}$$

Given these two inputs, define the vector fields $\mathbf{f}^- := \mathbf{f}(\mathbf{x}, u^-)$ and $\mathbf{f}^+ := \mathbf{f}(\mathbf{x}, u^+)$. Now, at a given point $\mathbf{x}$ on the switching surface, join the vectors $\mathbf{f}^-$ and $\mathbf{f}^+$. The resultant vector field at $\mathbf{x}$, denoted $\mathbf{f}^0$, is obtained by drawing the line tangent to the switching surface that intersects the line segment joining the velocity vectors $\mathbf{f}^-$ and $\mathbf{f}^+$ (see Figure 1.7):

$$\mathbf{f}^0 = \alpha\mathbf{f}^- + (1-\alpha)\mathbf{f}^+, \quad \alpha \in [0, 1]. \tag{1.93}$$

Clearly, $\mathbf{f}^0$ belongs to the smallest convex set containing $\mathbf{f}^-$ and $\mathbf{f}^+$.

This construction has a simple interpretation [Itk76, Utk78]. If one considers that sliding motion actually occurs as a sort of "limiting process" of an oscillation in a neighborhood of the switching surface $s = 0$ in a "small" time period Δt and interprets the number α as the fraction of this time that the trajectory spends below the switching surface, then $(1 - \alpha)$ is the time spent above it. The average vector field in this time period is then given by $\mathbf{f}^0$ as in (1.93).

The following definition from nonsmooth analysis is useful in order to give an alternative description of Filippov solutions.

Definition 1.33. *Consider a convex function* $F : \mathbb{R}^n \to \mathbb{R}$. *Then the* subdifferential *of* F *in* $\mathbf{x}_0 \in \mathbb{R}^n$ *is the set defined by*

$$\partial F(\mathbf{x}_0) := \{\mathbf{f} : F(\mathbf{x}) - F(\mathbf{x}_0) \geq \mathbf{f}^T(\mathbf{x} - \mathbf{x}_0)\},$$

and any vector $\mathbf{f} \in \partial F(\mathbf{x}_0)$ *is called a* subgradient *of* F *at the point* $\mathbf{x}_0$.

The set $\partial F(\mathbf{x}_0)$ is compact, closed, and convex [DV85]. Moreover, if $F(\cdot)$ is continuous in $\mathbf{x}_0$, then the set $\partial F(\mathbf{x}_0)$ has only one element, which is the gradient of F at $\mathbf{x}_0$.

Further details and properties of subdifferentials and subgradients of convex functions can be found, for instance, in [Cla83, DV85, SP94, CLSW98].

According to Filippov's solution concept, when the trajectories of (4.46) are not confined to the surface of discontinuity, the usual definition of solutions of differential equations holds. Otherwise, the solutions of (4.46) are absolutely continuous vector functions $\mathbf{x}(t)$, where its components $x_i(t)$ are defined in intervals $\mathcal{I}_i$, such that for almost all t in $\mathcal{I}_i$ the *differential inclusion* $\dot{\mathbf{x}} \in -\partial E(\mathbf{x})$ is satisfied.

Equivalent control

A simple alternative approach to that of Filippov, introduced by Utkin [Utk92], is that of equivalent control. Broadly speaking, equivalent control is the control input required to maintain an ideal sliding motion on the *sliding manifold* $\mathcal{S}$. In order to illustrate this in a simple case, consider a linear system

$$\dot{\mathbf{x}}(t) = \mathbf{F}\mathbf{x}(t) + \mathbf{G}\mathbf{u}(t).$$

Suppose that at time t_r (the reaching time), the state vector $\mathbf{x}(t_r)$ reaches the surface $\mathcal{S}$. This is expressed mathematically as $\mathbf{S}\mathbf{x}(t) = 0$ for all $t \geq t_r$. Differentiating this expression and substituting the system dynamics leads to

$$\mathbf{S}\dot{\mathbf{x}} = \mathbf{S}(\mathbf{F}\mathbf{x}(t) + \mathbf{G}\mathbf{u}(t)) \equiv 0 \text{ for all } t \geq t_r.$$

Assuming that the matrix $\mathbf{SG}$ is square and nonsingular, the unique solution to the above equation defines the equivalent control as follows:

$$\mathbf{u}_{\mathrm{eq}} = -(\mathbf{SG})^{-1}\mathbf{SF}\mathbf{x}(t). \tag{1.94}$$

Similar formal procedures allow the calculation of equivalent controls in the general nonlinear case; the reader should consult [Utk92] for further details.

1.6 Gradient Dynamical Systems

This section collects some of the basic results on the particular class of dynamical systems known as gradient dynamical systems (GDSs).

Smooth GDSs

GDSs have special properties that make their analysis simple, and furthermore, they occur in many applications. This book is no exception.

A GDS, defined on an open set $\mathcal{W} \subset \mathbb{R}^n$, is defined as

$$\dot{\mathbf{x}} = -\nabla V(\mathbf{x}), \tag{1.95}$$

where

$$V : \mathcal{W} \rightarrow \mathbb{R}$$

is a C^2 function and

$$\nabla V = \left(\frac{\partial V}{\partial x_1}, \ldots, \frac{\partial V}{\partial x_n} \right)$$

is the *gradient vector field* (also sometimes written as grad V)

$$\nabla V : \mathcal{W} \to \mathbb{R}^n.$$

The main property of the flow of a gradient vector field is its simplicity, in a sense to be made precise. Let the time derivative of V along the trajectories of (1.95) be denoted as $\dot{V}(\mathbf{x})$; i.e.,

$$\dot{V}(\mathbf{x}) = \frac{\mathrm{d}}{\mathrm{d}t} V(\mathbf{x}(t)).$$

The first basic result about a GDS is expressed in the following theorem.

Theorem 1.34. *The time derivative of V along the trajectories of* (1.95) *is nonpositive for all* $\mathbf{x} \in \mathcal{W}$, *and* $\dot{V}(\mathbf{x}) = 0$ *if and only if* $\mathbf{x}$ *is an equilibrium of* (1.95).

Proof. By the chain rule,

$$\dot{V}(\mathbf{x}) = \nabla V(\mathbf{x})^T \dot{\mathbf{x}} = -\nabla V(\mathbf{x})^T \nabla V(\mathbf{x}) = -\|\nabla V(\mathbf{x})\|_2^2 < 0,$$

which is negative definite, proving the theorem. □

An important corollary is as follows.

Corollary 1.35. *Let* $\overline{\mathbf{x}}$ *be a minimum of the real-valued function V and furthermore suppose that it is an isolated zero of* ∇V. *Then* $\overline{\mathbf{x}}$ *is an asymptotically stable equilibrium point of the GDS* (1.95).

Proof. It is straightforward to check that, in some neighborhood $\mathcal{N}$ of $\overline{\mathbf{x}}$, the function

$$V_1(\mathbf{x}) := V(\mathbf{x}) - V(\overline{\mathbf{x}}) \tag{1.96}$$

is a Liapunov function for $\overline{\mathbf{x}}$, strictly positive for all $\mathbf{x} \in \mathcal{N}$, such that $\mathbf{x} \neq \overline{\mathbf{x}}$. □

Thus the function V, modulo the constant value $V(\overline{\mathbf{x}})$, is a natural Liapunov function for the GDS (1.95) that it defines. It is also referred to as a *potential function* or as an *energy function*.

From a geometrical point of view, GDSs are also easily described. *Level sets* of the function $V : \mathcal{W} \to \mathbb{R}$ are defined as the subsets $\{V^{-1}(c), c \in \mathbb{R}\}$. If $\mathbf{w} \in V^{-1}(c)$ is such that $\nabla V(\mathbf{w}) \neq \mathbf{0}$, then $\mathbf{w}$ is referred to as a *regular point*; otherwise, if $\nabla V(\mathbf{w}) = \mathbf{0}$, then $\mathbf{w}$ is called a *critical point*. Critical points are clearly the equilibria of the GDS (1.95). By the implicit function theorem, it can be seen that, near a regular point, $V^{-1}(c)$ looks like the graph of a function, and moreover, the tangent plane to this graph has $\nabla V(\mathbf{w})$ as its normal vector [HS74]. This geometric information can be summarized in the following theorem (which sums up most of what will be needed in what follows regarding GDSs).

Theorem 1.36. *At regular points, the trajectories of the GDS* (1.95) *cross level surfaces orthogonally. Critical points are equilibria of the GDS* (1.95). *Minima that are isolated as critical points are asymptotically stable.*

To introduce one more important property of gradient flows, the notions of α- and ω-limit sets, from the general theory of dynamical systems, are now defined.

Consider the dynamical system (1.74) and let the *ω-limit set* of the trajectory $\mathbf{x}(t)$ (or any point on the trajectory) be denoted L_ω and be defined as the following subset of $\mathbb{R}^n$:

$$L_\omega := \{\mathbf{p} \in \mathbb{R}^n : \text{there exists a sequence } t_n \to \infty, \text{ such that } \mathbf{x}(t_n) \to \mathbf{p}\}. \tag{1.97}$$

Similarly the *α-limit set*, denoted L_α, of a trajectory $\mathbf{x}(t)$ is the set of all points $\mathbf{q}$ such that $\lim_{n\to\infty} \mathbf{x}(t_n) = \mathbf{q}$, for some sequence $t_n \to -\infty$. A set A in the domain of a dynamical system is called *invariant* if, for every $\mathbf{x} \in A$, the trajectory that starts from $\mathbf{x}$ remains in A for all $t \in \mathbb{R}$.

A fundamental fact about dynamical systems is that the α- and ω-limit sets of a trajectory are closed invariant sets. In terms of these definitions, the following theorem can be stated.

Theorem 1.37. *Let $\mathbf{z}$ be an α- or ω-limit point of a trajectory of a GDS* (1.95). *Then $\mathbf{z}$ is an equilibrium of the GDS.*

Extensions of gradient systems

A central theme in many recent developments in optimization and numerical analysis is that a given function has a variety of gradients with very different numerical and analytical properties, depending on the choice of a metric (see [Neu05, Neu97] and references therein).

We limit ourselves here to the elementary case which will occur in several places in the book. Recall that, given a function $\mathbf{f} : \mathbb{R}^n \to \mathbb{R}$, a property (or equivalent definition) of the gradient of $\mathbf{f}$, denoted $\nabla\mathbf{f}$, is that $\nabla\mathbf{f}(\mathbf{x})$ is the element of $\mathbb{R}^n$ such that

$$\mathbf{f}'(\mathbf{x})\mathbf{h} = \langle \mathbf{h}, \nabla\mathbf{f}(\mathbf{x})\rangle, \quad \mathbf{h} \in \mathbb{R}^n, \tag{1.98}$$

where

$$\mathbf{f}'(\mathbf{x})\mathbf{h} := \lim_{t\to 0}(\mathbf{f}(\mathbf{x} + t\mathbf{h}) - \mathbf{f}(\mathbf{x}))/t$$

and $\langle\cdot,\cdot\rangle$ denotes the standard inner product on $\mathbb{R}^n$. Recall also that a symmetric positive definite matrix $\mathbf{A} \in \mathbb{R}^{n\times n}$ defines an inner product, denoted $\langle \mathbf{x}, \mathbf{y}\rangle_{\mathbf{A}}$, as follows:

$$\langle \mathbf{x}, \mathbf{y}\rangle_{\mathbf{A}} := \langle \mathbf{A}\mathbf{x}, \mathbf{y}\rangle, \quad \mathbf{x}, \mathbf{y} \in \mathbb{R}^n. \tag{1.99}$$

The following natural question then leads to the concept of a *Sobolev gradient*. What is the gradient that results if the standard inner product in (1.98) is substituted by the inner product in (1.99)? In other words, given $\mathbf{x} \in \mathbb{R}^n$, it is necessary to find the element $\nabla_{\mathbf{A}}\mathbf{f}(\mathbf{x})$ that gives the identity

$$\mathbf{f}'(\mathbf{x})\mathbf{h} = \langle \mathbf{h}, \nabla_{\mathbf{A}}\mathbf{f}(\mathbf{x})\rangle_{\mathbf{A}}, \quad \mathbf{h} \in \mathbb{R}^n. \tag{1.100}$$

By calculation, for $\mathbf{x} \in \mathbb{R}^n$, $\mathbf{f}'(\mathbf{x})\mathbf{h} = \langle \mathbf{h}, \nabla_{\mathbf{A}}\mathbf{f}(\mathbf{x})\rangle_{\mathbf{A}} = \langle \mathbf{h}, \mathbf{A}\nabla_{\mathbf{A}}\mathbf{f}(\mathbf{x})\rangle$ for all $\mathbf{h} \in \mathbb{R}^n$. This implies that $\mathbf{A}\nabla_{\mathbf{A}}\mathbf{f}(\mathbf{x}) = \nabla\mathbf{f}(\mathbf{x})$, and hence

$$\nabla_{\mathbf{A}}\mathbf{f}(\mathbf{x}) = \mathbf{A}^{-1}\nabla\mathbf{f}(\mathbf{x}), \quad \mathbf{x} \in \mathbb{R}^n, \tag{1.101}$$

since $\mathbf{f}'(\mathbf{x})\mathbf{h} = \langle \mathbf{h}, \nabla\mathbf{f}(\mathbf{x})\rangle$, $\mathbf{x}, \mathbf{h} \in \mathbb{R}^n$. Neuberger [Neu05] refers to the gradient $\nabla_{\mathbf{A}}\mathbf{f}(\mathbf{x})$ as a *Sobolev gradient*.

Observe that, now taking $\mathbf{f}$ to be V as in (1.95), the dynamical system

$$\dot{\mathbf{x}} = -\mathbf{A}^{-1}\nabla V(\mathbf{x}) \tag{1.102}$$

can be written as a GDS that uses the Sobolev gradient, instead of the standard gradient, i.e.,

$$\dot{\mathbf{x}} = -\nabla_{\mathbf{A}} V(\mathbf{x}). \tag{1.103}$$

Note that the function $V(\cdot)$, possibly modulo a constant as in (1.96), is still a Liapunov function for the system (1.102). In fact, some authors [Ryb74] refer to (1.102) as a *quasigradient system*, while others [MQR98] call it a *linear gradient system*, but we will not use this terminology, and instead refer to both (1.102) and (1.103) as GDSs, with the understanding that, in the latter case, we mean that a Sobolev gradient is being thought of implicitly.

Gradient stabilization of a control system

Given a control system

$$\dot{\mathbf{x}} = \mathbf{f}(\mathbf{x}, \mathbf{u}), \tag{1.104}$$

one could ask for a feedback control $\mathbf{u}(\mathbf{x})$ that, when substituted into (1.104), leads to a gradient system of the form (1.95) or (1.102) for some choice of Liapunov function $V(\cdot)$ and, possibly, matrix $\mathbf{A}$ in the latter case. If this can be done, then from the preceding discussion, it is clear that such a feedback control stabilizes system (1.104) and that the function $V(\cdot)$ is a Liapunov function stability certificate for the resulting closed-loop system.

To fix ideas, consider the special case of a linear system

$$\dot{\mathbf{x}} = \mathbf{F}\mathbf{x} + \mathbf{G}\mathbf{u} \tag{1.105}$$

in which both $\mathbf{F}$ and $\mathbf{G}$ are real, square matrices of dimension n, and furthermore, $\mathbf{G}$ is invertible. The choice of state feedback

$$\mathbf{u} = -\mathbf{K}\mathbf{x} \tag{1.106}$$

results in the closed-loop system

$$\dot{\mathbf{x}} = (\mathbf{F} - \mathbf{G}\mathbf{K})\mathbf{x}. \tag{1.107}$$

Suppose that we want to choose $\mathbf{u}$, equivalently $\mathbf{K}$, such that the closed-loop system (1.107) becomes a gradient system for which the potential or Liapunov function is prespecified as

$$V(\mathbf{x}) = \frac{1}{2}\mathbf{x}^T\mathbf{P}\mathbf{x}, \tag{1.108}$$

where $\mathbf{P}$ is a symmetric positive definite matrix. Observe that this can always be done, since the choice

$$\mathbf{K} = \mathbf{G}^{-1}(\mathbf{F} + \mathbf{P}) \tag{1.109}$$

results in the GDS

$$\dot{\mathbf{x}} = -\mathbf{P}\mathbf{x} = -\nabla V(\mathbf{x}), \tag{1.110}$$

which is clearly stable, justifying the terminology gradient stabilizing control for the state feedback control defined above. The abbreviated term *gradient control* will also be used in Chapter 2. Gradient stabilization is discussed further in [VG97], on which this subsection is based.

1.6.1 Nonsmooth GDSs: Persidskii-type results

A generalized Persidskii-like theorem is outlined for applications to the stability analysis of a class of GDSs with discontinuous right-hand sides. These dynamical systems arise from the steepest descent technique applied to a variety of problems suitably formulated as constrained minimization problems, as will be seen in Chapter 4.

A class of Persidskii systems with discontinuous right-hand sides is analyzed in [HKB00], and a generalization of this class, with the corresponding stability result, is given in this section. Consider the generalized Persidskii-type system

$$\dot{\mathbf{x}}(t) = -\mathbf{A}\mathbf{f}(\mathbf{x}(t)), \tag{1.111}$$

where $\mathbf{x} = (\bar{\mathbf{x}}^T, \hat{\mathbf{x}}^T)^T$, $\bar{\mathbf{x}} \in \mathbb{R}^p$, $\hat{\mathbf{x}} \in \mathbb{R}^q$, $\mathbf{f}(\mathbf{x}) = (\bar{\mathbf{x}}^T, \mathbf{g}^T(\hat{\mathbf{x}}))^T$, $\mathbf{A} \in \mathbb{R}^{n\times n}$, $n = p + q$, and the vector function $\mathbf{g} : \mathbb{R}^q \to \mathbb{R}^q$ satisfies the following assumptions:

(i) $\mathbf{g}(\hat{\mathbf{x}})$ is a piecewise continuous diagonal-type function,

$$\mathbf{g}(\hat{\mathbf{x}}) = (g_1(\hat{x}_1), \ldots, g_q(\hat{x}_q))^T; \tag{1.112}$$

(ii) $\hat{x}_i g_i(\hat{x}_i) \geq 0$, $i = 1, \ldots, q$;

(iii) $\mathbf{g}$ is continuous almost everywhere (i.e., the points at which it is discontinuous form a set M of Lebesgue measure zero).

Furthermore, when the g_i's are chosen as hsgn, uhsgn, and sgn functions, the set

$$M := \bigcap_i \{\mathbf{x} : g_i(\hat{x}_i) = 0 \forall\, i\} \tag{1.113}$$

is described by the intersection of surfaces, which is referred to as a *surface of discontinuity*. Since system (1.111) has discontinuous right-hand side, its solutions must be considered in the sense of Filippov [Fil88]. According to Filippov's theory, when trajectories are not confined to the surface of discontinuity, the solutions are considered in the usual sense; otherwise the solutions of (1.111) are the solutions of the following differential inclusion:

$$\dot{\mathbf{x}}(t) \in G(\mathbf{x}(t)), \tag{1.114}$$

where $\mathbf{x}$ is absolutely continuous, defined almost everywhere within an interval, and the set G is described as the convex hull containing all the limiting values of $g(\mathbf{x})$, when $\mathbf{x} \to \mathbf{x}' \in M$. The set G can also be defined using the equivalent control method [Utk92].

With these preliminaries, the main stability result for the generalized Persidskii-type system can be formulated as follows.

Theorem 1.38. *Consider the Persidskii-type system* (1.111). *If there exist a symmetric positive semidefinite matrix* $\mathbf{S}$ *and a positive definite block diagonal matrix* $\mathbf{K}$ *such that*

$$\mathbf{K} = \left[\begin{array}{c|c} \mathbf{K}_{11} & \mathbf{0} \\ \hline \mathbf{0} & \mathbf{K}_{22} \end{array}\right], \tag{1.115}$$

where the block $\mathbf{K}_{11}$ *is symmetric positive definite and* $\mathbf{K}_{22}$ *is positive diagonal, such that* $\mathbf{A} = \mathbf{SK}$, *then the trajectories of* (1.111) *converge to the invariant set* $\mathcal{A} := \{\mathbf{x} : \mathbf{f}(\mathbf{x}) \in \mathcal{N}(\mathbf{A})\}$, *where* $\mathcal{N}(\mathbf{A})$ *denotes the null space of the matrix* $\mathbf{A}$.

Proof. (Outline.) Consider the nonsmooth candidate Liapunov function of Lur'e–Persidskii type, associated to the Persidskii-type system (1.111):

$$V(\mathbf{x}) = \frac{1}{2}\bar{\mathbf{x}}^T \mathbf{K}_{11}\,\bar{\mathbf{x}} + \sum_{i=p+1}^{q} k_{ii}^{(2)} \int_0^{\hat{x}_i} g_i(\tau)\mathrm{d}\tau - \left(\frac{1}{2}\bar{\mathbf{x}}^{*T} \mathbf{K}_{11}\,\bar{\mathbf{x}}^* + \sum_{i=p+1}^{q} k_{ii}^{(2)} \int_0^{\hat{x}_i^*} g_i(\tau)\mathrm{d}\tau\right), \tag{1.116}$$

where $\mathbf{x}^{*T}$ belongs to the set $\mathcal{A} := \{\mathbf{x} : \mathbf{f}(\mathbf{x}) \in \mathcal{N}(\mathbf{A})\}$ and $k_{ii}^{(2)}$ are elements of the positive diagonal matrix $\mathbf{K}_{22}$. The time derivative of (1.116) along the trajectories of (1.111) is

$$\dot{V} = \nabla V^T \dot{\mathbf{x}} = -\mathbf{f}^T(\mathbf{x})\mathbf{KSKf}(\mathbf{x}) \leq 0. \tag{1.117}$$

Since $\dot{\mathbf{x}} = \mathbf{0}$ for $\mathbf{f}(\mathbf{x}) \in \mathcal{N}(\mathbf{A})$, it follows that $\mathcal{A}$ is an invariant set; observe that $\dot{V} = 0$ if and only if $\mathbf{x}$ belongs to this invariant set. Since $\mathbf{f}(\mathbf{x})$ is discontinuous in the set M, $\dot{V}$ is analyzed further and there are two possibilities as follows:

(i) $\mathbf{x}(t)$ "off" M; i.e., the trajectories of system (1.111) are not in any of the sets $\{\mathbf{x} : g_i(\hat{x}_i) = 0\}$. In this case, the solutions of (1.111) exist in the usual sense, thus since $\mathbf{S}$ is positive semidefinite, it is immediate that $\dot{V}(\mathbf{x}) \leq 0$;

(ii) $\mathbf{x}(t)$ in one or more of the sets $\{\mathbf{x} : g_i(\hat{x}_i) = 0\}$, for $t \in [t_0, t_f]$. In this case, the vectors $\mathbf{SKf}(\mathbf{x})$ are described by some vector $\mathbf{e}$ such that $\dot{\mathbf{x}} = -\mathbf{e} \in G(\mathbf{x})$ and we have $\dot{V} = -\mathbf{e}^T\mathbf{e} \leq 0$.

To complete the proof, additional arguments or conditions are needed to show that, in both cases listed above, in fact, $\dot{V}(\mathbf{x}) < 0$, and this implies that the reaching phase takes place, either asymptotically or in finite time. This is done for specific situations that arise in the applications of this "theorem" in Chapters 3 and 4, rather than increasing the complexity of the theorem statement in this introductory chapter. □

Corollary 1.39. *If matrices* $\mathbf{S}$ *and* $\mathbf{K}$ *are symmetric positive definite and* $f_i(x_i) \in [a_i, b_i]$ *for all* $x_i \in \mathbb{R}^n$, *then the trajectories of system* (1.111) *converge in finite time to the invariant set* $\mathcal{A}$ *and remain in it thereafter.*

Proof. Consider the candidate Liapunov function (1.116) and its time derivative along the trajectories of system (1.111). If $\mathbf{S}$ is symmetric positive definite, using Rayleigh's quotient, it is immediate that

$$\begin{aligned}\dot{V}(\mathbf{x}) &\leq -\lambda_{\min}(\mathbf{S})\lambda_{\min}(\mathbf{K}^2)\mathbf{f}^T(\mathbf{x})\mathbf{f}(\mathbf{x}) \\ &\leq -\lambda_{\min}(\mathbf{S})\lambda_{\min}(\mathbf{K}^2)\mathbf{a}^T\mathbf{a}\end{aligned}$$

if $\mathbf{f}(\mathbf{x}) \neq \mathbf{0}$, where $\lambda_{\min}(\mathbf{S})$ and $\lambda_{\min}(\mathbf{K}^2)$ are the smallest eigenvalues of $\mathbf{S}$ and $\mathbf{K}^2$, respectively. From the positive-definiteness of $\mathbf{S}$ and $\mathbf{K}$ we have $\lambda_{\min}(\mathbf{S}) > 0$ and $\lambda_{\min}(\mathbf{K}^2) > 0$. Thus, the trajectories converge to $\mathcal{A}$ in finite time and remain in this set thereafter. □

Theorem 1.38 is an extension of the result in [HKB00] and it provides a general convergence result for Persidskii systems with discontinuous right-hand sides. An extension of Theorem 1.38 is crucial in applications to linear programming. Consider the modification of the dynamical system (1.111),

$$\dot{\mathbf{x}}(t) = -\mathbf{A}\mathbf{f}(\mathbf{x}(t)) - \mathbf{c}, \tag{1.118}$$

where $\mathbf{c}$ is a constant vector and the other symbols are as defined above. The following corollary extends the result of Theorem 1.38.

Corollary 1.40. *If*

$$\min_{\mathbf{x}}\{\mathbf{f}(\mathbf{x})^T\mathbf{A}\mathbf{f}(\mathbf{x})\} > \max_{\mathbf{x}}\left\{\left|\mathbf{f}(\mathbf{x})^T\mathbf{K}\mathbf{c}\right|\right\}, \tag{1.119}$$

then trajectories of (1.118) *converge to the set $\mathcal{A}$ and remain in this set thereafter.*

Proof. Consider the candidate Liapunov function (1.116). Clearly its time derivative along the trajectories of (1.118) can be written as

$$\dot{V} = -\mathbf{f}^T\mathbf{K}\mathbf{c} - \mathbf{f}^T\mathbf{A}\mathbf{f}. \tag{1.120}$$

The expression for $\dot{V}$ shows immediately that if (1.119) holds, then $\dot{V} < 0$, and the corollary is proved. □

In the specific applications discussed in Chapters 3 and 4, it will be shown how the general condition (1.119) results in checkable conditions for the reaching phase to occur in finite time.

Speed gradient algorithms

This section briefly describes the speed gradient (SG) method [FP98], emphasizing the class of affine nonlinear systems and, within this class the special system $\dot{\mathbf{x}} = \mathbf{u}$, $\mathbf{y} = \mathbf{f}(\mathbf{x})$. The objective is to show that some of the CLF/LOC methods described in section 2.1 can be systematized within the SG framework and have an interesting interpretation in terms of passivity. In addition, this opens up the possibility of a new class of algorithms, due to additional (gradient) dynamics introduced in the control.

Consider the plant

$$\dot{\mathbf{x}} = \mathbf{f}(\mathbf{x}, \mathbf{u}, t) \tag{1.121}$$

and assume that a general CLF is specified in terms of a nonnegative function

$$V := V(\mathbf{x}(t), t). \tag{1.122}$$

The time derivative $\dot{V}(\mathbf{x}, \mathbf{u}, t)$ along the trajectories of (1.121) can be interpreted as the speed of $V(\cdot)$ and is evaluated as

$$\dot{V}(\mathbf{x}, \mathbf{u}, t) = \frac{\partial V(\mathbf{x}, t)}{\partial t} + (\nabla_{\mathbf{x}} V(\mathbf{x}(t), t))^T \mathbf{f}(\mathbf{x}, \mathbf{u}, t). \tag{1.123}$$

The speed gradient is then defined as the gradient of the speed with respect to $\mathbf{u}$:

$$\nabla_{\mathbf{u}} \dot{V}(\mathbf{x}, \mathbf{u}, t) = \left(\frac{\partial \mathbf{f}(\mathbf{x}, \mathbf{u}, t)}{\partial \mathbf{u}}\right)^T \nabla_{\mathbf{x}} V(\mathbf{x}(t), t). \tag{1.124}$$

There are several types of speed gradient control laws defined in [FP98]; we confine ourselves to two. The first is given as

$$\frac{\mathrm{d}\mathbf{u}}{\mathrm{d}t} = \dot{\mathbf{u}} = -\mathbf{\Gamma} \nabla_{\mathbf{u}} \dot{V}(\mathbf{x}, \mathbf{u}, t), \tag{1.125}$$

where $\mathbf{\Gamma}$ is a symmetric positive definite matrix. The second type of speed gradient law is

$$\mathbf{u} = \mathbf{u}_0 - \boldsymbol{\psi}(\mathbf{x}, \mathbf{u}, t), \tag{1.126}$$

for some constant $\mathbf{u}_0$ and where $\boldsymbol{\psi}$ satisfies the so-called pseudogradient condition

$$\boldsymbol{\psi}(\mathbf{x}, \mathbf{u}, t)^T \nabla_{\mathbf{u}} \dot{V}(\mathbf{x}, \mathbf{u}, t) \geq 0. \tag{1.127}$$

For each choice of law, there is a corresponding stability theorem enunciated and proved in [FP98, p. 90–108], so the details are not presented here.

Note the following choices that clearly satisfy (1.127):

$$\boldsymbol{\psi}(\mathbf{x}, \mathbf{u}, t) = \nabla_{\mathbf{u}} \dot{V}(\mathbf{x}, \mathbf{u}, t), \tag{1.128}$$

$$\boldsymbol{\psi}(\mathbf{x}, \mathbf{u}, t) = \alpha \operatorname{sgn}(\nabla_{\mathbf{u}} \dot{V}(\mathbf{x}, \mathbf{u}, t)), \alpha > 0. \tag{1.129}$$

Finally, consider the affine time-invariant system

$$\dot{\mathbf{x}} = \mathbf{f}(\mathbf{x}) + \mathbf{g}(\mathbf{x})\mathbf{u} \tag{1.130}$$

and assume furthermore that V is time invariant. The speed gradient then becomes

$$\nabla_{\mathbf{u}} \dot{V} = (\nabla_{\mathbf{x}} V)^T \mathbf{g}(\mathbf{x}) = (L_{\mathbf{g}} V(\mathbf{x}))^T. \tag{1.131}$$

This has the following interpretation. If (1.130) has a Liapunov stable zero solution for $\mathbf{u}(t) \equiv \mathbf{0}$, with Liapunov function $V : \mathbb{R}^n \to \mathbb{R}$, then it is passive with respect to the output $\mathbf{y} = \nabla_{\mathbf{u}} \dot{V}$, which is the SG vector [FP98].

1.7 Notes and References

The notes and references sections throughout the book are organized by topic. They are merely pointers to the literature on the topics in question and make no attempt to be exhaustive or establish precedence.

Control and system theory

For a compact introduction to system theory, see [Des70]. The reader without a control background is referred to [Son98] for a mathematically sophisticated introduction or to [Kai80, Del88, CD91, Ter99, Che99] for more accessible approaches. Persidskii systems and their generalizations and stability theory are covered in [KB00].

Stability theory

Basic stability theory is covered in [Kha02, SL91, Vid93, Ela96]. At a more advanced level, see the classic [Hah67], as well as [SP94, BR05], which cover stability aspects of nonsmooth systems. A good discussion of the relationships between various convergence and stability concepts and their uses in the theory of iterative methods can be found in [Hur67, Ort73]. Matrix stability theory is covered comprehensively in the books [HJ88, HJ91]. Diagonal stability and applications can be found in [KB00].

Variable structure system theory

The standard references are [Utk78, Utk92]. An accessible introduction for applications to linear systems is [ES98].

Nonsmooth analysis

An elegant and authoritative source is [CLSW98].

GDSs

Gradient systems are treated in [HS74, Ryb74, KH95]. The extension of a gradient system, known variously as a quasi-gradient, Sobolev gradient, or linear gradient system, is treated, respectively, in [Ryb74], [Neu05, Neu97], and [MQR98].

Optimal control

An insightful and idiosyncratic treatment of optimal control is given in [You80]. The classic reference for optimal control is [PBGM62].

Chapter 2
Algorithms as Dynamical Systems with Feedback

Algorithms are inventions which very often appear to have little or nothing in common with one another. As a result, it was held for a long time that a coherent theory of algorithms could not be constructed. The last few years have shown that this belief was incorrect, that most convergent algorithms share certain basic properties, and hence a unified approach to algorithms is possible.

—E. L. Polak [Pol71]

At the risk of oversimplification, it can be said that the design of a successful numerical algorithm usually involves the choice of some parameters in such a way that a suitable measure of some residue or error decreases to a reasonably small value as fast as possible. Although this is the case with most numerical algorithms, they are usually analyzed on a case by case basis: there is no general framework to guide the beginner, or even the expert, in the choice of these parameters. At a more fundamental level, one can even say that the very choice of strategy that results in the introduction of the parameters to be chosen is not usually discussed. Thus, the intention of this chapter, and of this book, is to revisit the question raised in the above quote, suggesting that control theory provides a framework for the design or discovery of algorithms in a systematic way.

Control theory, once again oversimplifying considerably, is concerned with the problem of *regulation*. Given a system model, generally referred to as a *plant*, that describes the behavior of some variables to be controlled, the problem of regulation is that of finding a mechanism that either keeps or regulates these variables at constant values, despite changes or disturbances that may act on the system as a whole. A fundamental idea is that of *feedback*: The variable to be controlled is compared with the constant value that is desired, and a difference (error, or residue) variable is generated. This error variable is used (*fed back*) by a parameter-dependent control mechanism to influence the plant in such a way that the controlled variable is driven (or converges) to the desired value. This results in zero error and, consequently, zero control action, as long as no disturbance occurs. The point to be emphasized here is that, in the six decades or so of development of mathematical control theory, several approaches have been developed to the systematic introduction and choice of the so-called feedback control parameters in the regulation problem. One objective of this chapter is to show that one of these approaches—the control Liapunov function approach—

can be used, in a simple and systematic manner, to motivate and derive several iterative methods, both standard as well as new ones, by viewing them as dynamical systems with feedback control.

2.1 Continuous-Time Dynamical Systems that Find Zeros

We start out with a discussion of how one might arrive at a continuous-time dynamical system that finds a simple zero of a given nonlinear vector function $\mathbf{f} : \mathbb{R}^n \to \mathbb{R}^n$ from a feedback control perspective. In other words, the problem is to find a vector $\mathbf{x}^* \in \mathbb{R}^n$ such that

$$\mathbf{f}(\mathbf{x}^*) = \mathbf{0}. \tag{2.1}$$

For a general nonlinear function $\mathbf{f}(\cdot)$, several solutions will, in general, exist. For the moment, we will content ourselves with finding a simple (i.e., nonrepeated) zero. Let $\mathbf{x}, \mathbf{r} \in \mathbb{R}^n$ such that

$$\mathbf{r} = -\mathbf{f}(\mathbf{x}). \tag{2.2}$$

The variable $\mathbf{r}$ is, in fact, the familiar *residue* of numerical analysis, since its norm can be interpreted as a measure of how far the current guess $\mathbf{x}$ is from a zero of $\mathbf{f}(\cdot)$, i.e., $\mathbf{r} := \mathbf{0} - \mathbf{f}(\mathbf{x})$. The other names that it goes by are *error* and *deviation*. Note that if $\mathbf{f} = \mathbf{Ax} - \mathbf{b}$, then zeroing the residue $\mathbf{r} := \mathbf{b} - \mathbf{Ax}$ corresponds to solving the classical linear system $\mathbf{Ax} = \mathbf{b}$.

In order to introduce control concepts, the first step is to observe that, if the residue is thought of as a time-dependent variable $\mathbf{r}(t)$ that is to be driven to zero, then the variable $\mathbf{x}(t)$ is correspondingly driven to a solution of (2.1). The second step is to assume that this will be done using a suitably defined control variable $\mathbf{u}(t)$, acting directly on the variable $\mathbf{x}(t)$. In control terms, this is written as the following simple nonlinear dynamical system: The first equation is referred to as the *state equation* and the second as the *output equation*, for reasons that are clear from Figure 2.1.

$$\frac{d\mathbf{x}(t)}{dt} = \mathbf{u}(t), \quad \text{state equation}, \tag{2.3}$$

$$\mathbf{y}(t) = \mathbf{f}(\mathbf{x}(t)), \quad \text{output equation}. \tag{2.4}$$

Furthermore, from (2.2) and (2.4) the output $\mathbf{y}(t)$ is the negative of the residue $\mathbf{r}(t)$:

$$\mathbf{y}(t) = -\mathbf{r}(t). \tag{2.5}$$

The problem of finding a zero of $\mathbf{f}(\cdot)$ can now be formulated in control terms as follows. Find a control $\mathbf{u}(t)$ that will drive the output (i.e., negative of the residue) to zero and, consequently, the state variable $\mathbf{x}(t)$ to the desired solution. In terms of standard control jargon, this is a *regulation problem* where the output must regulate to (i.e., become equal to) a reference signal, which in this case is zero: a glance at Figure 2.1 makes this description clear.

If the input is regarded as arbitrary and denoted as $\mathbf{v}$, and the method (dynamical system) is now required to find $\mathbf{x}$ such that $\mathbf{f}(\mathbf{x}) = \mathbf{s}$, i.e., a trajectory $\mathbf{x}(t)$ such that $\lim_{t\to\infty} \mathbf{x}(t) = \mathbf{v}$, then the problem is referred to, in control terminology, as an asymptotic tracking problem, because the system output $\mathbf{y}$ is required to track the input $\mathbf{v}$ (see Figure 2.1). In this case, if it is also required that the method work in spite of perturbations (i.e.,

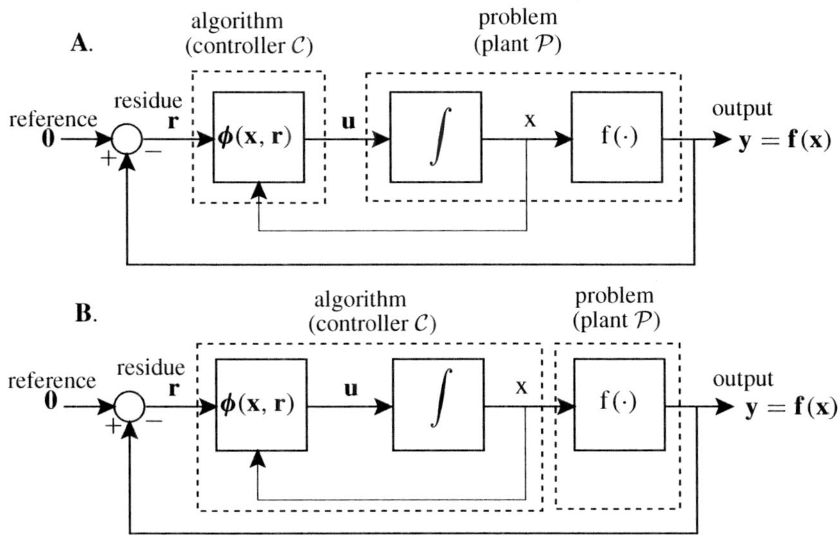

Figure 2.1. A: *A continuous realization of a general iterative method to solve the equation* $\mathbf{f}(\mathbf{x}) = \mathbf{0}$ *represented as a feedback control system. The plant, object of the control, represents the problem to be solved, while the controller* $\boldsymbol{\phi}(\mathbf{x}, \mathbf{r})$, *a function of* $\mathbf{x}$ *and* $\mathbf{r}$, *is a representation of the algorithm designed to solve it. Thus choice of an algorithm corresponds to the choice of a controller. As quadruples,* $\mathcal{P} = \{\mathbf{0}, \mathbf{I}, \mathbf{f}, \mathbf{0}\}$ *and* $\mathcal{C} = \{\mathbf{0}, \mathbf{0}, \mathbf{0}, \boldsymbol{\phi}(\mathbf{x}, \mathbf{r})\}$. B: *An alternative continuous realization of a general iterative method represented as a feedback control system. As quadruples,* $\mathcal{P} = \{\mathbf{0}, \mathbf{0}, \mathbf{0}, \mathbf{f}\}$ *and* $\mathcal{C} = \{\mathbf{0}, \boldsymbol{\phi}(\mathbf{x}, \mathbf{r}), \mathbf{I}, \mathbf{0}\}$. *Note that* $\mathbf{x}$ *is the state vector of the plant in part A, while it is the state vector of the controller in part B.*

errors in the input or output data), then the *internal model principle* (see Chapter 1) states that this so-called robust tracking property holds if and only if the feedback system is internally stable and there is a "model" of the disturbance (to be rejected) in the feedback loop. Since the input to be tracked can be viewed as a step function, i.e., one that goes from $\mathbf{0}$ to the constant value $\mathbf{v}$, this model is required to be an integrator (see Chapter 1 for a simple derivation in the linear case). This is depicted in Figures 2.1A and 2.1B, which show an integrator considered as part of the plant and part of the controller, respectively; in both cases, however, the integrator occurs in the feedback loop and the internal model principle is satisfied.

Having introduced the idea of time-dependence of the variables $\mathbf{x}(t)$, $\mathbf{u}(t)$, etc., in what follows, in order to lighten notation, the time variable will be dropped whenever possible and the variables written simply as $\mathbf{x}$, $\mathbf{u}$, etc.

General feedback control perspective on zero finding dynamical systems

From the point of view of the regulation problem in control, a natural idea is to feed back the output variable $\mathbf{y}$ in order to drive it to the reference value of zero. This is also referred to as *closing the loop*. It is also reasonable to expect a negative *feedback gain* that will be a function of the present "guess" of the state variable $\mathbf{x}$, as well as of the present value $\mathbf{f}(\mathbf{x}) = \mathbf{y}$. A simple approach is to choose a *feedback law* of the following type:

$$\mathbf{u} = \boldsymbol{\phi}(\mathbf{x}, \mathbf{r}), \tag{2.6}$$

leading to a so-called closed-loop system of the form

$$\frac{d\mathbf{x}}{dt} = \boldsymbol{\phi}(\mathbf{x}, \mathbf{r}). \tag{2.7}$$

Thus, the problem to be solved now is that of choosing the feedback law $\boldsymbol{\phi}(\mathbf{x}, \mathbf{r})$ in such a way as to make the closed-loop system asymptotically stable, driving the residue $\mathbf{r}$ to zero as fast as possible, thus solving the original problem of finding a solution to (2.1). In terms of Figure 2.1A, the choice (2.6) corresponds to choosing a controller

$$\mathcal{C}_s = \{\mathbf{0}, \mathbf{0}, \mathbf{0}, \boldsymbol{\phi}(\mathbf{x}, \mathbf{r})\} \tag{2.8}$$

for a plant $\mathcal{P} = \{\mathbf{0}, \mathbf{I}, \mathbf{f}, \mathbf{0}\}$. In control terminology, controller $\mathcal{C}_s$ is referred to as a *static state-dependent controller*, since the "gain" $\boldsymbol{\phi}(\mathbf{x}, \mathbf{r})$ between controller input and output depends on the controller input $\mathbf{r}$ as well as on the plant state $\mathbf{x}$.

More generally, in the configuration of Figure 2.1A, the problem is to choose a controller $\mathcal{C}_d$ (not necessarily of the form $\{\mathbf{0}, \mathbf{0}, \mathbf{0}, \boldsymbol{\phi}(\mathbf{x}, \mathbf{r})\}$) that regulates the plant output to zero, thus finding a zero of the function $\mathbf{f}(\cdot)$. In particular, the choice

$$\mathcal{C}_d = \{\boldsymbol{\Gamma}, \boldsymbol{\varphi}(\mathbf{x}, \mathbf{r}), \mathbf{I}, \mathbf{0}\}, \tag{2.9}$$

which is a *dynamic controller*, will be considered in what follows (see Figure 2.3), so that the combined plant-controller system is described by the equations

$$\dot{\mathbf{x}} = \mathbf{u}, \tag{2.10}$$
$$\dot{\mathbf{u}} = -\boldsymbol{\Gamma}\mathbf{u} + \boldsymbol{\varphi}(\mathbf{x}, \mathbf{r}). \tag{2.11}$$

The reader will observe that $\mathbf{u}$, in (2.11), is the controller state vector, which is also chosen as the controller output and, in turn, equal to the plant input in (2.10). The control problem for the plant-controller pair (2.10)–(2.11) is to choose the matrix $\boldsymbol{\Gamma}$ and the function $\boldsymbol{\varphi}(\mathbf{x}, \mathbf{r})$ in order to regulate the plant output to zero, thus finding a zero of the function $\mathbf{f}(\cdot)$.

In summary, it is desired to solve the problem of regulating the output to zero for a plant $\mathcal{P} = \{\mathbf{0}, \mathbf{0}, \mathbf{0}, \mathbf{f}\}$ and the following choices of controller: (i) $\mathcal{C}_s = \{\mathbf{0}, \mathbf{0}, \mathbf{0}, \boldsymbol{\phi}(\mathbf{x}, \mathbf{r})\}$ and (ii) $\mathcal{C}_d = \{\boldsymbol{\Gamma}, \boldsymbol{\varphi}(\mathbf{x}, \mathbf{r}), \mathbf{I}, \mathbf{0}\}$.

The remainder of this section shows how the CLF approach can be utilized to design both controllers $\mathcal{C}_s$ and $\mathcal{C}_d$. In fact, the CLF approach does more: the structure of the controllers $\mathcal{C}_s$ and $\mathcal{C}_d$, as well as the particular choices of $\boldsymbol{\phi}(\mathbf{x}, \mathbf{r})$, $\boldsymbol{\Gamma}$, and $\boldsymbol{\varphi}(\mathbf{x}, \mathbf{r})$, emerge naturally from the CLF approach.

The reader will observe that similar controller design problems can be formulated for the plant-controller partition presented in Figure 2.1B. This book, for the most part, will consider the controller design problem for the configuration of Figure 2.1A.

CLF/LOC approach to design continuous zero finding algorithms

The objective of this section is to show how control ideas can be used in the systematic design of algorithms, and thus the problems just posed above will be solved using the CLF and *Liapunov optimizing control* (LOC) approaches. Indeed, the specific form of the feedback law (2.7) will not be assumed, but rather derived from the CLF/LOC approach which results in the choice of the control $\mathbf{u}$.

The general scheme is as follows. A candidate Liapunov function V is chosen. The time derivative $\dot{V}$ along the trajectories of the dynamical system is calculated and is a function of the state vector $\mathbf{x}$, the output $\mathbf{y} = -\mathbf{r}$, and the input $\mathbf{u}$. The CLF approach can be described as the choice of *any* $\mathbf{u}$ that makes $\dot{V}$ negative definite, while the LOC approach goes a step further in using the degrees of freedom available and demands that the input $\mathbf{u}$ be chosen in such a way as to make $\dot{V}$ as negative as possible. The attempt to do the latter involves minimizing $\dot{V}$ with respect to the (free) control variable: This leads in one direction to a connection with the so-called speed-gradient method when there are no restrictions on the control. In another direction, when a bounded control is to be applied, it is often true that the optimizing control involves a signum function of the control, and this leads to a connection with variable structure control. There are, in fact, close connections between Liapunov optimizing control, speed-gradient control, variable structure control, and optimal control, and these will be examined in more detail in this chapter as well as in Chapter 3. Another way of thinking about LOC is to regard it as a greedy approach, in the sense that it makes the best local choice of control. The controllers $\mathcal{C}_s$ and $\mathcal{C}_d$ are designed using the CLF/LOC approach.

General stability result for quadratic CLF in residue coordinates

In order to use Liapunov theory, the control design is done in terms of the residual vector $\mathbf{r}$, so that the stability analysis is carried out with respect to the equilibrium $\mathbf{r} = \mathbf{0}$. Thus, it is assumed that a local change of coordinates is possible from the variable $\mathbf{x}$ to the variable $\mathbf{r}$. Since $\mathbf{r} = -\mathbf{f}(\mathbf{x})$, by the inverse function theorem, if it is assumed that the Jacobian of $\mathbf{f}(\cdot)$ is invertible, then $\mathbf{f}$ itself is locally invertible; i.e., the desired change of coordinates exists. Accordingly, taking the time derivative of (2.2) leads to

$$\frac{\mathrm{d}\mathbf{r}}{\mathrm{d}t} = -\frac{\partial \mathbf{f}(\mathbf{x})}{\partial \mathbf{x}} \frac{\mathrm{d}\mathbf{x}}{\mathrm{d}t} = -\mathbf{D}_{\mathbf{f}}(\mathbf{x})\dot{\mathbf{x}}, \tag{2.12}$$

where $\mathbf{D}_{\mathbf{f}}(\mathbf{x})$ denotes the Jacobian matrix of $\mathbf{f}$ at $\mathbf{x}$ and $\dot{\mathbf{x}}$ denotes the time derivative of $\mathbf{x}$. Notice, from (2.3), that (2.12) can also be written as

$$\frac{\mathrm{d}\mathbf{r}}{\mathrm{d}t} = -\mathbf{D}_{\mathbf{f}}(\mathbf{x})\mathbf{u}. \tag{2.13}$$

One simple choice of CLF for design of the static controller $\mathcal{C}_s$ is based on the 2-norm of the residue vector $\mathbf{r}$:

$$V(\mathbf{r}) := \frac{1}{2}\|\mathbf{r}\|_2^2 = \frac{1}{2}\mathbf{r}^T\mathbf{r}, \tag{2.14}$$

which is evidently positive definite and only assumes the value 0 at the desired equilibrium $\mathbf{r} = \mathbf{0}$. The time derivative of V along the trajectories of (2.13), denoted $\dot{V}$, is given by

$$\dot{V}(\mathbf{r}) = \mathbf{r}^T\dot{\mathbf{r}}. \tag{2.15}$$

Substituting (2.13) in (2.15) yields

$$\dot{V}(\mathbf{r}) = -\mathbf{r}^T \mathbf{D_f}(\mathbf{x})\mathbf{u}, \tag{2.16}$$

which is one possible starting point for a CLF/LOC approach to the design of the control variable $\mathbf{u}$. More precisely, in order for the system to be asymptotically stable, it is necessary to choose $\mathbf{u} = \mathbf{u}(\mathbf{r})$ in such a way that $\dot{V}$ becomes negative definite. Such a choice will prove asymptotic stability of the zero solution $\mathbf{r} \equiv \mathbf{0}$ of the closed-loop system

$$\frac{\mathrm{d}\mathbf{r}}{\mathrm{d}t} = -\mathbf{D_f}(\mathbf{x})\mathbf{u}(\mathbf{r}), \tag{2.17}$$

leading to a prototypical stability result, for the static controller case, of the following type.

Theorem 2.1. *Given a function* $\mathbf{f} : \mathbb{R}^n \to \mathbb{R}^n$, *suppose that* $\mathbf{x}^*$ *is a simple zero of* $\mathbf{f}$ *such that the Jacobian matrix* $\mathbf{D_f}(\mathbf{x})$ *of* $\mathbf{f}$ *is invertible in some neighborhood* $\mathcal{N}(\mathbf{x}^*)$ *of* $\mathbf{x}^*$. *Then, for all initial conditions* $\mathbf{x}_0 \in \mathcal{N}(\mathbf{x}^*)$, *the corresponding trajectories of the dynamical system*

$$\dot{\mathbf{x}} = \mathbf{u}(\mathbf{r}), \quad \mathbf{x}(0) = \mathbf{x}_0 \in \mathcal{N}(\mathbf{x}^*), \tag{2.18}$$

where $\mathbf{u}(\mathbf{r})$ *is a feedback control chosen by the CLF/LOC method, in the manner specified in Table* 2.1, *converge to the zero* $\mathbf{x}^*$ *of* $\mathbf{f}(\cdot)$.

Note that Theorem 2.1 yields a local stability result, since its proof depends on the stability properties of the residue system (2.17), which is obtained after a local change of coordinates. The stability type—exponential, asymptotic, or finite time—depends on the particular control that is chosen and will be discussed further below.

CLF/LOC design of continuous algorithms with static controllers $\mathcal{C}_s$

The CLF/LOC approach, applied to (2.16), is now used in order to derive the choices of $\mathbf{u}(\mathbf{r})$, leading to a proof of Theorem 2.1. Note that if

$$\mathbf{u} = \mathbf{D_f}^{-1}\mathbf{P}\mathbf{r},$$

then, from (2.16),

$$\dot{V} = -\mathbf{r}^T\mathbf{P}\mathbf{r},$$

which is clearly negative definite, for any choice of positive definite matrix $\mathbf{P}$. The resulting continuous algorithm is written as follows:

$$\dot{\mathbf{x}} = -\mathbf{D_f}^{-1}\mathbf{P}\mathbf{f}(\mathbf{x}). \tag{2.19}$$

This dynamical system, for $\mathbf{P} = \mathbf{I}$, is known as the *continuous Newton (CN) algorithm* and is well known in the literature (see [Alb71, Neu99] and the references therein).

We now show how the CLF/LOC approach can be exploited to go beyond the CN algorithm. Taking another look at (2.16), it is clear that if we choose

$$\mathbf{u} = \mathbf{D_f}^{-1}\mathrm{sgn}(\mathbf{r}), \tag{2.20}$$

where sgn is the signum function, then from (2.16),

$$\dot{V} = -\mathbf{r}^T \text{sgn}(\mathbf{r}) = -\|\mathbf{r}\|_1,$$

which is clearly negative definite and leads to a new algorithm, written as follows:

$$\dot{\mathbf{x}} = -\mathbf{D}_{\mathbf{f}}^{-1}\text{sgn}(\mathbf{f}(\mathbf{x})). \tag{2.21}$$

The above dynamical system is called a *Newton variable (NV) structure algorithm*, and its notable feature is that the right-hand side of (2.21) is discontinuous, so that the Filippov solution concept and associated stability theory discussed in section 1.5 must be used.

A third choice that suggests itself is

$$\mathbf{u} = \mathbf{P}\mathbf{D}_{\mathbf{f}}^T\mathbf{r}$$

leading, from (2.16), to

$$\dot{V} = -\mathbf{r}^T\mathbf{D}_{\mathbf{f}}\mathbf{P}\mathbf{D}_{\mathbf{f}}^T\mathbf{r},$$

which is clearly negative definite, under the hypotheses of Theorem 2.1. The resulting continuous algorithm is written as follows:

$$\dot{\mathbf{x}} = -\mathbf{P}\mathbf{D}_{\mathbf{f}}^T\mathbf{f}(\mathbf{x}). \tag{2.22}$$

This is again a new algorithm, which will be referred to as the *continuous Jacobian matrix transpose (CJT) algorithm.*

A fourth choice is

$$\mathbf{u} = \text{sgn}(\mathbf{D}_{\mathbf{f}}^T\mathbf{r}), \tag{2.23}$$

which yields, from (2.16),

$$\dot{V} = -(\mathbf{D}_{\mathbf{f}}^T\mathbf{r})^T\text{sgn}(\mathbf{D}_{\mathbf{f}}^T\mathbf{r}) = -\|\mathbf{D}_{\mathbf{f}}^T\mathbf{r}\|_1,$$

which again is negative definite under the nonsingularity hypothesis on $\mathbf{D}_{\mathbf{f}}$. Note that this choice is a Liapunov optimizing control, since it makes $\dot{V}$ as negative as possible, under the constraint that the 1-norm of the control should not exceed unity. The corresponding algorithm is written as follows:

$$\dot{\mathbf{x}} = -\text{sgn}(\mathbf{D}_{\mathbf{f}}^T\mathbf{f}(\mathbf{x})). \tag{2.24}$$

This algorithm will be referred to as a *variable structure Jacobian transpose (VJT) algorithm.* Once again, the right-hand side of (2.24) is discontinuous, so the Filippov solution concept and associated stability theory (section 1.5) should be used.

Finally, as an illustration of how the choice of Liapunov function influences the resulting continuous algorithm, instead of using the 2-norm as a CLF, let the 1-norm

$$W(\mathbf{r}) = \|\mathbf{r}\|_1 = \mathbf{r}^T\text{sgn}(\mathbf{r}) = \text{sgn}(\mathbf{r})^T\mathbf{r} \tag{2.25}$$

be chosen as a nonsmooth CLF. In this case, the signum function is constant except at the origin, so that, formally speaking [Utk92], its derivative is zero, except at the origin, and we can write

$$\dot{W}(\mathbf{r}) = \text{sgn}(\mathbf{r})^T\dot{\mathbf{r}} = \text{sgn}(\mathbf{r})^T(-\mathbf{D}_{\mathbf{f}}\mathbf{u}),$$

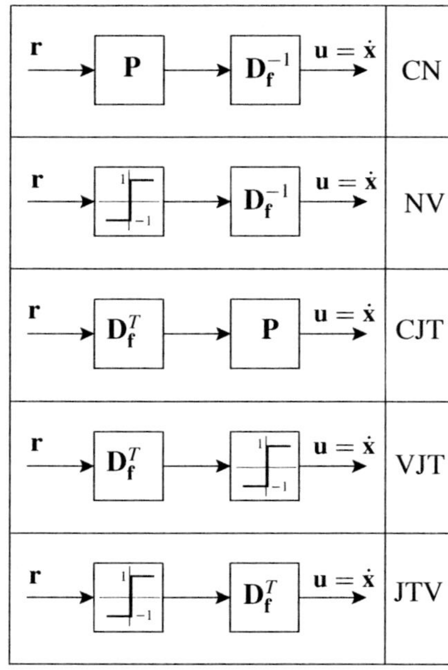

Figure 2.2. *The structure of the CLF/LOC controllers* $\boldsymbol{\phi}(\mathbf{x}, \mathbf{r})$: *The block labeled* $\mathbf{P}$ *corresponds to multiplication by a positive definite matrix* $\mathbf{P}$, *the blocks labeled* $\mathbf{D}_\mathbf{f}^T$ *and* $\mathbf{D}_\mathbf{f}^{-1}$ *depend on* $\mathbf{x}$ *(see Figure* 2.1A *and Table* 2.1).

using (2.13). Thus, the choice

$$\mathbf{u} = \mathbf{D}_\mathbf{f}^T \operatorname{sgn}(\mathbf{r}) \tag{2.26}$$

gives

$$\dot{W}(\mathbf{r}) = -\operatorname{sgn}(\mathbf{r})^T (\mathbf{D}_\mathbf{f} \mathbf{D}_\mathbf{f}^T) \operatorname{sgn}(\mathbf{r}) = -\|\mathbf{D}_\mathbf{f}^T \operatorname{sgn}(\mathbf{r})\|_2^2,$$

which is clearly negative definite under the nonsingularity hypothesis on $\mathbf{D}_\mathbf{f}$. The resulting algorithm is written as follows:

$$\dot{\mathbf{x}} = -\mathbf{D}_\mathbf{f}^T \operatorname{sgn}(\mathbf{f}(\mathbf{x})). \tag{2.27}$$

This algorithm is also new and will be referred to as a *Jacobian matrix transpose variable structure (JTV) algorithm* and, once again, the right-hand side is discontinuous, so the Filippov theory of section 1.5 should be used. The controller block, $\boldsymbol{\phi}(\mathbf{x}, \mathbf{r})$ (Figure 2.1), corresponding to each of the five algorithms, is shown in Figure 2.2.

At this point the observant reader might well ask where the LOC idea is being used, specifically in the design of the CN, NV, CJT, and JTV algorithms, since so far, LOC has been used only in the VJT method.

In order to answer this question, observe that there is flexibility in the choice of the plant. For example, suppose that, instead of (2.3), we choose the plant

$$\dot{\mathbf{x}} = \mathbf{D}_\mathbf{f}^{-1} \mathbf{u}, \tag{2.28}$$

which is easily seen to be the system

$$\dot{\mathbf{r}} = -\mathbf{u}, \tag{2.29}$$

Table 2.1. *Choices of* $\mathbf{u}(\mathbf{r})$ *in Theorem* 2.1 *that result in stable zero finding dynamical systems.*

$\mathbf{u}(\mathbf{r})$	$\dot{V}$ (rows 1–4) $\dot{W}$ (row 5)	Continuous algorithm	Acronym	Equation number
$\mathbf{D_f}^{-1}\mathbf{Pr}$,	$-\mathbf{r}^T\mathbf{Pr}$	$\dot{\mathbf{x}} = -\mathbf{D_f}^{-1}\mathbf{Pf}(\mathbf{x})$	CN	(2.19)
$\mathbf{D_f}^{-1}\mathrm{sgn}(\mathbf{r})$,	$-\Vert\mathbf{r}\Vert_1$	$\dot{\mathbf{x}} = -\mathbf{D_f}^{-1}\mathrm{sgn}(\mathbf{f}(\mathbf{x}))$	NV	(2.21)
$\mathbf{PD_f}^T\mathbf{r}$	$-\mathbf{r}^T\mathbf{D_f}\mathbf{PD_f}^T\mathbf{r}$	$\dot{\mathbf{x}} = -\mathbf{PD_f}^T\mathbf{f}(\mathbf{x})$	CJT	(2.22)
$\mathrm{sgn}(\mathbf{D_f}^T\mathbf{r})$	$-\Vert\mathbf{D_f}^T\mathbf{r}\Vert_1$	$\dot{\mathbf{x}} = -\mathrm{sgn}(\mathbf{D_f}^T\mathbf{f}(\mathbf{x}))$	VJT	(2.24)
$\mathbf{D_f}^T\mathrm{sgn}(\mathbf{r})$	$-\Vert\mathbf{D_f}^T\mathrm{sgn}(\mathbf{r})\Vert_2^2$	$\dot{\mathbf{x}} = -\mathbf{D_f}^T\mathrm{sgn}(\mathbf{f}(\mathbf{x}))$	JTV	(2.27)

in $\mathbf{r}$-coordinates. Maintaining the output equation (2.4) and the 2-norm Liapunov function V in (2.14), the time derivative of V along the trajectories of (2.28) is given by

$$\dot{V} = -\mathbf{r}^T\mathbf{D_f}\dot{\mathbf{x}} = -\mathbf{r}^T\mathbf{D_f}\mathbf{D_f}^{-1}\mathbf{u} = -\mathbf{r}^T\mathbf{u}. \tag{2.30}$$

Now, for (2.30), the choice $\mathbf{u} = \mathrm{sgn}(\mathbf{r})$ in (2.20) is clearly an LOC, under the constraint that all components of the control vector should be less than or equal to unity in absolute value (i.e., $\Vert\mathbf{u}\Vert_\infty = 1$). This yields the Newton variable structure method (2.21). Furthermore, the choice $\mathbf{u} = \mathbf{r}$ in (2.30), which leads to the CN method (2.19), is also optimal in the sense of being the simplest choice (of state feedback) that leads to an exponentially stable system in $\mathbf{r}$-coordinates,

$$\dot{\mathbf{r}} = -\mathbf{r}. \tag{2.31}$$

Similarly, consider the alternative choice of plant,

$$\dot{\mathbf{x}} = \mathbf{D_f}^T\mathbf{u}. \tag{2.32}$$

Then, maintaining the output equation (2.4) and using the 1-norm Liapunov function W in (2.25), the time derivative of W along the trajectories of (2.32) is given by

$$\dot{W} = \mathrm{sgn}(\mathbf{r})^T\dot{\mathbf{r}} = -\mathrm{sgn}(\mathbf{r})^T\mathbf{D_f}\dot{\mathbf{x}} = -\mathrm{sgn}(\mathbf{r})^T\mathbf{D_f}\mathbf{D_f}^T\mathbf{u}. \tag{2.33}$$

Now the choice $\mathbf{u} = \mathrm{sgn}(\mathbf{r})$ is clearly a possible LOC choice that gives the JTV algorithm in (2.27). If, instead of W, V is used, then CJT is seen to result from the simple choice $\mathbf{u} = \mathbf{r}$.

In summary, the CLF/LOC approach has been shown to lead, at least, to the five different algorithms considered above. For ease of reference, these zero finding dynamical systems are organized in Table 2.1. Each system is given an acronym formed using the following conventions: N = Newton, C = continuous-time, V = variable structure, JT = Jacobian matrix transpose.

The first row, for $\mathbf{P} = \mathbf{I}$, corresponds to the well-known and much studied CN method, so called because it is clear that a discretization of $\dot{\mathbf{x}} = -\mathbf{D_f}^{-1}\mathbf{f}(\mathbf{x})$, by the forward Euler method, will lead to the classical Newton–Raphson method, discussed further in section

2.2. It has several notable properties [HN05], one of which is singled out here for mention: Convergence to the zero is exponential. This is easily seen in the closed-loop residue system, where choosing $\mathbf{P} = \alpha\mathbf{I}$, $\alpha > 0$ gives

$$\dot{\mathbf{r}} = -\alpha\mathbf{r}, \quad \mathbf{r}_0 = -\mathbf{f}(\mathbf{x}_0), \tag{2.34}$$

which has the solution $\mathbf{r}(t) = e^{-\alpha t}\mathbf{r}_0$. The choice of α clearly affects the speed of convergence of the residue to zero and can be regarded as an additional design parameter. Having said this, from now on, unless otherwise stated, we will take $\mathbf{P}$ to be $\mathbf{I}$, to simplify matters. Another way of looking at the exponential stability of the zero solution of (2.19) is to observe that it can be written in the so-called integrated form by avoiding the inversion of the Jacobian matrix [Neu99]. In other words, (2.19) is first written as $\mathbf{D_f}\dot{\mathbf{x}} = -\mathbf{f}(\mathbf{x})$ and then, since $\dot{\mathbf{f}} = \mathbf{D_f}\dot{\mathbf{x}}$, finally

$$\dot{\mathbf{f}} = -\mathbf{f}, \quad \mathbf{f}_0 = \mathbf{f}(\mathbf{x}_0). \tag{2.35}$$

It should also be noted that, as far as the dynamical systems for zero finding are concerned, the variables $\mathbf{r}$ and $\mathbf{f}$ are equivalent: note the equivalence of (2.34) with $\alpha = 1$ and (2.35). Said differently, it is equivalent to work in $\mathbf{r}$-coordinates or $\mathbf{f}$-coordinates, up to a change of sign.

The second, fourth, and fifth rows correspond to *variable structure methods*, with some new features, the most notable one being that they lead to differential equation with discontinuous right-hand sides. It is necessary to be careful about existence, uniqueness, and stability issues for differential equations with discontinuous right-hand sides, but for reasons of organization and brevity (of this chapter), formal manipulations of Liapunov functions, smooth and nonsmooth, have been used to motivate the choices made. The reader should be aware that these formal manipulations can be made rigorous, and the basic ideas of one approach, due to Filippov, are outlined in Chapter 1, where further references are also supplied.

As pointed out at the beginning of this section, a general observation that should be made about the CLF and LOC approaches is that they lead to the form of feedback law that one expects intuitively, without the need to prespecify the form of the feedback law as (2.6), thus showing the power of the Liapunov approach. An additional demonstration of this power lies in the fact that it serves to unify several different types of algorithms, as well as providing a method of generating new algorithms. One way of doing the latter is to use different Liapunov functions, and an example of a choice different from (2.14) occurs in the fifth row of Table 2.1, as well as in the design of dynamic controllers, discussed further ahead in this section.

Gradient control perspective on zero finding algorithms

The static controller-based algorithms in Table 2.1 can also be regarded as examples of gradient control. To do this, we recall (2.17):

$$\dot{\mathbf{r}} = -\mathbf{D_f}\mathbf{u}. \tag{2.36}$$

For the Liapunov function (2.14),

$$\nabla V(\mathbf{r}) = \mathbf{r}. \tag{2.37}$$

Now, substituting the choices of $\mathbf{u}$ from rows 1 and 3 of Table 2.1 in (2.36), and using (2.37), yields, respectively,

$$\dot{\mathbf{r}} = -\mathbf{P}\nabla V(\mathbf{r}) \tag{2.38}$$

and

$$\dot{\mathbf{r}} = -\mathbf{D_f}\mathbf{P}\mathbf{D_f}^T\nabla V(\mathbf{r}), \tag{2.39}$$

which are both gradient systems in the terminology of Chapter 1 (for which V is a Liapunov function) since the matrices $\mathbf{P}$ and $\mathbf{D_f}\mathbf{P}\mathbf{D_f}^T$ are positive definite, by choice and by hypothesis, respectively. Thus the controls chosen in rows 1 and 3 can be viewed as *gradient controls* of the system (2.36).

Substitution of the control choices in rows 2 and 5 of Table 2.1 in (2.36) leads, respectively, to the following systems:

$$\dot{\mathbf{r}} = -\text{sgn}(\mathbf{r}) \tag{2.40}$$

and

$$\dot{\mathbf{r}} = -\mathbf{D_f}\mathbf{D_f}^T\text{sgn}(\mathbf{r}). \tag{2.41}$$

These systems are both in Persidskii-type form with discontinuous right-hand side, for which asymptotic stability of the zero solution $\mathbf{r} \equiv \mathbf{0}$ is ensured. Furthermore, as pointed out in Chapter 1, these particular Persidskii systems can also be written as GDSs with discontinuous right-hand sides, so that, once again, these entries in Table 2.1 can be viewed as examples of gradient control.

The observation that (2.40) is in Persidskii-type form permits a slight generalization of the second row, by making the choice $\mathbf{u} = \mathbf{D_f}^{-1}\mathbf{A}\text{sgn}(\mathbf{r})$, where $\mathbf{A}$ is a diagonally stable matrix, since this will lead to the asymptotically stable Persidskii-type system (see Chapter 1)

$$\dot{\mathbf{r}} = -\mathbf{A}\text{sgn}(\mathbf{r}), \tag{2.42}$$

with the corresponding generalized variable structure Newton method being

$$\dot{\mathbf{x}} = -\mathbf{D_f}^{-1}\mathbf{A}\,\text{sgn}(\mathbf{f}). \tag{2.43}$$

Continuous algorithms derived from a speed gradient perspective

The speed gradient method [FP98], reviewed in Chapter 1, can be used to systematize Liapunov optimizing control design in the context of the zero finding problem, as well as lead to the design of new algorithms.

In order to do this, consider the special affine nonlinear system given by (2.3) and (2.4) and let the CLF be chosen as in (2.14). Then, clearly

$$\dot{V}(\mathbf{r}) = \mathbf{r}^T\dot{\mathbf{r}} = -\mathbf{r}^T\mathbf{D_f}\mathbf{u}$$

and

$$\nabla_{\mathbf{u}}\dot{V}(\mathbf{r}) = -\mathbf{D_f}^T\mathbf{r}.$$

Using (1.126) and (1.128), with $\Gamma = \gamma\mathbf{I}$, yields CJT in Table 2.1, with $\mathbf{P} = \gamma\mathbf{I}$. Similarly, using (1.126) and (1.129), we arrive at VJT.

Starting with the dynamical system (2.28), and using the 2-norm CLF $V(\mathbf{r})$, it is easy to check that $\nabla_{\mathbf{u}}\dot{V} = \mathbf{r}$, so that using (1.126) and (1.128) yields CN, while using (1.126) and (1.129) leads to NV. Finally, using the 1-norm W, it is possible to recover JTV. A disadvantage of using the speed gradient method is that it requires some stringent hypotheses (see [FP98]). However, given the close connections between the CLF/LOC and the speed gradient approach, it is of interest to point out this possible route to the algorithms in Table 2.1.

Benchmark examples for zero finding algorithms

Two examples that are used as benchmarks to test the qualitative behavior of the various numerical algorithms proposed in this chapter, as well as elsewhere in the book, are given below.

The Rosenbrock function [Ros60] is a famous example of a difficult optimization problem and is defined as follows:

$$\mathbf{f}(\mathbf{x}) : \mathbb{R}^2 \to \mathbb{R}^2 : (x_1, x_2) \mapsto a(x_1^2 - x_2)^2 + (x_1 - b)^2, \tag{2.44}$$

where the parameters a and b are to be specified. The problem of finding a minimum of this function can be expressed as the zero finding problem for the gradient of $\mathbf{f}$, i.e., by finding the zeros of $\mathbf{g} = \nabla\mathbf{f}$, where

$$\mathbf{g}(\mathbf{x}) = \begin{bmatrix} g_1(\mathbf{x}) \\ g_2(\mathbf{x}) \end{bmatrix} = \begin{bmatrix} \partial\mathbf{f}/\partial x_1 \\ \partial\mathbf{f}/\partial x_2 \end{bmatrix} = \begin{bmatrix} 4ax_1(x_1^2 - x_2) + 2(x_1 - b) \\ -2a(x_1^2 - x_2) \end{bmatrix}. \tag{2.45}$$

The second example, due to Branin [Bra72], inspired by a tunnel diode circuit and used as a benchmark example for many zero finding algorithms, is the problem of finding the zeros of functions $f_1(x_1, x_2)$ and $f_2(x_1, x_2)$, where

$$\begin{aligned} f_1(x_1, x_2) &= a - bx_2 + c\sin(dx_2) - x_1, \\ f_2(x_1, x_2) &= x_2 - e\sin(fx_1). \end{aligned} \tag{2.46}$$

The values of the parameters, used in [Bra72] and several other papers that cite it, are $a = 1$, $b = 2$, $d = 4\pi$, $e = 0.5$, $f = 2\pi$; the value of c determines the number of zeros of the system $f_1(x_1, x_2) = f_2(x_1, x_2) = 0$. For example, for $c = 0$, the system has 5 zeros; for $c = 0.5$, the system has 7 zeros; for $c = 1$, the system has 15 zeros.

Changing the singularities of the Newton vector field

An important feature of the continuous Newton method, defined by the Newton vector field associated to a function $\mathbf{f}(\cdot)$,

$$\mathbf{N}_\mathbf{f} = -\mathbf{D}_\mathbf{f}^{-1}\mathbf{f}(\mathbf{x}), \tag{2.47}$$

is that the singularities of $\mathbf{N}_\mathbf{f}$ may occur in locations different from the desired zeros of $\mathbf{f}$. More specifically, convergence of trajectories of the CN method to the desired zeros (of $\mathbf{f}$) could be adversely affected by the presence of the so-called extraneous singularities [Bra72, Gom75, RZ99, ZG02], defined below.

Let the *set of zeros* of the function $\mathbf{f}$ be denoted as

$$\mathcal{Z} := \{\mathbf{x} : \mathbf{f}(\mathbf{x}) = \mathbf{0}\}. \tag{2.48}$$

The set of *singular points* is defined as the set of points

$$\mathcal{S} := \{\mathbf{x} : \det \mathbf{D_f}(\mathbf{x}) = 0\}. \tag{2.49}$$

Singular zeros are points in $\mathcal{S} \cap \mathcal{Z}$. The remaining zeros are called *regular*.

In particular, Zufiria, Guttalu, and coworkers [RZ99, ZG02] define a taxonomy of singularities of the Newton vector field and analyze most of the types that could occur. Rather than detail these results here, we point out that the analysis is based on the fact that the Newton vector field can be written as

$$\mathbf{N_f} = -\frac{(\operatorname{adj} \mathbf{D_f})\mathbf{f}(\mathbf{x})}{\det \mathbf{D_f}} =: \frac{1}{h(\mathbf{x})}\mathbf{g}(\mathbf{x}), \tag{2.50}$$

where $\operatorname{adj} \mathbf{D_f}$ is the classical adjoint of the Jacobian matrix $\mathbf{D_f}$. *Essential singularities* occur when the denominator $h(\mathbf{x})$ becomes zero, but the numerator $\mathbf{g}(\mathbf{x})$ does not become a zero vector. *Nonessential singularities* occur when $h(\mathbf{x}) = 0$ and $\mathbf{g}(\mathbf{x}) = \mathbf{0}$. Within the class of nonessential singularities, there is the class of *extraneous singularities* defined as

$$\mathcal{E} := \{\mathbf{x} \in \mathcal{S} : \mathbf{f}(\mathbf{x}) \neq \mathbf{0}, (\operatorname{adj} \mathbf{D_f})\mathbf{f}(\mathbf{x}) = \mathbf{0}\}. \tag{2.51}$$

Although the Newton vector field is constructed for locating the regular and singular zeros of $\mathbf{f}$, extraneous singularities have been shown to possess great importance in different methods designed for root finding [Bra72, Gom75, RZ99, ZG02].

From this discussion, it becomes clear that for the new methods introduced and the corresponding vector fields (given by the right-hand sides of the systems in column 3, rows 2 through 5 of Table 2.1), the structure of the singularities of the vector fields changes with respect to the Newton vector field. In particular, it becomes clear that although much research has focused on "desingularizing" and continuous extensions of the Newton vector field in the presence of the different types of singularities [DS75, DK80a, DK80b, Die93, RZ99, ZG02], control theory provides an alternative approach in which the flexibility in the choice of feedback control allows the algorithm designer to change the singularity structure of the vector field.

Example 2.2. Branin [Bra72] used Rosenbrock's example (2.44) to illustrate the problem of extraneous singularities of the Newton vector field. We use this example here to show that, while the Newton vector field has an extraneous singularity that affects trajectories of the CN dynamical system, on the other hand, the NV vector field does not possess this extraneous singularity. As a consequence, NV trajectories, for some initial conditions, are better behaved than CN trajectories, exhibiting faster and more consistent progress to the solution.

It is easy to check that $(x_1, x_2) = (b, b^2)$ is the solution of $\mathbf{g} = \mathbf{0}$ in (2.45). Furthermore, the Jacobian matrix $\mathbf{D_g}$ is easily calculated as follows:

$$\mathbf{D_g}(\mathbf{x}) = \begin{bmatrix} 4a(3x_1^2 - x_2) + 2 & -4ax_1 \\ -4ax_1 & 2a \end{bmatrix}. \tag{2.52}$$

Its classical adjoint adj $\mathbf{D_g}$ is

$$\operatorname{adj}\mathbf{D_g}(\mathbf{x}) = \begin{bmatrix} 2a & 4ax_1 \\ 4ax_1 & 4a(3x_1^2 - x_2) + 2 \end{bmatrix}. \tag{2.53}$$

It is now easy to verify that

$$\mathbf{x}_{\text{ex}}^{\text{ctn}} = \begin{bmatrix} b \\ b^2 + \frac{1}{2a} \end{bmatrix}$$

satisfies all the conditions of an extraneous singularity, namely:

$$\operatorname{adj}\mathbf{D_g}(\mathbf{x}_{\text{ex}}^{\text{ctn}}) \text{ is singular,}$$

$$\mathbf{g}(\mathbf{x}_{\text{ex}}^{\text{ctn}}) = \begin{bmatrix} -2b \\ 1 \end{bmatrix} \neq 0,$$

and

$$\operatorname{adj}\mathbf{D_g}(\mathbf{x}_{\text{ex}}^{\text{ctn}})\mathbf{g}(\mathbf{x}_{\text{ex}}^{\text{ctn}}) = \mathbf{0}.$$

Now, if the extraneous singularity analysis is carried out for the NV vector field $\mathbf{D_g}^{-1}\text{sgn}(\mathbf{g})$, all the above equations hold, except for the last one, i.e.,

$$\operatorname{adj}\mathbf{D_g}(\mathbf{x}_{\text{ex}}^{\text{ctn}})\text{sgn}(\mathbf{g}(\mathbf{x}_{\text{ex}}^{\text{ctn}})) = \operatorname{adj}\mathbf{D_g}(\mathbf{x}_{\text{ex}}^{\text{ctn}})\text{sgn}\begin{bmatrix} -2b \\ 1 \end{bmatrix} = \operatorname{adj}\mathbf{D_g}(\mathbf{x}_{\text{ex}}^{\text{ctn}})\begin{bmatrix} -1 \\ 1 \end{bmatrix} \neq \mathbf{0},$$

showing that $\mathbf{x}_{\text{ex}}^{\text{ctn}}$ is not an extraneous singularity for the NV vector field. In other words, it is indeed possible to change the singularity structure of the Newton vector field by making the alternative choices in rows 2 through 5 of Table 2.1. Trajectories of the CN and NV methods are compared in Figure 2.4 for the parameter values $a = 0.5$, $b = 1$.

Branin [Bra72] showed that, as c increases from zero, the Newton vector field (2.47) for the function defined in (2.46) correspondingly possesses an increasing number of extraneous singularities.

CLF design of continuous algorithms with dynamic controllers $\mathcal{C}_d$

Starting once again from (2.3), (2.4), and (2.13), a new quadratic CLF is chosen in order to design dynamic controllers of the type $\mathcal{C}_d$ presented in (2.11). Defining

$$V_2 := \frac{1}{2}\mathbf{r}^T\mathbf{r} + \frac{1}{2}\mathbf{u}^T\mathbf{u}, \tag{2.54}$$

it is clear that $V_2(\cdot)$ is a positive definite function of the variables $\mathbf{u}$ and $\mathbf{r}$. Taking the time derivative of V_2 yields

$$\dot{V}_2 = \mathbf{r}^T\dot{\mathbf{r}} + \mathbf{u}^T\dot{\mathbf{u}} = \mathbf{u}^T(\mathbf{D_f}^T\mathbf{f} + \dot{\mathbf{u}}). \tag{2.55}$$

This simple calculation motivates some choices of "stabilizing" dynamics for $\mathbf{u}$ which are analyzed below. The most straightforward choice to make $\dot{V}_2 \leq 0$ is

$$\dot{\mathbf{u}} = -\mathbf{\Gamma}\mathbf{u} - \mathbf{D_f}^T\mathbf{f}, \qquad \mathbf{\Gamma} = \mathbf{\Gamma}^T > 0, \tag{2.56}$$

which can be justified as follows. Substituting the choice (2.56) in (2.55) gives $\dot{V}_2 = -\mathbf{u}^T\mathbf{\Gamma}\mathbf{u} \leq 0$, which is negative semidefinite, since both $\mathbf{u}$ and $\mathbf{r}$ are now state variables and

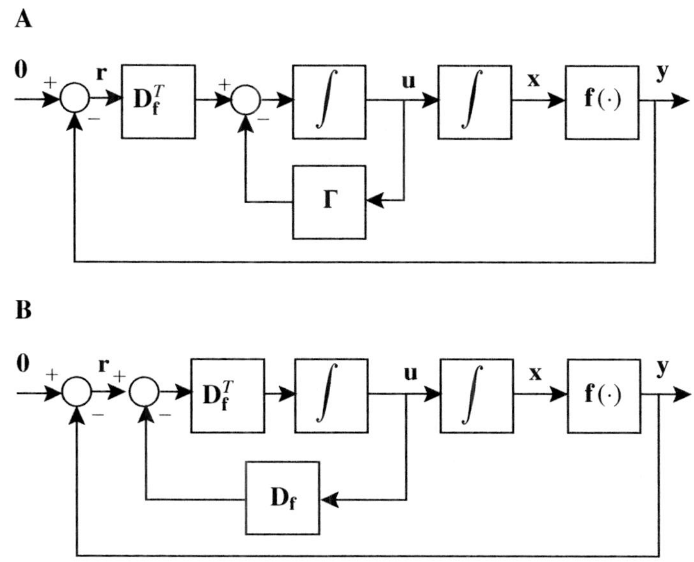

Figure 2.3. *A: Block diagram representations of continuous algorithms for the zero finding problem, using the dynamic controller defined by* (2.56). *B: With the particular choice* $\mathbf{\Gamma} = \mathbf{D}_\mathbf{f}^T \mathbf{D}_\mathbf{f}$.

Table 2.2. *Zero finding continuous algorithms with dynamic controllers designed using the quadratic CLF* (2.54) *(p.d. = positive definite).*

Plant	Dynamic controller	Choice of $\mathbf{\Gamma}$	$\varphi(\mathbf{x}, \mathbf{r})$	Acronym
$\dot{\mathbf{x}} = \mathbf{u}$	$\dot{\mathbf{u}} = -\mathbf{\Gamma u} - \mathbf{D}_\mathbf{f}^T \mathbf{f}(\mathbf{x})$	Arbitrary p.d.	$\mathbf{D}_\mathbf{f}^T \mathbf{r}$	DC1
$\dot{\mathbf{x}} = \mathbf{u}$	$\dot{\mathbf{u}} = -\mathbf{D}_\mathbf{f}^T(\mathbf{D}_\mathbf{f}\mathbf{u} + \mathbf{f}(\mathbf{x}))$	$\mathbf{D}_\mathbf{f}^T \mathbf{D}_\mathbf{f}$	$\mathbf{D}_\mathbf{f}^T \mathbf{r}$	DC2
$\dot{\mathbf{x}} = \mathbf{D}_\mathbf{f}^{-1}\mathbf{u}$	$\dot{\mathbf{u}} = -\mathbf{\Gamma u} - \mathbf{f}(\mathbf{x})$	Arbitrary p.d.	$\mathbf{r}$	DC3
$\dot{\mathbf{x}} = \mathbf{D}_\mathbf{f}^T \mathbf{u}$	$\dot{\mathbf{u}} = -\mathbf{\Gamma u} - \mathbf{D}_\mathbf{f}\mathbf{D}_\mathbf{f}^T \mathbf{f}(\mathbf{x})$	Arbitrary p.d.	$\mathbf{D}_\mathbf{f}^T \mathbf{D}_\mathbf{f} \mathbf{r}$	DC4

$\dot{V}_2$ does not depend on the state variable $\mathbf{r}$. In this situation, LaSalle's theorem (Theorem 1.27) can be applied. In fact, $\dot{V}_2 = 0$ implies that $\mathbf{u} = \mathbf{0}$ and thus $\dot{\mathbf{u}} = \mathbf{0}$. From (2.56), it follows that $\mathbf{D}_\mathbf{f}^T \mathbf{f} = \mathbf{0}$. Now, under the usual assumption that the matrix $\mathbf{D}_\mathbf{f}$ is invertible, it follows that $\mathbf{f}(\mathbf{x}) = \mathbf{0}$, as desired.

One choice of $\mathbf{\Gamma}$ is $\mathbf{D}_\mathbf{f}^T \mathbf{D}_\mathbf{f}$, which also allows a simple implementation of the resulting dynamic controller (see Figure 2.3B).

The block diagrams for these continuous algorithms are given in Figure 2.3.

Similarly, using (2.28) and (2.32), the quadratic CLF (2.54), and LaSalle's theorem, two other continuous algorithms for zero finding are easily derived. For convenience, all these algorithms are displayed in Table 2.2. Once again, it should be observed that both the controller structure and the choices that define the specific controller emerge naturally from the CLF approach and do not need to be specified a priori.

The prototypical stability result for the class of dynamical controllers (2.11) can be stated as follows.

Theorem 2.3. *Given a function* $\mathbf{f} : \mathbb{R}^n \to \mathbb{R}^n$, *suppose that* $\mathbf{x}^*$ *is a zero of* $\mathbf{f}$ *such that the Jacobian matrix* $\mathbf{D_f}(\mathbf{x})$ *of* $\mathbf{f}$ *is invertible in some neighborhood* $\mathcal{N}(\mathbf{x}^*)$ *of* $\mathbf{x}^*$. *Then, for all initial conditions* $\mathbf{x}_0 \in \mathcal{N}(\mathbf{x}^*)$, *the corresponding trajectories (in* $\mathbf{x}(\cdot)$*) of the dynamical system* (2.10)–(2.11) *(repeated here for ease of reference)*

$$\dot{\mathbf{x}} = \mathbf{u},$$
$$\dot{\mathbf{u}} = -\mathbf{\Gamma u} + \boldsymbol{\varphi}(\mathbf{x}, \mathbf{r}),$$

where $\mathbf{\Gamma}$ *and* $\boldsymbol{\varphi}(\mathbf{x}, \mathbf{r})$ *are chosen by the CLF approach in the manner specified in Table* 2.2, *converge to the zero* $\mathbf{x}^*$ *of* $\mathbf{f}(\cdot)$.

A final commentary on the class of dynamic controllers is that they lead to second-order dynamical systems whose trajectories converge to the zeros of a function. To see this, one can eliminate $\mathbf{u}$ from the pair of equations (2.10)–(2.11), which yields

$$\ddot{\mathbf{x}} + \mathbf{\Gamma}\dot{\mathbf{x}} + \boldsymbol{\varphi}(\mathbf{x}, \mathbf{r}) = \mathbf{0}, \tag{2.57}$$

where $\boldsymbol{\varphi}(\mathbf{x}, \mathbf{r})$ is chosen in the manner specified in Table 2.2.

This idea of using second-order dynamical systems has been proposed before, having been studied in [Pol64] in an optimization context, and subsequently in [Bog71, IPZ79, DS80] in a zero finding context. All these authors, however, arrived at second-order dynamical systems using classical mechanics analogies, such as that of a heavy ball moving in a force field and subject to friction (or some other dissipative force). The discussion above shows that second-order dynamical systems for zero finding can be derived in a natural way from the CLF approach. Further remarks are made below, in an optimization context, in section 2.3.3.

Numerical simulations of continuous algorithms

Some numerical simulations of the continuous algorithms derived above are presented in Figures 2.4 and 2.5 for the Rosenbrock function; in keeping with the "perspectives" nature of this book, these are illustrative examples, and further research needs to be done in order to determine the effectiveness of the new algorithms, beyond the fact that the singularity structure of the associated vector fields is different (from that of the Newton algorithm) and, as a consequence, the trajectories take different routes to the zeros to which they converge.

2.2 Iterative Zero Finding Algorithms as Discrete-Time Dynamical Systems

> *Increasingly often, it is not optimal to try to solve a problem exactly in one pass; instead, solve it approximately, then iterate ... iterative, infinite algorithms are sometimes better.... our central mission is to compute quantities that are typically uncomputable, from an analytical point of view, and to do it with lightning speed.*
>
> –L. N. Trefethen [Tre92]

We now turn to discrete or iterative algorithms, maintaining our feedback control point of view. From (2.4) and (2.5), given $\mathbf{x}_k$ at the kth iteration, we can define

$$\mathbf{r}_k := \mathbf{r}(\mathbf{x}_k) = -\mathbf{f}(\mathbf{x}_k). \tag{2.58}$$

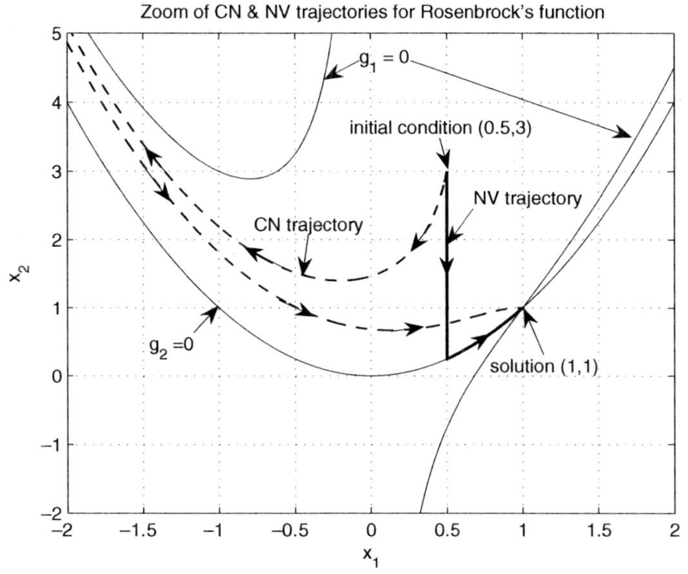

Figure 2.4. *Comparison of CN and NV trajectories for minimization of Rosenbrock's function* (2.44), *with* $a = 0.5$, $b = 1$, *or equivalently, finding the zeros of* $\mathbf{g}$ *in* (2.45).

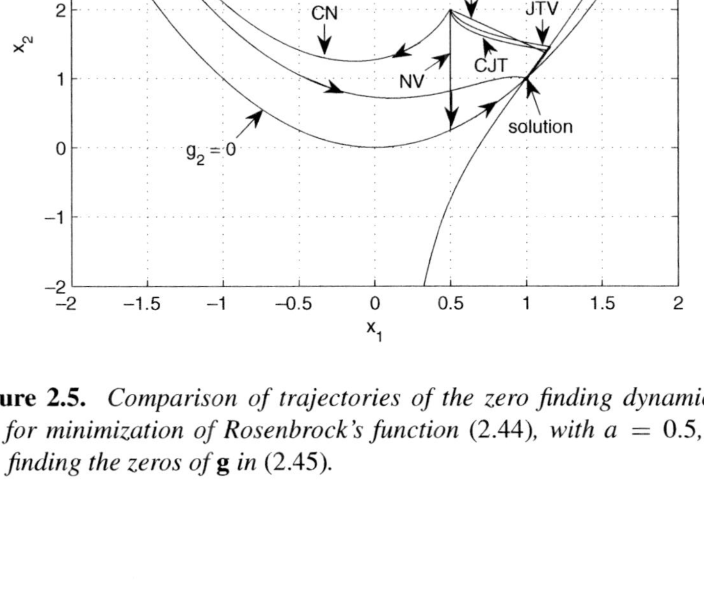

Figure 2.5. *Comparison of trajectories of the zero finding dynamical systems of Table* 2.1 *for minimization of Rosenbrock's function* (2.44), *with* $a = 0.5$, $b = 1$, *or equivalently, finding the zeros of* $\mathbf{g}$ *in* (2.45).

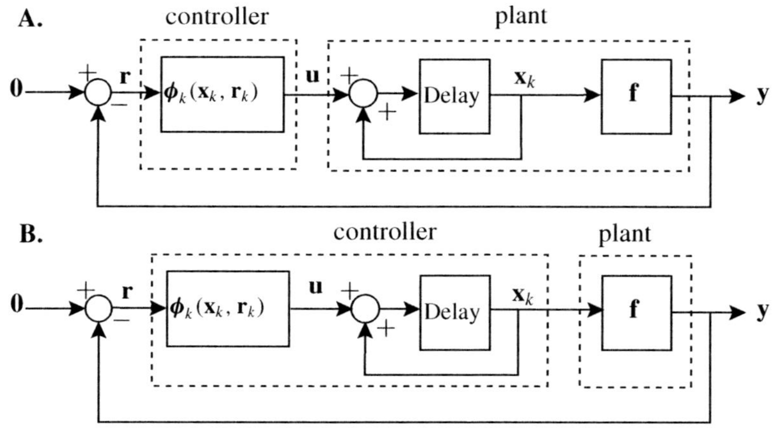

Figure 2.6. A: *A discrete-time dynamical system realization of a general iterative method represented as a feedback control system. The plant, object of the control, represents the problem to be solved, while the controller is a representation of the algorithm designed to solve it. As quadruples, plant* $\mathcal{P} = \{\mathbf{I}, \mathbf{I}, \mathbf{f}, \mathbf{0}\}$ *and controller* $\mathcal{C} = \{\mathbf{0}, \mathbf{0}, \mathbf{0}, \boldsymbol{\phi}_k(\mathbf{x}_k, \mathbf{r}_k)\}$. B: *An alternative discrete-time dynamical system realization of a general iterative method, represented as a feedback control system. As quadruples,* $\mathcal{P} = \{\mathbf{0}, \mathbf{0}, \mathbf{0}, \mathbf{f}\}$ *and* $\mathcal{C} = \{\mathbf{I}, \boldsymbol{\phi}_k(\mathbf{x}_k, \mathbf{r}_k), \mathbf{I}, \mathbf{0}\}$.

Now assume that we define an iteration in $\mathbf{x}$ as follows:

$$\mathbf{x}_{k+1} = \mathbf{x}_k + \mathbf{u}_k, \tag{2.59}$$

where $\mathbf{u}_k$ is a control to be chosen. From (2.59):

$$\mathbf{f}(\mathbf{x}_{k+1}) = \mathbf{f}(\mathbf{x}_k + \mathbf{u}_k),$$

and, using (2.58) and the Taylor expansion of the right-hand side around $\mathbf{x}_k$ and keeping only the first-order term, yields

$$\mathbf{r}_{k+1} = \mathbf{r}_k - \mathbf{D}_\mathbf{f}\mathbf{u}_k, \tag{2.60}$$

where $\mathbf{D}_\mathbf{f} := \mathbf{D}_\mathbf{f}(\mathbf{x}_k)$ is the Jacobian matrix of $\mathbf{f}$. Note that (2.60) defines $\mathbf{r}_{k+1}$ and our task is to choose $\mathbf{u}_k$ using the CLF/LOC approach in order that $\mathbf{r}_k = -\mathbf{f}(\mathbf{x}_k) \to \mathbf{0}$ as $k \to \infty$.

Note that an equivalent interpretation of (2.60) is that it is the forward Euler discretization of (2.13). In this interpretation, it is assumed that the choice of stepsize has been absorbed into the control $\mathbf{u}_k$. If the general expression $\boldsymbol{\phi}_k(\mathbf{x}_k, \mathbf{r}_k)$ defines the term $\mathbf{u}_k$ in (2.59), then the iterative method can be represented as a feedback control system as shown in Figure 2.6.

The quadratic CLF/LOC lemma

Section 2.1 introduced some simple choices of feedback laws, found by a CLF/LOC approach that led to corresponding continuous-time zero finding dynamical systems. A natural idea is to try and discretize the latter in order to come up with discrete-time dynamical systems (i.e., iterative algorithms in the conventional sense) that determine zeros. As the preceding discussion shows, this leads to (2.60). From this point of view, all that needs to be done is to choose an appropriate time-varying stepsize. In keeping with the philosophy of the previous section, this choice of time-varying stepsize will also be done using the CLF/LOC approach, by postulating that the control can be written as

$$\mathbf{u}_k = \alpha_k \boldsymbol{\phi}(\mathbf{r}_k), \tag{2.61}$$

where α_k represents the time-varying stepsize that is to be chosen, and $\boldsymbol{\phi}(\mathbf{r}_k)$ represents the choice of feedback law. Substituting (2.61) in (2.60) gives

$$\mathbf{r}_{k+1} = \mathbf{r}_k - \alpha_k \boldsymbol{\psi}(\mathbf{r}_k), \tag{2.62}$$

where

$$\boldsymbol{\psi}(\mathbf{r}_k) = \mathbf{D_f}\boldsymbol{\phi}(\mathbf{r}_k). \tag{2.63}$$

In order to choose α_k, (2.62) will be thought of as a discrete-time dynamical system, affine in the scalar control α_k. The following lemma uses a quadratic CLF and LOC to derive α_k and does not assume the particular form (2.63) so that it can be used repeatedly in what follows, in different situations.

Lemma 2.4 CLF/LOC lemma. *The zero solution of the discrete-time dynamical system* (2.62) *is asymptotically stable if α_k is chosen as*

$$\alpha_k = \frac{\langle \mathbf{r}_k, \boldsymbol{\psi}(\mathbf{r}_k)\rangle}{\langle \boldsymbol{\psi}(\mathbf{r}_k), \boldsymbol{\psi}(\mathbf{r}_k)\rangle}, \tag{2.64}$$

provided that

$$\boldsymbol{\psi}(\mathbf{r}_k) \neq \mathbf{0}, \ \textit{whenever } \mathbf{r}_k \neq \mathbf{0}. \tag{2.65}$$

Proof. Consider the quadratic CLF

$$V(\mathbf{r}_k) = \langle \mathbf{r}_k, \mathbf{r}_k\rangle = \mathbf{r}_k^T\mathbf{r}_k = \|\mathbf{r}_k\|^2.$$

Taking the inner product of each side of (2.62) with itself gives

$$\langle \mathbf{r}_{k+1}, \mathbf{r}_{k+1}\rangle = \langle \mathbf{r}_k - \alpha_k\boldsymbol{\psi}(\mathbf{r}_k), \mathbf{r}_k - \alpha_k\boldsymbol{\psi}(\mathbf{r}_k)\rangle,$$

which can be expanded and rearranged as

$$\Delta V := \langle \mathbf{r}_{k+1}, \mathbf{r}_{k+1}\rangle - \langle \mathbf{r}_k, \mathbf{r}_k\rangle = -2\alpha_k\langle \mathbf{r}_k, \boldsymbol{\psi}(\mathbf{r}_k)\rangle + \alpha_k^2\langle \boldsymbol{\psi}(\mathbf{r}_k), \boldsymbol{\psi}(\mathbf{r}_k)\rangle. \tag{2.66}$$

The LOC choice of α_k is the one that makes ΔV as negative as possible and is found by setting the partial derivative of ΔV with respect to α_k equal to zero.

$$\frac{\partial \Delta V}{\partial \alpha_k} = -2\langle \mathbf{r}_k, \boldsymbol{\psi}(\mathbf{r}_k)\rangle + 2\alpha_k\langle \boldsymbol{\psi}(\mathbf{r}_k), \boldsymbol{\psi}(\mathbf{r}_k)\rangle = 0$$

Table 2.3. *The entries of the first and fourth columns define an iterative zero finding algorithm* $\mathbf{x}_{k+1} = \mathbf{x}_k + \alpha_k \boldsymbol{\phi}(\mathbf{r}_k) = \mathbf{x}_k + \alpha_k \boldsymbol{\phi}(-\mathbf{f}(\mathbf{x}_k))$ *(see Theorem 2.5). The matrices in the third and fifth rows are defined, respectively, as* $\mathbf{M}_k = \mathbf{D_f P D_f}^T$, $\mathbf{W}_k = \mathbf{D_f D_f}^T$.

$\phi(\mathbf{r}_k)$	$\psi(\mathbf{r}_k)$	Acronym	LOC choice of α_k
$\mathbf{D_f}^{-1}\mathbf{r}_k$	$\mathbf{r}_k$	DN	1
$\mathbf{D_f}^{-1}\text{sgn}(\mathbf{r}_k)$	$\text{sgn}(\mathbf{r}_k)$	DNV	$\dfrac{\Vert\mathbf{r}_k\Vert_1}{\Vert\text{sgn}(\mathbf{r}_k)\Vert_2^2}$
$\mathbf{P D_f}^T\mathbf{r}_k$	$\mathbf{D_f P D_f}^T\mathbf{r}_k$	DJT	$\dfrac{\langle\mathbf{r}_k, \mathbf{M}_k\mathbf{r}_k\rangle}{\Vert\mathbf{M}_k\mathbf{r}_k\Vert_2^2}$
$\text{sgn}(\mathbf{D_f}^T\mathbf{r}_k)$	$\mathbf{D_f}\text{sgn}(\mathbf{D_f}^T\mathbf{r}_k)$	DVJT	$\dfrac{\Vert\mathbf{D_f}^T\mathbf{r}_k\Vert_1}{\Vert\mathbf{D_f}\text{sgn}(\mathbf{D_f}^T\mathbf{r}_k)\Vert_2^2}$
$\mathbf{D_f}^T\text{sgn}(\mathbf{r}_k)$	$\mathbf{D_f D_f}^T\text{sgn}(\mathbf{r}_k)$	DJTV	$\dfrac{\langle\mathbf{r}_k, \mathbf{W}_k\text{sgn}(\mathbf{r}_k)\rangle}{\Vert\mathbf{W}_k\text{sgn}(\mathbf{r}_k)\Vert_2^2}$

yields

$$\alpha_k = \frac{\langle \mathbf{r}_k, \boldsymbol{\psi}(\mathbf{r}_k)\rangle}{\langle \boldsymbol{\psi}(\mathbf{r}_k), \boldsymbol{\psi}(\mathbf{r}_k)\rangle},$$

which leads to

$$\Delta V = -\frac{\langle \mathbf{r}_k, \boldsymbol{\psi}(\mathbf{r}_k)\rangle^2}{\langle \boldsymbol{\psi}(\mathbf{r}_k), \boldsymbol{\psi}(\mathbf{r}_k)\rangle} < 0 \quad \text{for all } \mathbf{r}_k \neq \mathbf{0}, \tag{2.67}$$

since the numerator is a square and the denominator is a squared two-norm. □

From this lemma, the following theorem is immediate.

Theorem 2.5. *Given the function* $\mathbf{f} : \mathbb{R}^n \to \mathbb{R}^n$, *consider the iterative algorithm*

$$\mathbf{x}_{k+1} = \mathbf{x}_k + \alpha_k \boldsymbol{\phi}(-\mathbf{f}(\mathbf{x}_k)), \tag{2.68}$$

where α_k *satisfies* (2.64) *and the function* $\boldsymbol{\phi}(\cdot)$ *is well defined in some neighborhood* $\mathcal{N}(\mathbf{x}^*)$ *of a zero* $\mathbf{x}^*$ *of* $\mathbf{f}(\cdot)$ *and, in addition, satisfies* (2.65). *Then, for each initial condition* $\mathbf{x}_0 \in \mathcal{N}(\mathbf{x}^*)$, *the corresponding trajectory of* (2.68) *converges to the zero* $\mathbf{x}^*$ *of* $\mathbf{f}(\cdot)$.

It remains to substitute each of the five specific choices of control from section 2.1 (Table 2.1) into the formula (2.64), check (2.65), and calculate the resulting α_k's, to get discrete-time versions, of the form (2.68), of the continuous-time algorithms of section 2.1.

The results correspond to the different iterative methods presented below and tabulated in Table 2.3, which gives the specific LOC choices of stepsize α_k.

The discrete-time *Newton method (DN)*, also called the *Newton–Raphson* method, is

$$\mathbf{x}_{k+1} = \mathbf{x}_k - \mathbf{D_f}^{-1}\mathbf{f}(\mathbf{x}_k). \tag{2.69}$$

The discrete-time *Newton variable structure method (DNV)* is

$$\mathbf{x}_{k+1} = \mathbf{x}_k - \alpha_k \mathbf{D_f}^{-1}\text{sgn}(\mathbf{f}(\mathbf{x}_k)). \tag{2.70}$$

The discrete-time *Jacobian matrix transpose method (DJT)* is

$$\mathbf{x}_{k+1} = \mathbf{x}_k - \alpha_k \mathbf{P}\mathbf{D}_\mathbf{f}^T \mathbf{f}(\mathbf{x}_k). \tag{2.71}$$

The discrete-time *variable structure Jacobian matrix transpose method (DVJT)* is

$$\mathbf{x}_{k+1} = \mathbf{x}_k - \alpha_k \mathrm{sgn}(\mathbf{D}_\mathbf{f}^T \mathbf{f}(\mathbf{x}_k)). \tag{2.72}$$

The discrete-time *Jacobian matrix transpose variable structure method (DJTV)* is

$$\mathbf{x}_{k+1} = \mathbf{x}_k - \alpha_k \mathbf{D}_\mathbf{f}^T \mathrm{sgn}(\mathbf{f}(\mathbf{x}_k)). \tag{2.73}$$

Observe that the CLF/LOC lemma can also be used to derive discrete-time versions of the dynamic controller–based continuous-time algorithms in Table 2.2. This derivation is not carried out here (see [PB06]), but is similar to the one in section 2.3.2, in which two coupled discrete-time iterations, which represent a dynamic controller for a linear equation, are analyzed using the CLF/LOC lemma: The discrete-time algorithms corresponding to the dynamic controller-based continuous-time algorithms DC1, DC2, DC3, DC4 in Table 2.2 are given the acronyms DDC1, DDC2, etc.

Comparison of discrete-time methods in Table 2.3

Some numerical examples of the application of the various discrete-time algorithms are given. Once again, these are illustrative examples, and further research needs to be done in order to determine the effectiveness of the new algorithms, beyond the fact that the trajectories take different routes to the zeros to which they converge.

Example 2.6. Branin's example (2.46) is used here to compare the different discrete-time zero finding algorithms proposed in Table 2.3. In Figure 2.7, we show the behavior of the discrete-time algorithms with an LOC choice of stepsize (described in Table 2.3) for this example, with the parameter $c = 0.05$ (one of the values studied in [Bra72, p. 510ff]). Figures 2.9 and 2.10 show the behavior of the static controller–based DJT and DJTV algorithms, compared with the dynamic controller–based algorithms DDC1 and DDC2, with the parameter $c = 0.5$.

Continuous-time algorithms as a first step in the design of discrete-time algorithms

We put the simple but useful quadratic CLF/LOC lemma in perspective by pointing out that, in the discrete-time case, it is no longer as easy to use the 1-norm as it was in the continuous-time case. This is because the requisite manipulations to arrive at a majorization or equivalent expression for the decrement ΔV in the 1 norm are not easy to carry out.

Thus the strategy is to use the continuous-time algorithms to get at the form of the (feedback) controls. Then, thinking of the discrete-time algorithm as a (forward Euler) variable stepsize discretization of the continuous-time algorithm, the stepsize is interpreted as a control, whereas the (possibly nonlinear) feedback law is now interpreted as part of a (possibly nonlinear) system, affine in the control (stepsize). This is where the quadratic CLF/LOC lemma comes in, giving a simple way to choose the variable stepsize, using the familiar quadratic (2-norm) CLF and LOC strategy.

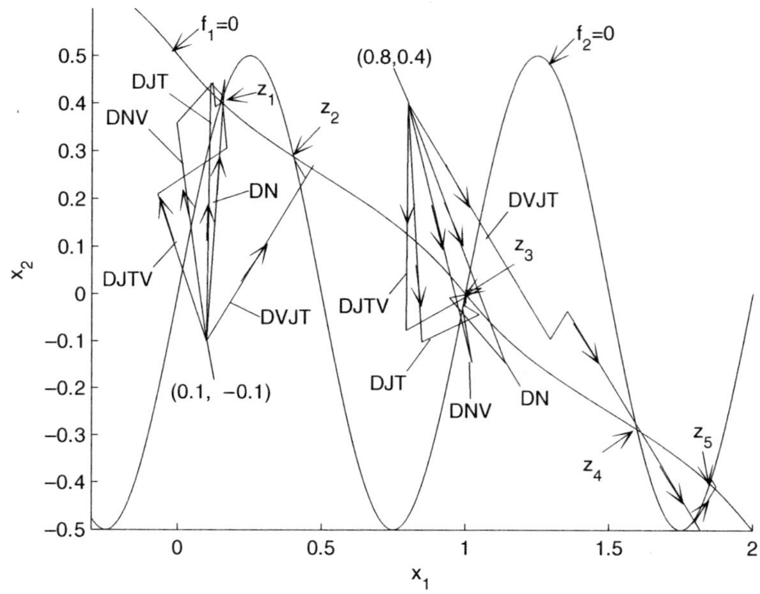

Figure 2.7. *Comparison of different discrete-time methods with LOC choice of stepsize (Table* 2.3) *for Branin's function with* $c = 0.05$. *Calling the zeros* z_1 *through* z_5 *(from left to right), observe that, from initial condition* $(0.1, -0.1)$, *the algorithms DN, DNV, DJT, DJTV converge to* z_1, *whereas DVJT converges to* z_2. *From initial condition* $(0.8, 0.4)$, *once again, DN, DNV, DJT, and DJTV converge to the same zero* (z_3), *whereas DVJT converges to* z_5.

Simplifying the adaptive stepsize formulas

From a computational point of view, it is simpler, whenever possible, to avoid the complex adaptation formulas for α_k in the last column of Table 2.3 and to use a simplified stepsize formula.

This is exemplified by taking another closer look at ΔV for the DNV:

$$\Delta V = \langle \mathbf{r}_{k+1}, \mathbf{r}_{k+1} \rangle - \langle \mathbf{r}_k, \mathbf{r}_k \rangle = -2\langle \alpha_k \text{sgn}(\mathbf{r}_k), \mathbf{r}_k \rangle + \langle \alpha_k \text{sgn}(\mathbf{r}_k), \alpha_k \text{sgn}(\mathbf{r}_k) \rangle. \quad (2.74)$$

From (2.74), clearly a sufficient condition for $\Delta V < 0$ is

$$\alpha_k \leq (2\|\mathbf{r}_k\|_1/n). \quad (2.75)$$

Use of (2.75) to choose α_k is not optimal (in the LOC sense), but could lead to some simplification. For example, if $\|\mathbf{r}_k\|_1$ is being monitored (to measure progress to convergence), then it is enough to keep α_k piecewise constant and change its value only when (2.75) is about to be violated.

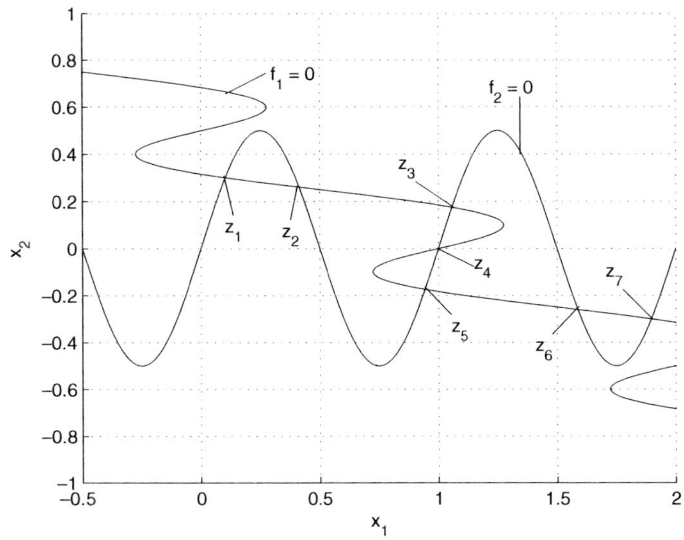

Figure 2.8. *Plot of Branin's function* (2.46), *with* $c = 0.5$, *showing its seven zeros,* z_1 *through* z_7. *This value of* c *is used in Figure* 2.9.

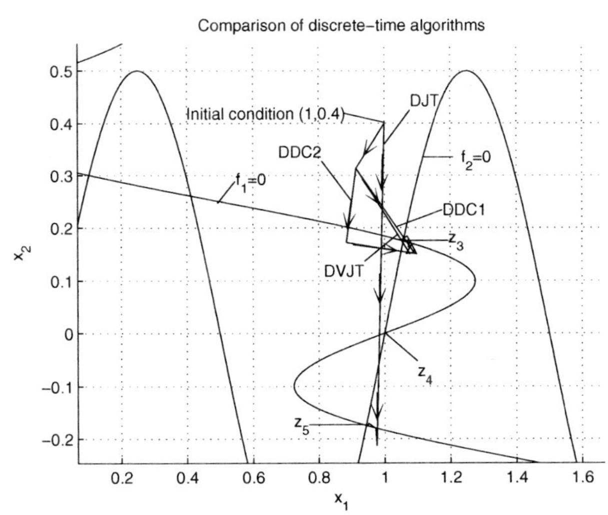

Figure 2.9. *Comparison of the trajectories of the DJT, DVJT, DDC*1, *and DDC*2 *algorithms from the initial condition* $(1, 0.4)$, *showing convergence to the zero* z_5 *of the DJT algorithm and to the zero* z_3 *of the other three algorithms. Note that this initial condition is outside the basin of attraction of the zero* z_3 *for all the other algorithms, including the Newton algorithm (DN). This figure is a zoom of the region around the zeros* z_3, z_4, *and* z_5 *in Figure* 2.8. *A further zoom is shown in Figure* 2.10.

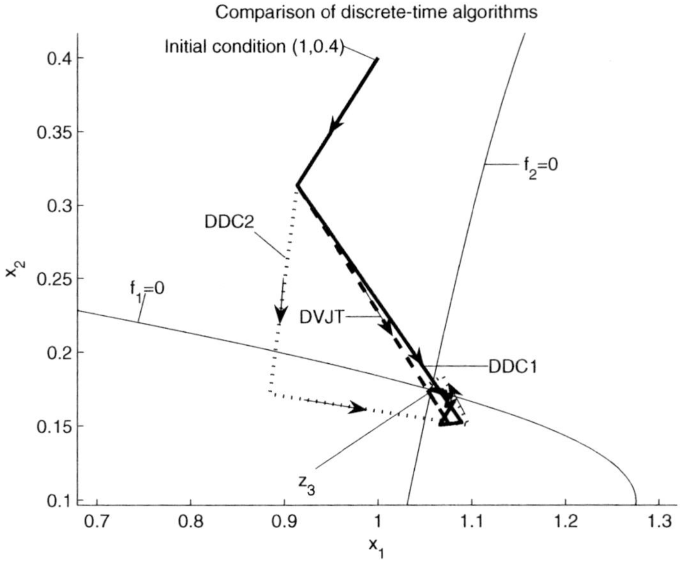

Figure 2.10. *Comparison of the trajectories of the DVJT, DDC*1*, and DDC*2 *algorithms from the initial condition* (1, 0.4)*, showing the paths to convergence to the zero* z_3*. This figure is a further zoom of the region around the zero* z_3 *in Figure* 2.9.

"Paradox" of one-step convergence of the discrete Newton method

From the expression for ΔV, it is clear that the choice $\alpha = 1$ is optimal as shown in the first row of Table 2.3, since it maximizes the decrease in the norm of $\mathbf{r}_k$. Indeed, rewriting ΔV as $V(\mathbf{r}_{k+1}) - V(\mathbf{r}_k) = (\alpha^2 - 2\alpha)V(\mathbf{r}_k)$, it appears that, for the LOC choice $\alpha = 1$, $V(\mathbf{r}_{k+1}) = 0$, implying that $\mathbf{r}_{k+1} = \mathbf{0}$. This, of course, is true only for the residue $\mathbf{r}$ in (2.60), which is a first-order approximation or linearization. The real residue, corresponding to the nonlinear Newton iteration (2.68), is given by

$$\mathbf{r}_{k+1} = -\mathbf{f}(\mathbf{x}_{k+1}) = -\mathbf{f}(\mathbf{x}_k - \mathbf{D}_\mathbf{f}^{-1}\mathbf{f}(\mathbf{x}_k)),$$

which is zero only up to first order because

$$\mathbf{f}(\mathbf{x}_k - \mathbf{D}_\mathbf{f}^{-1}\mathbf{f}(\mathbf{x}_k)) = \mathbf{f}(\mathbf{x}_k) - \mathbf{D}_\mathbf{f}\mathbf{D}_\mathbf{f}^{-1}\mathbf{f}(\mathbf{x}_k) + \text{h.o.t.} = \mathbf{0} + \text{h.o.t.},$$

where h.o.t. denotes higher order terms.

Should other discretization methods be used?

We have seen that the discretization of (2.19), using the forward Euler method, results in the standard discrete Newton iterative method. This raises the question of applying different approximation methods to the left-hand side of (2.19) in order to get corresponding discrete iterative methods that belong to the class of Newton methods (because they arise from discretizations of the continuous Newton method (2.19)), but have different convergence properties. Similar remarks apply to the other static controller–based algorithms in Table 2.3, as well as to the dynamic controller–based continuous algorithms in Table 2.2.

Deeper discussion of this point will take us too far afield, so we will refer the reader to Brezinski [Bre01] and earlier papers [Bog71, BD76, IPZ79] for details. Essentially, Brezinski [Bre01] shows that (i) the Euler method applied to (2.7) is "optimal" in the sense that explicit r-stage Runge–Kutta methods of order strictly greater than 1 cannot have a superlinear order of convergence; and (ii) suitable choice of a variable stepsize results in most of the known and popular methods. We have followed this line of reasoning, adopting the unifying view of the stepsize as a control input.

Condition for CLF of a continuous algorithm to work for its discrete version

A general result is available for variable stepsize Euler discretization. Consider the ODE

$$\dot{\mathbf{x}} = -\mathbf{g}(\mathbf{x}), \qquad \mathbf{x}(\mathbf{0}) = \mathbf{x}_0 \tag{2.76}$$

as well as

$$\mathbf{x}_{k+1} = \mathbf{x}_k - t_k \mathbf{g}(\mathbf{x}_k), \qquad \mathbf{x}_0 \text{ given,} \tag{2.77}$$

where $\mathbf{g} : D \subset \mathbb{R}^n \to \mathbb{R}^n$ is continuous and D is an open convex subset of $\mathbb{R}^n$.

Observe that Euler's method applied to (2.76) with a variable stepsize t_k yields (2.77). Since all iterative methods can be expressed in the form (2.77), (2.76) can be considered as the prototype *continuous analog* of (2.77), also referred to as a continuous algorithm; finally, it is often easier to work with (2.76) to obtain qualitative information on its behavior and then to use this to analyze the iterative method (2.77). Also, as Alber [Alb71] pointed out "theorems concerning the convergence of these (continuous) methods and theorems concerning the existence of solutions of equations and of minimum points of functionals are formulated under weaker assumptions than is the case for the analogous discrete processes."

Boggs [Bog76] observed that it is sometimes difficult to find an appropriate Liapunov function, but that it is often easier to find a Liapunov function for the continuous counterpart (2.76) and then use the same function for (2.77). His result and its simple proof are reproduced below.

Theorem 2.7 [Bog76]. *Let V be a Liapunov function for* (2.76) *at* $\mathbf{x}^*$. *Assume that ∇V is Lipschitz continuous with constant K on D. Suppose that there is a constant c independent of* $\mathbf{x}$ *such that* $\nabla V^T(\mathbf{x})\mathbf{g}(\mathbf{x}) \geq c\|\mathbf{g}(\mathbf{x})\|^2$. *Then there are constants $\underline{t}$ and $\bar{t}$ such that V is a Liapunov function for* (2.77) *at* $\mathbf{x}^*$ *for* $t_k \in [\underline{t}, \bar{t}]$. *Furthermore,* $\bar{t} < 2c/K$.

Proof. It only needs to be shown that $\Delta V < 0$ is satisfied along the trajectories of (2.77). Observe that

$$\begin{aligned}\Delta V &= V(\mathbf{x}_k - t_k\mathbf{g}(\mathbf{x}_k)) - V(\mathbf{x}_k)\\ &= \{V(\mathbf{x}_k - t_k\mathbf{g}(\mathbf{x}_k)) - V(\mathbf{x}_k) + t_k\nabla V(x)^T(\mathbf{x}_k)\mathbf{g}(\mathbf{x_k})\}\\ &\quad + [V(\mathbf{x}_k) - t_k\nabla V(x)^T(\mathbf{x}_k)\mathbf{g}(\mathbf{x}_k)] - V(\mathbf{x}_k).\end{aligned}$$

By the Lipschitz condition and by [OR70, Thm. 3, 2.12], the term in braces is bounded by $(1/2)Kt_k^2\|\mathbf{g}(\mathbf{x}_k)\|^2$. Therefore,

$$\begin{aligned}\Delta V &\leq -t_k \nabla V(\mathbf{x})^T \mathbf{g}(\mathbf{x}_k) + (1/2)Kt_k^2\|\mathbf{g}(\mathbf{x}_k)\|^2 \\ &\leq [-t_k c + (1/2)Kt_k^2]\|\mathbf{g}(\mathbf{x}_k)\|^2,\end{aligned}$$

which is strictly less than zero if $t_k c > (1/2)Kt_k^2$. Choose $\bar{t} < 2c/K$ and $\underline{t}$ such that $0 < \underline{t} < \bar{t} < 2c/K$, and therefore, for $t \in [\underline{t}, \bar{t}]$ the result follows. □

For the case of steepest descent, $\mathbf{g}(\mathbf{x}) = \nabla \mathbf{f}(\mathbf{x})$ and $\frac{\partial V}{\partial x}^T \mathbf{g}(\mathbf{x}) = \|\mathbf{g}(\mathbf{x})\|^2$, so that $c = 1$, and the steplengths are restricted to the interval $[\underline{t}, 2/K]$.

Clearly, the stepsize can be identified with the control input, and Theorem 2.7 is then seen as a result giving sufficient conditions under which a CLF for the continuous-time system (2.76) works for its discrete counterpart (2.77). Note that the control or stepsize (t_k) is restricted to lie in a bounded interval—a situation which is quite common in control as well. Boggs [Bog76] uses Theorem 2.7 to analyze the Ben-Israel iteration for nonlinear least squares problems; thus his analysis may be viewed as another application of the CLF approach.

Scalar iterative methods

Consider the scalar iterative method

$$x_{k+1} = x_k + u_k f(x_k). \tag{2.78}$$

Various well-known choices of u_k can be arrived at by analyzing the residual (linearized) iteration:

$$r_{k+1} = (1 - f'(x_k)u_k)r_k, \tag{2.79}$$

where $f'(x_k) = \mathrm{d}f(x_k)/\mathrm{d}x$ denotes the derivative (in x) of the function $f(\cdot)$, evaluated at x_k.

Some results for scalar iterative methods are presented in Table 2.4. More details on the order of convergence and choices of u_k for higher order methods that work when $f'(x)$ is not invertible (e.g., when f has a multiple zero), such as the Halley and Chebyshev methods, can be found in [Bre01], where u_k is regarded as a nonstationary stepsize for an Euler method (and accordingly denoted as h_k).

Scalar Newton method subject to disturbances

Using the control formulation of the Newton method discussed in this chapter and first published in [BK03], Kashima and coworkers [KAY05] develop an approach to the scalar Newton iteration with disturbances by treating it as a linear system with nonlinear feedback, known in control terminology as a Lur'e system.

We outline this approach here as an illustration of how control methods lead to further insights into the question of robustness to computational errors in zero finding problems.

Table 2.4. *Showing the choices of control u_k in (2.78) that lead to the common variants of the Newton method for scalar iterations.*

Choice of u_k	Name of method
$-f'(x_k)^{-1}$	Newton
$\dfrac{-(x_k - x_{k-1})}{[f(x_k) - f(x_{k-1})]}$	Secant
$\dfrac{f(x_k)}{[f(x_k + f(x_k)) - f(x_k)]}$	Steffensen

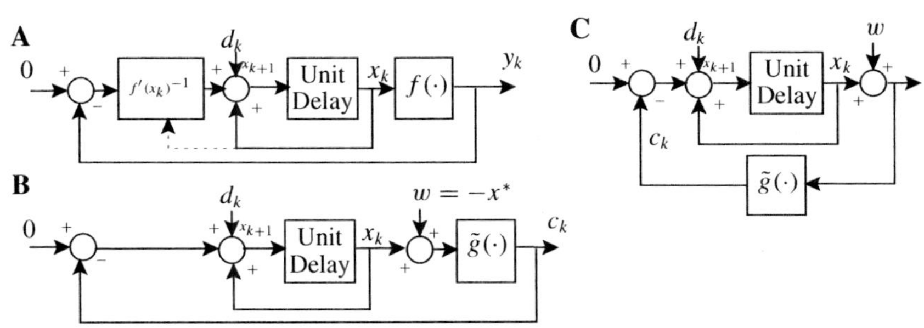

Figure 2.11. *Block diagram manipulations showing how the Newton method with disturbance d_k* (A) *can be redrawn in Lur'e form* (C). *The intermediate step* (B) *shows the introduction of a constant input w that shifts the input of the nonlinear function ${f'}^{-1}(\cdot)f(\cdot)$. The shifted function is named $\tilde{g}(\cdot)$. Note that part* A *is identical to Figure* 2.6A, *except for the additional disturbance input d_k.*

Suppose that $f : \mathbb{R} \to \mathbb{R}$ is a continuously differentiable function with nonzero derivative everywhere in $\mathbb{R}$. Assume that there exists x^* such that $f(x^*) = 0$. Then the Newton iteration subject to a disturbance or computational error d_k can be described as follows:

$$x_{k+1} = x_k - f'(x_k)^{-1} f(x_k) + d_k. \tag{2.80}$$

The feedback system description of (2.80) is shown in Figure 2.11A (which is a scalar version of Figure 2.5A, with the addition of a disturbance d_k). Convergence of x_k to x^*, in the absence of computational error ($d_k \equiv 0$), means that the output $y_k = f(x_k)$ converges to the reference input 0. The feedback system of Figure 2.11A may be equivalently redrawn as a system of the Lur'e type, in order to carry out the stability analysis under perturbations. The intermediate steps are explained with reference to Figure 2.11 as follows. In Figure 2.11 B, the function $(f')^{-1}f$ is denoted as g, and the shifted function $g(x + x^*)$ is denoted as $\tilde{g}$. Note that $\tilde{g}(x)$ is 0 for $x = 0$ (i.e., its graph goes through the origin). In addition, in order to use the so-called absolute stability theory, it is assumed that the function $\tilde{g}$ satisfies a *sector condition*; namely, there exist positive real numbers a and b such that

$$a \leq \frac{\tilde{g}(x)}{x} \leq b \quad \text{for all } x \neq 0. \tag{2.81}$$

This condition is referred to as a sector condition because it means that the graph of the function $\tilde{g}(x)$ is confined between straight lines of positive slope a and b through the origin.

In terms of the new function $\tilde{g}$, the block diagram of Figure 2.11A can be redrawn as in Figure 2.11B, in which the step disturbance w must be chosen as $-x^*$, so that, as before, the output $z_k = \tilde{g}(x_k + w) = \tilde{g}(x_k - x^*) = g(x_k)$ in Figure 2.11A.

This transforms the problem to the classical Lur'e form, since by defining the absolute error as $e_k := x_k - x^*$, the absolute error dynamics is as follows:

$$e_{k+1} = e_k - \tilde{g}(e_k) + d_k. \tag{2.82}$$

In this error system in Lur'e form, the objective is either to guarantee bounded error when disturbances are present or to make the error e_k converge to 0, in the absence of disturbances.

From absolute stability theory [Vid93], the following theorem can be proved.

Theorem 2.8 [KAY05]. *Consider the algorithm* (2.80) *and suppose that there exist constants a and b such that*

$$0 < a \leq \frac{f(x)}{(x - x^*)f'(x)} \leq b < 2 \tag{2.83}$$

for all $x \neq x^*$. *Then, for any initial value* x_0 *and any disturbance sequence* $d_k \in \ell^2$, *the error sequence* e_k *also belongs to* ℓ^2. *Furthermore, if* $d_k \equiv 0$, *then* x_k *converges to the exact solution* x^*.

Note that this theorem says that, if the disturbance sequence is square-summable, then so is the absolute error sequence.

For a proof of this result, some extensions, and further discussion on the relation of this result to the classical contraction mapping convergence condition, we refer the reader to [KAY05], which is based on the control formulation first given in [BK03].

Region of convergence for Newton's method via Liapunov theory

The Liapunov method can also be used to study the Newton method with disturbances (errors) in the vector case. The first step is to use Corollary 1.18 to find a region of convergence for iterative methods; the Newton method is used as an illustrative example [Hur67, p. 593ff].

Assuming, as usual, that $\mathbf{D_f}(\mathbf{x}_k)$ always has an inverse and that the desired zero $\mathbf{x}^*$ is simple, let the *absolute error* be defined as

$$\mathbf{e}_k := \mathbf{x}_k - \mathbf{x}^*. \tag{2.84}$$

Since $\mathbf{f}(\mathbf{x}^*) = \mathbf{0}$, the Taylor expansion of $\mathbf{f}$ around $\mathbf{x}^*$ becomes

$$\mathbf{f}(\mathbf{x}^* + \mathbf{e}_k) = \left[\mathbf{D_f}(\mathbf{x}^*)\right]\mathbf{e}_k + \mathbf{h}(\mathbf{e}_k), \tag{2.85}$$

where $\mathbf{h}(\cdot)$ denotes the higher order terms. Define the matrices

$$\mathbf{M}_1(\mathbf{e}_k) := \left[\mathbf{D_f}(\mathbf{x}^* + \mathbf{e}_k)\right]^{-1}, \quad \mathbf{M}_2(\mathbf{e}_k) := \left[\mathbf{D_f}(\mathbf{x}^* + \mathbf{e}_k) - \mathbf{D_f}(\mathbf{x}^*)\right]. \tag{2.86}$$

Subtracting $\mathbf{x}^*$ from both sides of the Newton iteration

$$\mathbf{x}_{k+1} = \mathbf{x}_k - [\mathbf{D_f}(\mathbf{x}_k)]^{-1}\,\mathbf{f}(\mathbf{x}_k), \tag{2.87}$$

the difference equation for the absolute error can be written as follows:

$$\mathbf{e}_{k+1} = \mathbf{M}_1(\mathbf{e}_k)\,[\mathbf{M}_2(\mathbf{e}_k)\mathbf{e}_k - \mathbf{h}(\mathbf{e}_k)]. \tag{2.88}$$

If $\mathbf{x}^*$ is a simple zero of $\mathbf{f}(\cdot)$ and $\mathbf{f}$ is twice continuously differentiable at $\mathbf{x}^*$, then for each $\eta > 0$, there exists a positive constant $c(\eta)$ such that for all $\mathbf{e}$ and for any vector norm $\|\cdot\|$:

$$\|\mathbf{e}\| < \eta \Rightarrow \|\mathbf{M}_1(\mathbf{e})\,[\mathbf{M}_2(\mathbf{e})\mathbf{e} - \mathbf{h}(\mathbf{e})]\,\| \le c(\eta)\|\mathbf{e}\|^2. \tag{2.89}$$

Choosing the Liapunov function as

$$V(\mathbf{e}) = \|\mathbf{e}\|, \tag{2.90}$$

it follows that

$$\Delta V(\mathbf{e}) \le -(1 - c(\eta)\|\mathbf{e}\|)\|\mathbf{e}\|, \tag{2.91}$$

which is nonpositive if

$$c(\eta)\|\mathbf{e}\| \le 1. \tag{2.92}$$

From Corollary 1.18, it follows that a region of convergence for the Newton method is

$$G(\eta_0) := \{\mathbf{x} : \|\mathbf{x} - \mathbf{x}^*\| < \eta_0\}, \tag{2.93}$$

where

$$\eta_0 := \min\{\eta, 1/c(\eta)\}. \tag{2.94}$$

The parameter η can then be chosen so as to maximize η_0, in order to obtain the largest region of convergence corresponding to the choice of Liapunov function (2.90).

Effect of disturbances on the Newton method via Liapunov theory

Corollary 1.20 is used to study the Newton method subject to a disturbance, modeled by the following discrete dynamical system, which is a vector version of (2.80):

$$\mathbf{x}_{k+1} = \mathbf{x}_k - [\mathbf{D_f}(\mathbf{x}_k)]^{-1}\,\mathbf{f}(\mathbf{x}_k) + \mathbf{d}_k, \tag{2.95}$$

where the error term $\mathbf{d}_k$ is only known in terms of an upper bound

$$\|\mathbf{d}_k\| \le \varepsilon, \tag{2.96}$$

in some norm $\|\cdot\|$, and for some $\varepsilon > 0$. Mimicking the calculations of a convergence region for the Newton method, the absolute error, defined as in (2.84), has the dynamics

$$\mathbf{e}_{k+1} = \mathbf{M}_1(\mathbf{e}_k)\,[\mathbf{M}_2(\mathbf{e}_k)\mathbf{e}_k - \mathbf{h}(\mathbf{e}_k)] + \overline{\mathbf{d}}_k(\mathbf{e}_k), \tag{2.97}$$

where $\mathbf{M}_i$, $i = 1, 2$, are defined as in (2.86). Using the Liapunov function (2.90), the decrement is (cf. (2.91))

$$\Delta V(\mathbf{e}) \le -(1 - c(\eta)\|\mathbf{e}\|)\|\mathbf{e}\| + \varepsilon =: W(\mathbf{e}) \le \varepsilon. \tag{2.98}$$

The set S in (1.72) is defined, in this case, as

$$S := \{\mathbf{e} : W(\mathbf{e}) \ge 0\} = \{\mathbf{e} : \|\mathbf{e}\| \le b\}, \tag{2.99}$$

where

$$b(\eta) = b = \frac{1 - \sqrt{1 - 4c(\eta)\varepsilon}}{2c(\eta)} = \varepsilon + 2c(\eta)\varepsilon^2 + \cdots, \tag{2.100}$$

provided that $4c(\eta)\varepsilon < 1$. If $4c(\eta)\varepsilon \ge 1$, then $W(\mathbf{e}) \ge 0$ everywhere, and the iterations may not converge. Finally, the set A in (1.72) is defined, in this case, as follows:

$$A := \{\mathbf{e} : V(\mathbf{e}) \le b + \varepsilon\} = \{\mathbf{e} : \|\mathbf{e}\| \le b + \varepsilon\}. \tag{2.101}$$

Clearly, for η small enough,

$$W(\mathbf{e}) \le -(b - c(\eta)(b + \varepsilon)^2) = -\delta. \tag{2.102}$$

From Corollary 1.20, if $\delta > 0$, then all solutions that start in $G(\eta_0)$ remain in $G(\eta_0)$, enter A in a finite number of iterations, and remain in A thereafter. The choice of η_0 is now a little more involved than in (2.94); it must be chosen such that

$$c(\eta_0)\eta_0 \le 1, \quad 4c(\eta_0)\varepsilon < 1, \quad b(\eta_0) - c(\eta_0)(b(\eta_0) + \varepsilon)^2 > 0.$$

If η_1 is chosen as the smallest positive solution of $\eta_1 c(\eta_1) = 1$, then choosing $\eta_0 < \eta_1$ will satisfy, simultaneously, the inequalities $c(\eta_0)\eta_0 < 1$ and $b - c(\eta_0)(b + \varepsilon)^2 > 0$. The remaining condition $4c(\eta_0)\varepsilon < 1$ can then be interpreted as a condition on the precision or accuracy required in the computation, clearly showing that disturbances reduce the region of convergence. If the disturbance is roundoff error, it is more realistic to replace $\mathbf{d}_k$ by $\mathbf{d}_k(\mathbf{x}_k)$, allowing for a roundoff error that depends on the vector $\mathbf{x}_k$, i.e., in control language, a state-dependent disturbance [Hur67]. Note that the analysis above is still valid, provided that the upper bound (2.96) holds for all $\mathbf{x}_k$. In light of this observation, we now refer to the disturbance as roundoff error. Another effect of roundoff errors that can be observed from this Liapunov analysis is that the error in the calculation of each $\mathbf{x}_k$ cannot, in general, be reduced below the value $b + \varepsilon = 2\varepsilon + 2c(\eta)\eta^2 + \cdots$, regardless of the number of iterations carried out. Thus this value is called the *ultimate accuracy* obtainable in the presence of roundoff errors. For small ε, the ultimate accuracy can be seen to be approximately 2ε, which is twice the roundoff error made at each step. The smaller the ultimate accuracy, the better the method is judged to be, in terms of sensitivity to roundoff errors, and, in this respect, the Newton method is a good method.

2.3 Iterative Methods for Linear Systems as Feedback Control Systems

This section specializes the discussion of the previous section, focusing on iterative methods to solve linear systems of equations of the form

$$\mathbf{A}\mathbf{x} = \mathbf{b}, \tag{2.103}$$

where $\mathbf{A} \in \mathbb{R}^{n\times n}$, $\mathbf{b} \in \mathbb{R}^n$. First, assuming that $\mathbf{A}$ is nonsingular, (2.103) has a unique solution $\mathbf{x}^* = \mathbf{A}^{-1}\mathbf{b} \in \mathbb{R}^n$, which it is desired to find, without explicitly inverting the matrix $\mathbf{A}$. A brief discussion of Krylov subspace iterative methods is followed by a control perspective on these and other methods.

Krylov subspace iterative methods

In order to solve the system of linear equations $\mathbf{Ax} = \mathbf{b}$, where $\mathbf{A}$ is an $n \times n$ nonsingular matrix and $\mathbf{b}$ is a given n-vector, one classical method is Gaussian elimination. It requires, for a dense matrix, storage of all n^2 entries of the matrix and approximately $2n^3/3$ arithmetic operations. Many matrices that arise in practical applications are, however, quite sparse and structured. A typical example is a matrix that arises in the numerical solution of a partial differential equation: Such a matrix has only a few nonzero entries per row. Other applications lead to banded matrices, in which the (i, j)-entry is zero whenever $|i - j| > m$. In general, Gaussian elimination can take only partial advantage of such structure and sparsity.

Matrix-vector multiplication can, on the other hand, take advantage of sparsity and structure. The reason for this is that, if a square $n \times n$ matrix has only k nonzero entries per row ($k \ll n$), then the product of this matrix with a vector will need just kn operations, compared to the $2n^2$ operations that would be required for a general dense matrix-vector multiplication. In addition, only the few nonzero entries of the matrix need to be stored.

The observations made in the preceding paragraphs lead to the question of whether it is possible to solve, at least to a good approximation, a linear system using mainly matrix-vector multiplications and a small number of additional operations. For, if this can be done, the corresponding iterative solution method should certainly be superior to Gaussian elimination, in terms of both computational effort and memory storage requirements.

For iterative methods, an initial guess for the solution is needed to start the iterative method. If no such guess is available, a natural first guess is some multiple of $\mathbf{b}$:

$$\mathbf{x}_1 \in \text{span}\,\{\mathbf{b}\}. \tag{2.104}$$

The next step is to compute the matrix-vector product $\mathbf{Ab}$ and take the next iterate to be some linear combination of the vectors $\mathbf{b}$ and $\mathbf{Ab}$:

$$\mathbf{x}_2 \in \text{span}\,\{\mathbf{b}, \mathbf{Ab}\}. \tag{2.105}$$

Proceeding in this way, at the kth step:

$$\mathbf{x}_k \in \text{span}\,\{\mathbf{b}, \mathbf{Ab}, \ldots, \mathbf{A}^{k-1}\mathbf{b}\}. \tag{2.106}$$

The subspace on the right-hand side of (2.106) is called, in numerical linear algebra, the *Krylov subspace* associated to the pair $(\mathbf{A}, \mathbf{b})$, denoted $\mathcal{K}_k(\mathbf{A}, \mathbf{b})$. In control theory, this subspace is a controllability subspace, as seen in Chapter 1.

Given that $\mathbf{x}_k$ is to be taken from the Krylov subspace $\mathcal{K}_k(\mathbf{A}, \mathbf{b})$, the two main questions that must be answered about such methods can be posed as follows [Gre97]: (i) How good an approximate solution to $\mathbf{Ax} = \mathbf{b}$ is contained in the Krylov subspace? (ii) How can a good (optimal) approximation from this subspace be computed with a moderate amount of work and storage?

Modifying Krylov subspaces by preconditioners

If the Krylov subspace $\mathcal{K}_k(\mathbf{A}, \mathbf{b})$ does not contain a good approximation of the solution for some reasonably small value of k, then a remedy might be to modify the original problem to obtain a better Krlyov subspace. One way to achieve this is to use a so-called preconditioner and solve the modified problem

$$\mathbf{P}^{-1}\mathbf{A}\mathbf{x} = \mathbf{P}^{-1}\mathbf{b} \tag{2.107}$$

by generating approximate solutions $\mathbf{x}_1, \mathbf{x}_2, \ldots$ that satisfy

$$\mathbf{x}_k = \text{Span}\,\{\mathbf{P}^{-1}\mathbf{b}, (\mathbf{P}^{-1}\mathbf{A})\mathbf{P}^{-1}\mathbf{b}, \ldots, (\mathbf{P}^{-1}\mathbf{A})^{k-1}\mathbf{P}^{-1}\mathbf{b}\}. \tag{2.108}$$

Note that at each step of the modified or preconditioned problem, it is necessary to compute the product of $\mathbf{P}^{-1}$ with a vector or, equivalently, to solve a linear system with coefficient matrix $\mathbf{P}$. This means that the matrix $\mathbf{P}$ should be chosen such that the linear system in $\mathbf{P}$ is easy to solve, and specifically, much easier to solve than the original system. The problem of finding a good preconditioner is a difficult one, and although much recent progress has been made, most preconditioners are designed for specific classes of problems, such as those arising in finite element and finite difference approximations of elliptic partial differential equations (PDEs) [Gre97]. In section 5.3, the preconditioning problem is approached from a control viewpoint.

Krylov subspace methods for symmetric matrices

Suppose that an approximation $\mathbf{x}_k$ is called optimal if its residual $\mathbf{b} - \mathbf{A}\mathbf{x}_k$ has minimal Euclidean norm. An algorithm that generates this optimal approximation is called *minimal residual*. If the matrix $\mathbf{A}$ is symmetric, there are known efficient (i.e., short) recurrences to find such an optimal approximation. If, in addition, $\mathbf{A}$ is also positive definite, then another possibility is to minimize the $\mathbf{A}$-norm of the error, $\|\mathbf{e}_k\|_{\mathbf{A}} := \langle \mathbf{A}^{-1}\mathbf{b} - \mathbf{x}_k, \mathbf{b} - \mathbf{A}\mathbf{x}_k \rangle^{1/2}$, and the conjugate gradient algorithm can be shown to do this [Gre97]. Each of these so-called Krylov subspace algorithms carries out one matrix-vector product, as well as a few vector inner products, implying that work and storage requirements are modest. Some Krylov subspace methods are approached from a control viewpoint in sections 2.3.1 and 2.3.2.

Simple iterative methods correspond to static controllers

Following the discussion of the previous section, a general linear iterative method, also called a simple iteration [Gre97], to solve (2.103) can be described by a recurrence of the form

$$\mathbf{x}_{k+1} = \mathbf{x}_k + \mathbf{K}\mathbf{r}_k, \quad k = 0, 1, 2, \ldots, \tag{2.109}$$

where $\mathbf{K}$ is a real $n \times n$ matrix and the residue $\mathbf{r}_k$, in each iteration, with respect to (2.103), is defined by

$$\mathbf{r}_k = \mathbf{b} - \mathbf{A}\mathbf{x}_k. \tag{2.110}$$

Exactly as in the previous section, it is possible to associate a discrete-time dynamical feedback system to the iterative method (2.109), and in consequence (2.111) can be viewed as a closed-loop dynamical system with a block diagram representation depicted in Figure

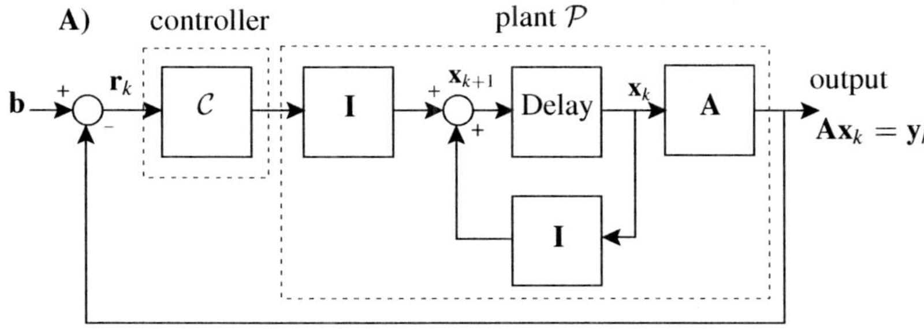

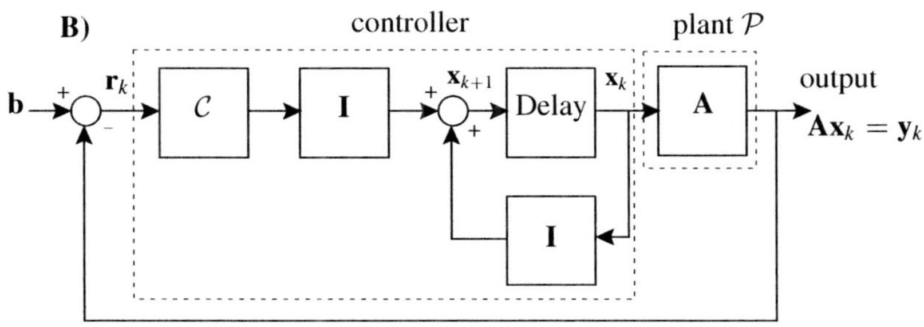

Figure 2.12. A: *A general linear iterative method to solve the linear system of equations* $\mathbf{Ax} = \mathbf{b}$ *represented in standard feedback control configuration. The plant is* $\mathcal{P} = \{\mathbf{I}, \mathbf{I}, \mathbf{A}, \mathbf{0}\}$*, whereas different choices of the linear controller* $\mathcal{C}$ *lead to different linear iterative methods.* B: *A general linear iterative method to solve the linear system of equations* $\mathbf{Ax} = \mathbf{b}$ *represented in an alternative feedback control configuration. The plant is* $\mathcal{P} = \{\mathbf{0}, \mathbf{0}, \mathbf{0}, \mathbf{A}\}$*, whereas different choices of the controller* $\mathcal{C}$ *lead to different iterative methods.*

2.12, where the controller $\mathcal{C}$ is given by $\{\mathbf{0}, \mathbf{0}, \mathbf{0}, \mathbf{K}\}$. The matrix $\mathbf{K}$ is often referred to as the *feedback gain matrix* and we will also use this term whenever appropriate. The observant reader will note that Figure 2.12 is a discrete version with linear plant of Figure 2.1.

Defining $\mathbf{y}_k := \mathbf{A}\mathbf{x}_k$ as the output vector of $S(\mathcal{P}, \mathcal{C})$, consider the constant vector $\mathbf{b}$ as the constant input to this system. The vector $\mathbf{r}_k$ represents the residue or error between the input $\mathbf{b}$ and the output $\mathbf{y}_k$ vectors. The numerical problem of solving the linear system $\mathbf{Ax} = \mathbf{b}$ is thus equivalent to the problem known in control terminology as the *servomechanism* or *regulator* problem of forcing the output $\mathbf{y}$ to regulate to the constant input $\mathbf{b}$, by a suitable choice of controller. When this is achieved, the state vector $\mathbf{x}$ reaches the desired solution of the linear system $\mathbf{Ax} = \mathbf{b}$.

Substituting the expression for $\mathbf{r}_k$ into (2.109), the iterative equation is obtained in the so-called output feedback form, i.e.,

$$\mathbf{x}_{k+1} = (\mathbf{I} - \mathbf{KA})\mathbf{x}_k + \mathbf{Kb}. \tag{2.111}$$

Notice that this corresponds to the choice of a static controller $\mathcal{C} = \{\mathbf{0}, \mathbf{0}, \mathbf{0}, \mathbf{K}\}$ and the iterative method (2.111) corresponds to this particular choice of controller $\mathcal{C}$. We exemplify this here by the classical *Jacobi iterative method*, described by the recurrence equation

$$\mathbf{x}_{k+1} = \mathbf{H}\mathbf{x}_k + \mathbf{D}^{-1}\mathbf{b}, \tag{2.112}$$

where $\mathbf{H} = \mathbf{D}^{-1}(\mathbf{E} + \mathbf{F})$ and the matrices $\mathbf{D}$, $\mathbf{E}$, and $\mathbf{F}$ are, respectively, strictly diagonal, lower, and upper triangular matrices obtained by splitting matrix $\mathbf{A}$ as $\mathbf{A} = -\mathbf{E} + \mathbf{D} - \mathbf{F}$.

Equating (2.111) and the classical Jacobi iterative equation (2.112), the relationship between the corresponding matrices is given by

$$\mathbf{H} = (\mathbf{I} - \mathbf{K}\mathbf{A}); \quad \mathbf{K} = \mathbf{D}^{-1}. \tag{2.113}$$

Other examples are as follows. If $\mathbf{K} = (\mathbf{D} - \mathbf{E})^{-1}$, then the recurrence (2.111) represents the *Gauss–Seidel* iterative method; if $\mathbf{K} = \left(\omega^{-1}\mathbf{D} - \mathbf{E}\right)^{-1}$, then it represents the *successive overrelaxation (SOR)* method; and, finally, if $\mathbf{K} = \omega\mathbf{D}^{-1}$, then it represents the *extrapolated Jacobi* method. This set of examples should make it clear that all these classical methods correspond to the choice of a static controller $\mathcal{C} = \{\mathbf{0}, \mathbf{0}, \mathbf{0}, \mathbf{K}\}$; the particular choice of $\mathbf{K}$ distinguishes one method from another. The formulation of iterative methods for linear systems as feedback control systems presented here was initiated in [SK01], where shooting methods for ODEs were also analyzed from this perspective. In order to complete the analysis, observe that, in all the cases considered above, the evolution of the residue $\mathbf{r}_k$ is given by the linear recurrence equation below, derived from (2.109) by multiplying both sides by $\mathbf{A}$ and subtracting each side from the vector $\mathbf{b}$:

$$\mathbf{r}_{k+1} = (\mathbf{I} - \mathbf{A}\mathbf{K})\mathbf{r}_k. \tag{2.114}$$

From (2.114) it is clear that convergence of the linear iterative method is ensured if the matrix $\mathbf{S}$ has all its eigenvalues within the unit disk (i.e., is Schur stable), where

$$\mathbf{S} = (\mathbf{I} - \mathbf{A}\mathbf{K}). \tag{2.115}$$

The eigenvalues of this matrix are given by the roots of the characteristic equation:

$$\det(\mathbf{I}\lambda - \mathbf{I} + \mathbf{A}\mathbf{K}) = 0. \tag{2.116}$$

Thus, designing an iterative method with an adequate rate of convergence corresponds to a certain choice of the feedback gain matrix $\mathbf{K}$. This is an instance of the well-studied *inverse eigenvalue problem* known in control as the *problem of pole assignment by state feedback* [Kai80, Son98]. More precisely, (2.114) can be viewed as the dynamical system $\{\mathbf{I}, \mathbf{A}, \mathbf{0}, \mathbf{0}\}$ subject to state feedback with gain matrix $\mathbf{K}$. From standard control theory, it is well known that there exists a state feedback gain $\mathbf{K}$ that results in arbitrary placement of the eigenvalues of the "closed-loop" matrix $\mathbf{S} = \mathbf{I} - \mathbf{A}\mathbf{K}$ if and only if the pair $\{\mathbf{I}, \mathbf{A}\}$ of the quadruple $\{\mathbf{I}, \mathbf{A}, \mathbf{0}, \mathbf{0}\}$ is controllable. Furthermore, the latter occurs if the rank of the controllability matrix is equal to n; i.e.,

$$\operatorname{rank} \mathbf{C} := \operatorname{rank} \begin{bmatrix} \mathbf{A} & \mathbf{I}\mathbf{A} & \mathbf{I}^2\mathbf{A} \cdots \mathbf{I}^{n-1}\mathbf{A} \end{bmatrix} = n. \tag{2.117}$$

This condition reduces to rank $\mathbf{A} = n$, i.e., $\mathbf{A}$ nonsingular. We thus have the following lemma.

Lemma 2.9. *Consider the matrices* $\mathbf{A}$ *and* $\mathbf{K}$ *both real and square of dimension* n. *The eigenvalues of the matrix* $\mathbf{S} := (\mathbf{I} - \mathbf{AK})$ *can be arbitrarily assigned by choice of* $\mathbf{K}$ *if and only if the matrix* $\mathbf{A}$ *is nonsingular.*

Actually, it is possible to state a slightly more general form of this lemma, showing that the less stringent requirement of stabilizability also implies that the matrix $\mathbf{A}$ must be nonsingular.

Lemma 2.10. *There exists a matrix* $\mathbf{K}$ *such that* $\rho(\mathbf{S}) = \rho(\mathbf{I} - \mathbf{AK}) < 1$ *if and only if the matrix* $\mathbf{A}$ *is nonsingular.*

Proof. ("If") Choose $\mathbf{K} = \mathbf{A}^{-1}$.

("Only if") The contrapositive is proved. Note first that if $\mathbf{A}$ is singular, then for all matrices $\mathbf{K}$, the product $\mathbf{AK}$ is also singular, and moreover, rank $\mathbf{AK} \leq$ rank $\mathbf{A}$. It now suffices to observe that a singular matrix $\mathbf{Z} \in \mathbb{R}^{n \times n}$ with rank $\mathbf{Z} = p$ has $n - p$ eigenvalues equal to zero. Thus the matrix $\mathbf{I} - \mathbf{Z}$ has $n - p$ eigenvalues equal to 1, and hence $\rho(\mathbf{I} - \mathbf{Z}) \geq 1$. Take $\mathbf{Z} = \mathbf{AK}$. □

Notice that the particular choice $\mathbf{K} = \mathbf{A}^{-1}$ makes all the eigenvalues of matrix $\mathbf{S}$ equal to zero, implying that the iterative scheme (2.114) will converge in one iteration. This is, of course, only a theoretical remark, since if the inverse of matrix $\mathbf{A}$ were in fact available, it would be enough to compute $\mathbf{A}^{-1}\mathbf{b}$ in order to solve the linear system $\mathbf{Ax} = \mathbf{b}$ and unnecessary to resort to any iteration. In fact, the problem of solving a linear system $\mathbf{Ax} = \mathbf{b}$ without inverting $\mathbf{A}$ can be stated in control terms as that of "emulating" $\mathbf{A}^{-1}$ without actually computing it, and this is exactly what iterative methods do. We also remark that the convergence in one iteration, or more generally in a finite number of iterations, is just a question of making the iteration matrix in (2.114) *nilpotent*, with the index of nilpotency representing an upper bound on the number of iterations required to zero the residue. This is another topic, called *deadbeat control*, that is well studied in the control literature in [ÅW84], to which we refer the reader.

Lemma 2.10 says that stabilizability of the pair $\{\mathbf{I}, \mathbf{A}\}$ implies that the matrix $\mathbf{A}$ must be nonsingular. Another result of this nature is that controllability of the pair $\{\mathbf{A}, \mathbf{b}\}$ implies that the system $\mathbf{Ax} = \mathbf{b}$ possesses a unique solution [dSB81]. Actually, there are deeper connections with Krylov subspaces that we will not dwell on here; however, see [IM98].

2.3.1 CLF/LOC derivation of minimal residual and Krylov subspace methods

The next natural question is whether it is possible to do better with more complicated controllers than the static constant controllers of the preceding discussion.

First, consider the case in which matrix $\mathbf{K}$ is no longer a constant and is, in fact, dependent on the state $\mathbf{x}$ or the iteration counter k:

$$\mathbf{x}_{k+1} = \mathbf{x}_k + \mathbf{K}_k \mathbf{r}_k. \tag{2.118}$$

In particular, in many iterative methods, the matrix $\mathbf{K}_k$ is chosen as $\alpha_k\mathbf{I}$, leading to

$$\mathbf{x}_{k+1} = (\mathbf{I} - \alpha_k\mathbf{A})\mathbf{x}_k + \alpha_k\mathbf{b}, \tag{2.119}$$

where α_k is a scalar sequence and $\mathbf{I}$ is an identity matrix of appropriate dimension. One method differs from another in the way in which the scalars α_k are chosen; e.g., if the α_k's are precomputed (from arguments involving clustering of eigenvalues of the iteration matrix), we get the class of Chebyshev-type "semi-iterative" methods; if the α_k are computed in terms of the current values of $\mathbf{r}_k$, the resulting class is referred to as adaptive Richardson, etc.)

The objective of this section is to show that this and some related classes of methods have a natural control interpretation that permits the analysis and design of this class, using the CLF/LOC approach.

Considering the matrix $\mathbf{K}_k$ in (2.118) given by $\mathbf{K}_k = \alpha_k\mathbf{I}$, it is convenient to rewrite (2.118) in terms of the residue $\mathbf{r}_k$ as follows:

$$\mathbf{r}_{k+1} = \mathbf{r}_k - \alpha_k\mathbf{A}\mathbf{r}_k. \tag{2.120}$$

In control language, now thinking of the parameter α_k as a control u_k and vector $\mathbf{r}_k$ as the state, (2.120) describes a dynamical system of the type known as *bilinear*, since it is linear in the state $\mathbf{r}$ if the control α is fixed and linear in the control α if the state is fixed. It is not linear if both state and control are allowed to vary, since the right-hand side contains a product term $\alpha_k\mathbf{A}\mathbf{r}_k$ involving the control input and the state $\mathbf{r}_k$.

Since the system is no longer linear or time invariant, straightforward eigenvalue analysis is no longer applicable. A control Liapunov function is used to design an asymptotically stabilizing state feedback control for (2.120) that drives $\mathbf{r}_k$ to the origin and thus solves the original problem (2.103).

Consider the control Liapunov function candidate $V(\mathbf{r}_k) := \langle\mathbf{r}_k, \mathbf{r}_k\rangle = \mathbf{r}_k^T\mathbf{r}_k$. Then, from (2.120),

$$\begin{aligned}\langle\mathbf{r}_{k+1}, \mathbf{r}_{k+1}\rangle &= \langle\mathbf{r}_k - \alpha_k\mathbf{A}\mathbf{r}_k, \mathbf{r}_k - \alpha_k\mathbf{A}\mathbf{r}_k\rangle &(2.121)\\ &= \langle\mathbf{r}_k, \mathbf{r}_k\rangle - 2\alpha_k\langle\mathbf{r}_k, \mathbf{A}\mathbf{r}_k\rangle + \alpha_k^2\langle\mathbf{A}\mathbf{r}_k, \mathbf{A}\mathbf{r}_k\rangle, &(2.122)\end{aligned}$$

from which it follows that

$$\Delta V := V(\mathbf{r}_{k+1}) - V(\mathbf{r}_k) = -\alpha_k\left(2\langle\mathbf{r}_k, \mathbf{A}\mathbf{r}_k\rangle - \alpha_k\langle\mathbf{A}\mathbf{r}_k, \mathbf{A}\mathbf{r}_k\rangle\right). \tag{2.123}$$

From this expression it is clear that the LOC choice

$$\alpha_k = \frac{\langle\mathbf{r}_k, \mathbf{A}\mathbf{r}_k\rangle}{\langle\mathbf{A}\mathbf{r}_k, \mathbf{A}\mathbf{r}_k\rangle} \tag{2.124}$$

leads to

$$\Delta V = -\frac{\langle\mathbf{r}_k, \mathbf{A}\mathbf{r}_k\rangle^2}{\langle\mathbf{A}\mathbf{r}_k, \mathbf{A}\mathbf{r}_k\rangle} < 0, \tag{2.125}$$

showing that the candidate control Liapunov function works and that (2.124) is the appropriate choice of feedback control. Furthermore, ΔV is strictly negative unless $\langle\mathbf{r}_k, \mathbf{A}\mathbf{r}_k\rangle = 0$. One way of saying that this possibility is excluded is to say that zero does not belong to the *field of values* of $\mathbf{A}$ [Gre97]. In other words, the control Liapunov function proves that the residual vector $\mathbf{r}_k$ decreases monotonically to the zero vector.

Note that the stabilizing feedback control α_k is a nonlinear function of the state, which is not surprising, since the system being stabilized is not linear, but bilinear. The choice (2.124) results in the so-called Orthomin(1) method [Gre97].

Richardson's method is an LOC/CLF steepest descent method

The observant reader will note that (2.124) corresponds to the LOC choice for the quadratic CLF V and that it could have been derived by the application of Lemma 2.4. Note, however, that it is not necessary to stick to the quadratic CLF $\mathbf{r}^T\mathbf{r}$ in order to use the LOC approach. In particular, a small change in the candidate Liapunov function, together with the assumption that the matrix $\mathbf{A}$ is positive definite, leads to another well-known method. Since $\mathbf{A}$ is positive definite, $\mathbf{A}^{-1}$ exists and the following choice is legitimate:

$$V(\mathbf{r}_k) := \langle \mathbf{r}_k, \mathbf{A}^{-1}\mathbf{r}_k \rangle. \tag{2.126}$$

Repeating the steps above, it is easy to arrive at

$$\Delta V = -\alpha_k \left(2\langle \mathbf{A}\mathbf{r}_k, \mathbf{A}^{-1}\mathbf{r}_k \rangle - \alpha_k \langle \mathbf{A}\mathbf{r}_k, \mathbf{A}^{-1}\mathbf{A}\mathbf{r}_k \rangle \right), \tag{2.127}$$

from which it follows, in exact analogy to the development above, that

$$\alpha_k = \frac{\langle \mathbf{A}\mathbf{r}_k, \mathbf{A}^{-1}\mathbf{r}_k \rangle}{\langle \mathbf{A}\mathbf{r}_k, \mathbf{A}^{-1}\mathbf{A}\mathbf{r}_k \rangle} = \frac{\langle \mathbf{r}_k, \mathbf{r}_k \rangle}{\langle \mathbf{r}_k, \mathbf{A}\mathbf{r}_k \rangle} \tag{2.128}$$

is the appropriate choice of feedback control that makes $\Delta V < 0$, which, in fact, corresponds to *Richardson's iterative method* for symmetric matrices [You89, Var00, SvdV00], sometimes also qualified with the adjectives *adaptive* and *parameter-free*, since the α_k's are calculated in feedback form.

From (2.118), Richardson's method can be written as

$$\mathbf{x}_{k+1} = \mathbf{x}_k + \alpha_k \mathbf{r}_k, \quad \text{where } \alpha_k = \frac{\langle \mathbf{r}_k, \mathbf{r}_k \rangle}{\langle \mathbf{r}_k, \mathbf{A}\mathbf{r}_k \rangle}. \tag{2.129}$$

Now observe that the problem of solving the linear system is equivalent to that of minimizing the quadratic form $\langle \mathbf{x}, \mathbf{A}\mathbf{x} \rangle - 2\langle \mathbf{b}, \mathbf{x} \rangle$ (since the latter attains its minimum where $\mathbf{A}\mathbf{x} = \mathbf{b}$).

Since the negative gradient of this function at $\mathbf{x} = \mathbf{x}_k$ is $\mathbf{r}_k = \mathbf{b} - \mathbf{A}\mathbf{x}_k$, clearly (2.129) can be viewed as a *steepest descent method* in which the stepsize α_k is chosen optimally in the LOC/CLF sense. From a control viewpoint, since the control action on the state is $\alpha_k \mathbf{r}_k$, i.e., proportional to the error or residue $\mathbf{r}_k$, Richardson's method can be viewed as the application of a proportional controller with a time-varying gain α_k.

2.3.2 The conjugate gradient method derived from a proportional-derivative controller

In a survey of the top ten algorithms of the century, Krylov subspace methods have a prominent place [DS00, vdV00]. Furthermore, as Trefethen says in his essay on numerical

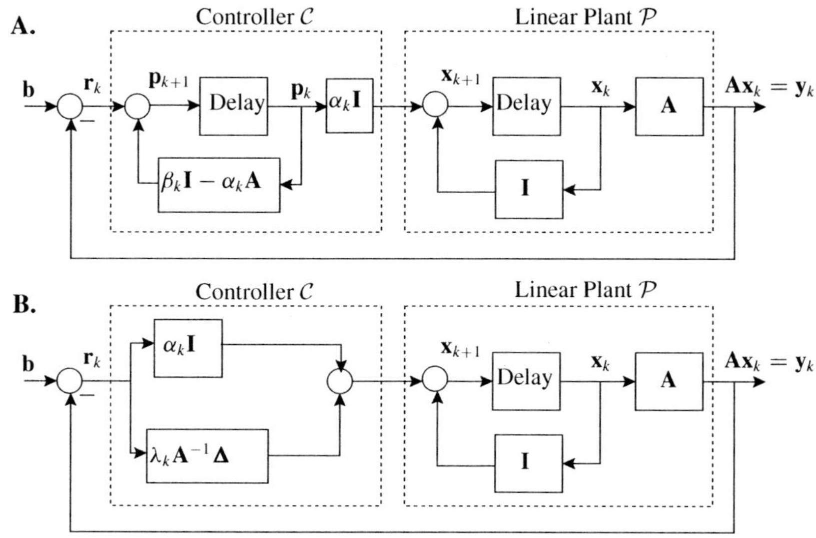

Figure 2.13. A: *The conjugate gradient method represented as the standard plant* $\mathcal{P} = \{\mathbf{I}, \mathbf{I}, \mathbf{A}, \mathbf{0}\}$ *with dynamic nonstationary controller* $\mathcal{C} = \{(\beta_k\mathbf{I} - \alpha_k\mathbf{A}), \mathbf{I}, \alpha_k\mathbf{I}, \mathbf{0}\}$ *in the variables* $\mathbf{p}_k$, $\mathbf{x}_k$. B: *The conjugate gradient method represented as the standard plant* $\mathcal{P}$ *with a nonstationary (time-varying) proportional-derivative (PD) controller in the variables* $\mathbf{r}_k$, $\mathbf{x}_k$, *where* $\lambda_k = \beta_{k-1}\alpha_k/\alpha_{k-1}$. *This block diagram emphasizes the conceptual proportional-derivative structure of the controller. Of course, the calculations represented by the derivative block,* $\lambda_k\mathbf{A}^{-1}\boldsymbol{\Delta}$, *are carried out using formulas* (2.143), (2.148) *that do not involve inversion of the matrix* $\mathbf{A}$.

analysis: "For guidance to the future we should study not Gaussian elimination and its beguiling stability properties, but the diabolically fast conjugate gradient iteration" [Tre92]. This section shows that the formal conjugate gradient method, one of the best known Krylov subspace methods, is also easily arrived at from a control viewpoint. This has the merit of providing a natural control motivation for the conjugate gradient method in addition to providing some insights as to why it has certain desirable properties, such as speed and robustness, in the face of certain types of errors.

In fact, the conjugate gradient method is conveniently viewed as an acceleration of Richardson's steepest descent method (2.129).

Note that the latter can be viewed as the standard feedback control system $S(\mathcal{P}, \mathcal{C})$ with the controller $\{\mathbf{0}, \mathbf{0}, \mathbf{0}, \alpha_k(\mathbf{r}_k)\mathbf{I}\}$, which is referred to as a *proportional controller*. The acceleration is achieved by using a discrete version of a classical control strategy for faster "closed-loop" response (i.e., acceleration of convergence to the solution). This strategy is known as *derivative action* in the controller. The development of this approach is as follows.

Suppose that a new method is to be derived from the steepest descent method by adding a new term that is proportional to a discrete derivative of the state vector $\mathbf{x}_k$. In other words, the new increment $\Delta\mathbf{x}_k := \mathbf{x}_{k+1} - \mathbf{x}_k$ is a linear combination of the steepest descent direction $\mathbf{r}_k$ and the previous increment or discrete derivative of the state $\mathbf{x}_k - \mathbf{x}_{k-1}$.

Denoting the scalar gains as α_k and γ_k, the new method, depcited in Figure 2.13, can be expressed mathematically as follows:

$$\mathbf{x}_{k+1} = \mathbf{x}_k + \alpha_k[\mathbf{r}_k + \gamma_k(\mathbf{x}_k - \mathbf{x}_{k-1})], \tag{2.130}$$

which can be rewritten as

$$\mathbf{x}_{k+1} = \mathbf{x}_k + \alpha_k \mathbf{p}_k, \tag{2.131}$$

where

$$\begin{aligned} \mathbf{p}_k &= \mathbf{r}_k + \gamma_k(\mathbf{x}_k - \mathbf{x}_{k-1}) = \mathbf{r}_k + \gamma_k \alpha_{k-1}\mathbf{p}_{k-1} \\ &= \mathbf{r}_k + \beta_{k-1}\mathbf{p}_{k-1}, \end{aligned} \tag{2.132}$$

and

$$\beta_{k-1} := \gamma_k \alpha_{k-1}. \tag{2.133}$$

Combining these equations leads to

$$\mathbf{x}_{k+1} = \mathbf{x}_k + \alpha_k \mathbf{p}_k, \tag{2.134}$$

$$\mathbf{r}_{k+1} = \mathbf{r}_k - \alpha_k \mathbf{A}\mathbf{p}_k, \tag{2.135}$$

$$\mathbf{p}_{k+1} = \mathbf{r}_{k+1} + \beta_k \mathbf{p}_k. \tag{2.136}$$

From a control viewpoint, it is natural to think of the "parameters" α_k and β_k as scalar control inputs. The motivation for doing this is the observation that the systems to be controlled then belong to the class of systems known as bilinear, similarly to system (2.120). More precisely, taking $\mathbf{r}_k$ and $\mathbf{p}_k$ as the state variables, it is necessary to analyze the following pair of coupled bilinear systems:

$$\mathbf{r}_{k+1} = \mathbf{r}_k - \alpha_k \mathbf{A}\mathbf{p}_k, \tag{2.137}$$

$$\mathbf{p}_{k+1} = \mathbf{r}_{k+1} + \beta_k \mathbf{p}_k. \tag{2.138}$$

Provided that α_k is not identically zero, it is easy to see that the equilibrium solution of this system is the zero solution $\mathbf{r}_k = \mathbf{p}_k = \mathbf{0}$ for all k. The control objective is to choose the scalar controls α_k, β_k so as to drive the state vectors $\mathbf{r}_k$ and $\mathbf{p}_k$ to zero. The analysis will be carried out in terms of the variables $\mathbf{r}_k$ and $\mathbf{p}_k$. Thus the objective is to show that the same control Liapunov function approach that has been successfully applied to other iterative methods above can also be used here to derive choices of α_k and β_k that result in stability of the zero solution. The analysis proceeds in two stages. In the first stage, a choice of α_k guided by a control Liapunov function is shown to result in a decrease of a suitable norm of $\mathbf{r}_k$. In the second stage, a second control Liapunov function orients the choice of β_k that results in a decrease of a suitable norm of $\mathbf{p}_k$. The conclusion is that $\mathbf{r}_k$ and $\mathbf{p}_k$ both converge to zero, as required.

Since $\mathbf{A}$ is a real positive definite matrix, so is $\mathbf{A}^{-1}$ and both matrices define weighted 2-norms. The control Liapunov method is used to choose the controls, using the $\mathbf{A}^{-1}$-norm for (2.137) and the $\mathbf{A}$-norm for (2.138). Before proceeding, it should be pointed out that these choices are arbitrary and that exactly the same control Liapunov argument with different choices of norms leads to different methods.

Thus the first step is to calculate the $\mathbf{A}^{-1}$-norm of both sides of (2.137) in order to choose a control α_k that will result in the reduction of this norm of $\mathbf{r}$ to zero:

$$\langle \mathbf{r}_{k+1}, \mathbf{A}^{-1}\mathbf{r}_{k+1}\rangle = \langle \mathbf{r}_k - \alpha_k \mathbf{A}\mathbf{p}_k, \mathbf{A}^{-1}(\mathbf{r}_k - \alpha_k \mathbf{A}\mathbf{p}_k)\rangle \tag{2.139}$$
$$= \langle \mathbf{r}_k, \mathbf{A}^{-1}\mathbf{r}_k\rangle - 2\alpha_k \langle \mathbf{r}_k, \mathbf{p}_k\rangle + \alpha_k^2 \langle \mathbf{A}\mathbf{p}_k, \mathbf{p}_k\rangle. \tag{2.140}$$

It follows that

$$\Delta V_{\mathbf{r}} := \langle \mathbf{r}_{k+1}, \mathbf{A}^{-1}\mathbf{r}_{k+1}\rangle - \langle \mathbf{r}_k, \mathbf{A}^{-1}\mathbf{r}_k\rangle = -2\alpha_k \langle \mathbf{r}_k, \mathbf{p}_k\rangle + \alpha_k^2 \langle \mathbf{A}\mathbf{p}_k, \mathbf{p}_k\rangle. \tag{2.141}$$

The LOC choice of α_k is found from the calculation

$$\frac{\partial \Delta V}{\partial \alpha_k} = -2\langle \mathbf{r}_k, \mathbf{p}_k\rangle + 2\alpha_k \langle \mathbf{A}\mathbf{p}_k, \mathbf{p}_k\rangle \tag{2.142}$$

so that $\frac{\partial \Delta V}{\partial \alpha_k} = 0$ when

$$\alpha_k = \frac{\langle \mathbf{r}_k, \mathbf{p}_k\rangle}{\langle \mathbf{A}\mathbf{p}_k, \mathbf{p}_k\rangle}. \tag{2.143}$$

This choice of α_k is the LOC choice, i.e., optimal in the sense that it makes ΔV as negative as possible. In other words, it makes the reduction in the $\mathbf{A}^{-1}$-norm of $\mathbf{r}$ as large as possible:

$$\Delta V = -\frac{\langle \mathbf{r}_k, \mathbf{p}_k\rangle^2}{\langle \mathbf{A}\mathbf{p}_k, \mathbf{p}_k\rangle}. \tag{2.144}$$

This derivation of α_k also gives a clue to the robustness of the conjugate gradient algorithm, since the argument so far has not used any information on the properties of the vectors $\mathbf{p}_k$ (such as $\mathbf{A}$-orthogonality). This indicates that, in a finite precision implementation, even when properties such as $\mathbf{A}$-orthogonality are lost, the choice of α_k in (2.143) ensures that the $\mathbf{A}^{-1}$-norm of $\mathbf{r}$ will decrease.

Proceeding with the analysis, consider the "$\mathbf{p}_k$-subsystem" subject to the control β_k. The $\mathbf{A}$-norm of both sides of (2.138) is calculated in order to choose an appropriate control input β_k:

$$\langle \mathbf{p}_{k+1}, \mathbf{A}\mathbf{p}_{k+1}\rangle = \langle \mathbf{r}_{k+1} + \beta_k \mathbf{p}_k, \mathbf{A}(\mathbf{r}_{k+1} + \beta_k \mathbf{p}_k)\rangle \tag{2.145}$$
$$= \langle \mathbf{r}_{k+1}, \mathbf{A}\mathbf{r}_{k+1}\rangle + 2\beta_k \langle \mathbf{p}_k, \mathbf{A}\mathbf{r}_{k+1}\rangle + \beta_k^2 \langle \mathbf{p}_k, \mathbf{A}\mathbf{p}_k\rangle. \tag{2.146}$$

Using the same line of argument as above, calculate

$$\frac{\partial \|\mathbf{p}_{k+1}\|_{\mathbf{A}}^2}{\partial \beta_k} = 2\langle \mathbf{p}_k, \mathbf{A}\mathbf{r}_{k+1}\rangle + 2\beta_k \langle \mathbf{p}_k, \mathbf{A}\mathbf{p}_k\rangle, \tag{2.147}$$

so that

$$\beta_k = -\frac{\langle \mathbf{p}_k, \mathbf{A}\mathbf{r}_{k+1}\rangle}{\langle \mathbf{p}_k, \mathbf{A}\mathbf{p}_k\rangle} \tag{2.148}$$

is the LOC choice, resulting in

$$\|\mathbf{p}_{k+1}\|_{\mathbf{A}}^2 = \|\mathbf{r}_{k+1}\|_{\mathbf{A}}^2 - \frac{\langle \mathbf{p}_k, \mathbf{A}\mathbf{r}_{k+1}\rangle^2}{\langle \mathbf{p}_k, \mathbf{A}\mathbf{p}_k\rangle}. \tag{2.149}$$

Since the second term is negative (except at the solution $\mathbf{p}_k = \mathbf{0}$), this results in the inequality

$$\|\mathbf{p}_{k+1}\|_{\mathbf{A}} < \|\mathbf{r}_{k+1}\|_{\mathbf{A}}. \tag{2.150}$$

From (2.144) and the equivalence of norms, it can be concluded that $\mathbf{r}_{k+1}$ decreases in any induced norm (in particular, in the $\mathbf{A}$-norm). Thus (2.150) implies that $\mathbf{p}_{k+1}$ decreases in the $\mathbf{A}$-norm, although not necessarily monotonically.

The above derivations may be summarized in the form of an algorithm, which we will refer to as the CLF/LOC version of the conjugate gradient algorithm.

The conjugate gradient algorithm: CLF/LOC version.
Compute $\mathbf{r}_0 := \mathbf{b} - \mathbf{A}\mathbf{x}_0$, $\mathbf{p}_0 := \mathbf{r}_0$.
For $k = 0, 1, \ldots,$ until convergence
Do:
$\alpha_k = \langle \mathbf{r}_k, \mathbf{p}_k \rangle / \langle \mathbf{A}\mathbf{p}_k, \mathbf{p}_k \rangle$
$\mathbf{x}_{k+1} := \mathbf{x}_k + \alpha_k \mathbf{p}_k$
$\mathbf{r}_{k+1} := \mathbf{r}_k - \alpha_k \mathbf{A}\mathbf{p}_k$
$\beta_k = -\langle \mathbf{p}_k, \mathbf{A}\mathbf{r}_{k+1} \rangle / \langle \mathbf{p}_k, \mathbf{A}\mathbf{p}_k \rangle$
$\mathbf{p}_{k+1} := \mathbf{r}_{k+1} + \beta_k \mathbf{p}_k$
EndDo

Under the assumption that the conjugate gradient method is initialized choosing $\mathbf{r}_0 = \mathbf{p}_0$, (2.143) and (2.148) are equivalent to the more commonly used but less obvious forms [Gre97]

$$\alpha_k = \langle \mathbf{r}_k, \mathbf{r}_k \rangle / \langle \mathbf{p}_k, \mathbf{A}\mathbf{p}_k \rangle, \tag{2.151}$$

$$\beta_k = \langle \mathbf{r}_{k+1}, \mathbf{r}_{k+1} \rangle / \langle \mathbf{r}_k, \mathbf{r}_k \rangle. \tag{2.152}$$

With these choices of α_k and β_k the CLF/LOC version becomes the standard textbook [Saa96, Alg. 6.17, p. 179] version of the conjugate gradient algorithm, given below for comparison.

The standard conjugate gradient algorithm.
Compute $\mathbf{r}_0 := \mathbf{b} - \mathbf{A}\mathbf{x}_0$, $\mathbf{p}_0 := \mathbf{r}_0$.
For $k = 0, 1, \ldots,$ until convergence
Do:
$\alpha_k := \langle \mathbf{r}_k, \mathbf{r}_k \rangle / \langle \mathbf{A}\mathbf{p}_k, \mathbf{p}_k \rangle$
$\mathbf{x}_{k+1} := \mathbf{x}_k + \alpha_k \mathbf{p}_k$
$\mathbf{r}_{k+1} := \mathbf{r}_k - \alpha_k \mathbf{A}\mathbf{p}_k$
$\beta_k := \langle \mathbf{r}_{k+1}, \mathbf{r}_{k+1} \rangle / \langle \mathbf{r}_k, \mathbf{r}_k \rangle$
$\mathbf{p}_{k+1} := \mathbf{r}_{k+1} + \beta_k \mathbf{p}_k$
EndDo

The difference between these two algorithms appears exactly when the assumption $\mathbf{r}_0 = \mathbf{p}_0$ is violated; in this case, it can happen that, for some choices of $\mathbf{r}_0 \neq \mathbf{p}_0$, the standard version of the conjugate gradient algorithm diverges, whereas the CLF/LOC version does not. A practical application in which such a situation occurs is in *adaptive filtering*, which, in the current context, can be described as using an "on-line" conjugate gradient-type algorithm to solve a system of the type $\mathbf{A}_k\mathbf{x}_k = \mathbf{b}_k$. The term "on-line" refers to the updating of the matrix $\mathbf{A}_k$ and the right-hand side $\mathbf{b}_k$ at each iteration and means, in practice, that the assumption $\mathbf{r}_0 = \mathbf{p}_0$ does not hold after the first iteration. The paper [CW00] discusses the nonconvergent behavior of an "on-line" version of the standard conjugate gradient algorithm in this situation and proposes a solution based on choices of α_k and β_k that involve a heuristic choice of an additional parameter and a line search. In contrast, numerical experiments show that, for this class of problems, an appropriately modified on-line CLF/LOC variant of the conjugate gradient algorithm works well, without the need for line search and heuristically chosen parameters [DB06].

Variants of the conjugate gradient algorithm

The Orthomin(2) algorithm [Gre97] differs from the standard conjugate gradient algorithm only in the choice of the controls α_k and β_k. From the viewpoint adopted here, it can be said that the difference lies in the choice of the norms used for the control Liapunov functions for the $\mathbf{r}$ and $\mathbf{p}$ subsystems. More precisely, consider the algorithm (coupled bilinear systems)

$$\mathbf{r}_{k+1} = \mathbf{r}_k - \alpha_k \mathbf{A}\mathbf{p}_k, \tag{2.153}$$

$$\mathbf{p}_{k+1} = \mathbf{r}_{k+1} - \beta_k \mathbf{p}_k. \tag{2.154}$$

Suppose that the 2-norm is used as the control Liapunov function for the $\mathbf{r}$ subsystem and that the 2-norm of $\mathbf{A}\mathbf{p}$ (recall that for the Orthomin(2) method it is not assumed that the matrix $\mathbf{A}$ is symmetric) is the control Liapunov function for the $\mathbf{p}$ subsystem. A calculation that is strictly analogous to the one above for the conjugate gradient method shows that this choice of norms results in

$$\alpha_k = \frac{\langle \mathbf{r}_k, \mathbf{A}\mathbf{p}_k \rangle}{\langle \mathbf{A}\mathbf{p}_k, \mathbf{A}\mathbf{p}_k \rangle}, \quad \beta_k = \frac{\langle \mathbf{A}\mathbf{p}_k, \mathbf{A}\mathbf{r}_{k+1} \rangle}{\langle \mathbf{A}\mathbf{p}_k, \mathbf{A}\mathbf{p}_k \rangle}, \tag{2.155}$$

which is exactly the Orthomin(2) choice of α_k and β_k (see [Gre97]).

The CLF proof of the conjugate gradient choices of α_k, β_k allows another observation that, to the authors' knowledge, has not been made in the literature. Consider the following variant of the conjugate gradient algorithm:

$$\mathbf{r}_{k+1} = \mathbf{r}_k - \alpha_k \mathbf{A}\mathbf{p}_k, \tag{2.156}$$

$$\mathbf{p}_{k+1} = \mathbf{r}_k + \beta_k \mathbf{p}_k. \tag{2.157}$$

In this version of conjugate gradient, the second equation (in $\mathbf{p}$) has been modified and does not make use of the iterate $\mathbf{r}_{k+1}$ computed (sequentially) "before" it, but instead uses the iterate $\mathbf{r}_k$. In this sense, this version may be thought of as a "Jacobi" version of the standard "Gauss–Seidel-like" conjugate gradient algorithm. The analysis of the standard conjugate gradient algorithm made above may be repeated almost verbatim, leading to the conclusion that the choices

$$\alpha_k = \frac{\langle \mathbf{r}_k, \mathbf{p}_k \rangle}{\langle \mathbf{A}\mathbf{p}_k, \mathbf{p}_k \rangle}, \quad \beta_k = -\frac{\langle \mathbf{p}_k, \mathbf{A}\mathbf{r}_k \rangle}{\langle \mathbf{p}_k, \mathbf{A}\mathbf{p}_k \rangle} \tag{2.158}$$

(the only difference is in the numerator of β_k) ensure that $\mathbf{r}_k$ is a decreasing sequence in $\mathbf{A}$-norm, and furthermore that $\|\mathbf{p}_{k+1}\|_{\mathbf{A}}^2 < \|\mathbf{r}_k\|_{\mathbf{A}}^2$, implying that $\mathbf{p}_k$ is also a decreasing sequence, although it decreases more slowly than it would in the standard conjugate gradient method (for which the inequality $\|\mathbf{p}_{k+1}\|_{\mathbf{A}}^2 < \|\mathbf{r}_{k+1}\|_{\mathbf{A}}^2$ was obtained). This confirms the conventional wisdom that "Gauss–Seidelization" is conducive to faster convergence, and indeed, numerical experiments confirm this.

The conjugate gradient algorithm interpreted as a dynamic controller

A block diagram representation is helpful in order to interpret what has just been done, in terms of the taxonomy of iterative methods proposed as well as making the controller structure explicit. Comparing the block diagrams of Figures 2.12 and 2.13, it becomes clear that, although the box representing the plant (i.e., problem or equation to be solved) has remained the same, the box representing the controller (i.e., solution method) is considerably more sophisticated with respect to the simple controllers studied in section 2.3. It is, in fact, a dynamic time-varying or nonstationary controller, in the spirit of the dynamic controllers considered for nonlinear equations in section 2.1. The upshot of the increased sophistication is that the method (conjugate gradient) is more efficient. In fact, it is well known that, in infinite precision, conjugate gradient is actually a direct method (i.e., it converges in n steps for an $n \times n$ matrix $\mathbf{A}$) [Kel95]. In control terms, this last observation can be rephrased by saying that the "conjugate gradient controller" achieves so-called dead beat control in n steps.

The backpropagation with momentum algorithm is the conjugate gradient algorithm

The well-known backpropagation with momentum (BPM) method (resp., algorithm) is a variant of the gradient or steepest descent method that is popular in the neural network literature [PS94, YC97]. In light of the control formulation of the conjugate gradient method presented in section 2.3.2, the BPM method for quadratic functions is now shown to be exactly equivalent to the conjugate gradient method, allowing derivation of a so-called optimally tuned learning rate and momentum parameters for the former method, for the nonstationary or time-varying case (as opposed to most of the literature in the field of neural networks, which treats only the time-invariant case).

Consider the problem of determining a vector $\mathbf{x}$ (thought of as a "set of network weights" in the neural network context) that minimizes a quadratic error function

$$f(\mathbf{x}) = \frac{1}{2}\langle \mathbf{x}, \mathbf{A}\mathbf{x}\rangle - \langle \mathbf{b}, \mathbf{x}\rangle + c, \tag{2.159}$$

where $\mathbf{A}$ is a symmetric positive definite matrix. The gradient $\nabla f(\mathbf{x}) = \mathbf{A}\mathbf{x} - \mathbf{b} =: -\mathbf{r}$ is also called the residue (in the context of solution of the linear system $\mathbf{A}\mathbf{x} = \mathbf{b}$). In this notation, the BPM algorithm can be written as

$$\mathbf{x}_{k+1} - \mathbf{x}_k = \mu_k(\mathbf{x}_k - \mathbf{x}_{k-1}) + (1 - \mu_k)\lambda_k \mathbf{r}_k. \tag{2.160}$$

Clearly, the idea behind the BPM algorithm is to define the new increment $(\mathbf{x}_{k+1} - \mathbf{x}_k)$ as a (time-varying) convex combination of the old increment $(\mathbf{x}_k - \mathbf{x}_{k-1})$ and a multiple (λ_k) of the residue $(\mathbf{r}_k)$. The parameter λ_k is called the *learning rate*, while the parameter μ_k is called the *momentum factor* [TH02].

Notice that (2.160) is the same as (2.130), so that the BPM method for quadratic functions is exactly equivalent to the conjugate gradient method, for an adequate choice of the parameters μ_k and λ_k.

Indeed, the optimal CLF/LOC choice of the controls α_k, β_k immediately yields the optimal learning rate λ_k and momentum factors μ_k:

$$\begin{aligned} \alpha_k &= (1 - \mu_k)\lambda_k, \\ \alpha_k \gamma_k &= \mu_k. \end{aligned} \tag{2.161}$$

One can solve for the optimal learning rate and momentum factor in terms of the optimal choices of CG parameters α_k (2.151) and β_k (2.152):

$$\begin{aligned} \mu_k &= (\alpha_k/\alpha_{k-1})\beta_{k-1}, \\ \lambda_k &= \alpha_k\alpha_{k-1}/(\alpha_{k-1} - \alpha_k\beta_{k-1}). \end{aligned} \tag{2.162}$$

Equations (2.162) and (2.151) show that the optimal learning rate and momentum factor can be calculated in terms of the state variables $(\mathbf{r}, \mathbf{p})$, although this involves calculation of more inner products than the conjugate gradient method.

The overall conclusion is that it is easier just to use the standard conjugate gradient method, which has tried and tested variants, both linear and nonlinear [NW99], rather than use the equivalent "optimally tuned" BPM algorithm. Of course, there are many practical issues involved in determining ultimate performance, and the complexity and cost of each iteration of an optimal algorithm may offset its faster rate of convergence.

In section 2.3.3, a continuous version of the BPM or conjugate gradient method is proposed and analyzed and shown to be a version of the so-called heavy ball with friction method for continuous optimization.

Taxonomy of linear iterative methods from a control perspective

The block diagram representation has the virtue of allowing us to make a clear separation between the problem and the algorithm, making it easy to classify as well as generalize the strategies used in the algorithms. Taking the example of linear iterative methods, we see a progression of successively more complex controllers: constant (α), nonstationary or time varying $(\alpha_k\mathbf{I})$, multivariable $(\mathbf{K})$, multivariable time varying $(\alpha_k\mathbf{K}_k)$, and dynamic, leading to most of the standard iterative methods in a natural manner. For linear iterations, the results of this section lead to a "dictionary" relating controller choice to numerical algorithm that we present in the form of Table 2.5, which makes reference to Figure 2.1 and uses the terminology of [Kel95, SvdV00]. The standardized CLF analysis technique leads to the conventional choices of control parameters. It is worth noting that 2-norms, possibly weighted with a diagonal or positive definite matrix, usually work as CLFs. This is in sharp contrast with the situation for an arbitrary nonlinear dynamical system, for which, as a rule, considerable ingenuity is required to find a suitable CLF. Another consequence of the relative ease in finding quadratic CLFs is that each of these leads to a different algorithm, so that there is scope for devising new algorithms, showing that the CLF approach has an inherent richness. Moreover, the control Liapunov approach is easily generalizable to a Hilbert space setting, following the work on iterative methods for operators by Kantorovich [KA82], Krasnosel'skii [KLS89], and others.

Some disclaimers should also be made here. Although the control approach provides guidelines for algorithm design, it does not free the designer of the need for a careful analysis of issues such as roundoff error (robustness), computational complexity, order of convergence, etc. It should also be noted that many standard solutions of control problems are infeasible in numerical analysis because they would involve more computation for their implementation than the original problem. Here the challenge is for control theorists to develop limited complexity controllers, which, to some extent, driven by technological needs such as miniaturization and low energy consumption, are now being researched in control theory.

Table 2.5. *Taxonomy of linear iterative methods from a control perspective, with reference to Figure* 2.1. *Note that* $\mathcal{P} = \{\mathbf{I}, \mathbf{I}, \mathbf{A}, \mathbf{0}\}$ *in all cases.*

Controller $\mathcal{C}$	Controller type	Class of method	Specific methods
$\{\mathbf{0}, \mathbf{0}, \mathbf{0}, \mathbf{I}\}$	Static, stationary	Richardson	
$\{\mathbf{0}, \mathbf{0}, \mathbf{0}, \alpha_k\mathbf{I}\}$	Static, nonstationary	Adaptive Richardson	Chebyshev
$\{\mathbf{0}, \mathbf{0}, \mathbf{0}, \mathbf{K}\}$	Static, stationary	Preconditioned Richardson	Jacobi, Gauss–Seidel, SOR, extrap. Jacobi
$\{\mathbf{0}, \mathbf{0}, \mathbf{0}, \alpha_k\mathbf{K}_k\}$	Static, nonstationary	Adaptive preconditioned Richardson	
$\{\mathbf{0}, \mathbf{0}, \mathbf{0}, \alpha_k(\mathbf{r}_k)\mathbf{I}\}$	Static, nonstationary	Steepest descent	Orthomin(1)
$\{\beta_k\mathbf{I} - \alpha_k\mathbf{A}, \mathbf{I}, \mathbf{I}, \mathbf{0}\}$, in variables $\mathbf{p}_k$, $\mathbf{x}_k$	Dynamic, nonstationary	Conjugate gradient	Conjugate gradient, Orthomin(2), Orthodir
α_k, β_k, in variables $\mathbf{r}_k$, $\mathbf{x}_k$	Proportional-derivative, nonstationary	Conjugate gradient	Conjugate gradient, Orthomin, Orthodir
$\{\alpha_k\mathbf{I}, \mathbf{I}, \beta_k\mathbf{I}, \mathbf{0}\}$	Dynamic, nonstationary	Second-order Richardson	Frankel

2.3.3 Continuous algorithms for finding optima and the continuous conjugate gradient algorithm

The problem of unconstrained minimization of a differentiable function $f : \mathbb{R}^n \to \mathbb{R}$ can be viewed as the zero finding problem of the gradient of f, denoted $\nabla f : \mathbb{R}^n \to \mathbb{R}^n$. From this perspective, the zero finding methods studied in this chapter can be used for unconstrained optimization. In this section, continuous methods based on dynamical systems will be studied. In particular, a continuous-time analog of the conjugate gradient method will be developed and studied using a CLF approach, and will be related to existing dynamical system methods that are referred to collectively as *continuous optimization methods*.

There are many different ways to write a continuous version of the discrete conjugate gradient iteration. One natural approach is to write continuous versions of (2.137) and (2.138) as follows:

$$\dot{\mathbf{r}} = -\alpha\mathbf{A}\mathbf{p}, \tag{2.163}$$

$$\dot{\mathbf{p}} = \mathbf{r} - \beta\mathbf{p}. \tag{2.164}$$

Elimination of the vector $\mathbf{p}$ yields the conjugate gradient ODE:

$$\ddot{\mathbf{r}} + \beta\dot{\mathbf{r}} + \alpha\mathbf{A}\mathbf{r} = \mathbf{0}. \tag{2.165}$$

Introducing the quadratic potential function

$$\Phi = \frac{1}{2}\mathbf{x}^T\mathbf{A}\mathbf{x} - \mathbf{b}^T\mathbf{x}, \tag{2.166}$$

for which

$$\nabla\Phi(\mathbf{x}) = \mathbf{A}\mathbf{x} - \mathbf{b} =: -\mathbf{r},$$

allows (2.165) to be rewritten as

$$\ddot{\mathbf{r}} + \beta\dot{\mathbf{r}} - \alpha\mathbf{A}\nabla\Phi(\mathbf{x}) = \mathbf{0}. \tag{2.167}$$

Since $\mathbf{A} = \nabla^2\Phi(\mathbf{x})$, (2.167) can be written as

$$\ddot{\mathbf{r}} + \beta\dot{\mathbf{r}} - \alpha[\nabla^2\Phi(\mathbf{x})]\nabla\Phi(\mathbf{x}) = \mathbf{0}. \tag{2.168}$$

Observing that $\dot{\mathbf{r}} = -\mathbf{A}\dot{\mathbf{x}}$, $\ddot{\mathbf{r}} = -\mathbf{A}\ddot{\mathbf{x}}$, in $\mathbf{x}$-coordinates, (2.165) becomes

$$\ddot{\mathbf{x}} + \beta\dot{\mathbf{x}} + \alpha\nabla\Phi(\mathbf{x}) = \mathbf{0}. \tag{2.169}$$

In [Pol64], the idea of using a dynamical system that represents a heavy ball with friction (HBF) moving under Newtonian dynamics in a conservative force field is investigated. Specifically, the *HBF ODE* is

$$\ddot{\mathbf{x}}(t) + \theta\dot{\mathbf{x}}(t) + \nabla\Phi(\mathbf{x}(t)) = \mathbf{0}. \tag{2.170}$$

Clearly, (2.169) is an instance of the HBF method, where the parameters β (friction coefficient) and α (related to the spring constant) need to be chosen in order to make the trajectories of (2.165) tend to zero asymptotically.

The *steepest descent ODE* is defined as follows:

$$\dot{\mathbf{x}}(t) = -\epsilon\nabla\Phi(\mathbf{x}(t)). \tag{2.171}$$

As pointed out in [AGR00], the damping term $\theta\dot{\mathbf{x}}(t)$ confers optimizing properties on (2.170), but it is isotropic and ignores the geometry of Φ. Another connection pointed out in [AGR00] is that the second derivative term $\ddot{\mathbf{x}}(t)$, which induces inertial effects, is a singular perturbation or regularization of the classical *Newton ODE*, which may be written as follows:

$$\nabla^2\Phi(\mathbf{x}(t))\dot{\mathbf{x}}(t) + \nabla\Phi(\mathbf{x}(t)) = \mathbf{0}. \tag{2.172}$$

In the neural network context, $\mathbf{x}$ is the weight vector, usually denoted $\mathbf{w}$, and the potential energy function Φ is the error function, usually denoted $E(\mathbf{w})$ (as in [Qia99]). In fact, with these changes of notation, it is clear that the HBF equation (2.170) is exactly the equation that has been proposed [Qia99] as the continuous analog of BPM, using a similar physical model (point mass moving in a viscous medium with friction under the influence of a conservative force field and with Newtonian dynamics). Thus, the continuous version of BPM is the HBF ODE and may be regarded either as a regularization of the steepest descent ODE or as the classical Newton ODE.

Allowing variable coefficients into (2.165) gives

$$\ddot{\mathbf{x}} + \beta(\mathbf{x})\dot{\mathbf{x}} + \alpha(\mathbf{x})\nabla\Phi(\mathbf{x}) = \mathbf{0}, \tag{2.173}$$

where $\alpha(\cdot)$ and $\beta(\cdot)$ are nonnegative parameters to be chosen adequately in order that the trajectories of (2.173) converge to an equilibrium (i.e., a minimum of the potential or energy function).

In order to view this problem as a control problem amenable to treatment using a CLF, observe that (2.173) can be written as a first-order differential equation in the standard fashion, by introducing the state vector $\mathbf{z} = (\mathbf{z}_1, \mathbf{z}_2) := (\mathbf{x}, \dot{\mathbf{x}})$ as

$$\dot{\mathbf{z}} = \begin{bmatrix} \mathbf{z}_2 \\ -\alpha(\mathbf{z})\nabla\Phi(\mathbf{z}_1) - \beta(\mathbf{z})\mathbf{z}_2 \end{bmatrix} =: \mathbf{f}(\mathbf{z}, \mathbf{u}(\mathbf{z})), \tag{2.174}$$

where $\mathbf{u}(\mathbf{z}) := (\alpha(\mathbf{z}), \beta(\mathbf{z}))$ is the control input to be designed, using a CLF, such that the zero solution of (2.174) is asymptotically stable.

Choice of continuous conjugate gradient algorithm parameters using LOC/CLF

From the discussion in section 2.3.2, it is natural to consider the discrete conjugate gradient iteration (2.137), (2.138) as a pair of coupled bilinear systems that are the starting point in the derivation of the parameters α and β, regarded as control inputs. We will repeat this approach in the analysis of (2.163) and (2.164), rather than simply analyzing stability properties of the second-order vector ODE (2.165) with variable parameters α and β. A control Liapunov argument similar to that in section 2.3.2 is used. The continuous-time analog of the conjugate gradient algorithm (section 2.3.2) is described in the following theorem.

Theorem 2.11. *Given the symmetric, positive definite matrix* $\mathbf{A}$ *and a quadratic function* $\Phi = \frac{1}{2}\mathbf{x}^T\mathbf{A}\mathbf{x} - \mathbf{b}^T\mathbf{x}$, *the trajectories of the conjugate gradient dynamical system, dependent on the positive parameters* α *and* β, *defined as*

$$\begin{aligned} \dot{\mathbf{r}} &= \alpha\mathbf{A}\mathbf{p}, \\ \dot{\mathbf{p}} &= \mathbf{r} - \beta\mathbf{p}, \end{aligned}$$

converge globally to the minimum of the quadratic function Φ *(i.e., to the solution of the linear system* $\mathbf{A}\mathbf{x} = \mathbf{b}$*) if the parameters* α *and* β *are chosen as follows:*

$$\begin{aligned} &\text{if } \langle \mathbf{r}, \mathbf{p}\rangle \neq 0, \quad \alpha = \langle \mathbf{r}, \mathbf{A}\mathbf{p}\rangle / \langle \mathbf{r}, \mathbf{p}\rangle, \quad \beta > 0, \\ &\text{if } \langle \mathbf{r}, \mathbf{p}\rangle = 0, \quad \beta \text{ such that } \langle \mathbf{r}, \mathbf{A}\mathbf{p}\rangle - \beta\langle \mathbf{p}, \mathbf{A}\mathbf{p}\rangle < 0, \end{aligned}$$

where $\mathbf{r} := \mathbf{b} - \mathbf{A}\mathbf{x}$. *The parameter* β *is chosen as follows: If the inner product* $\langle \mathbf{r}, \mathbf{A}\mathbf{p}\rangle$ *is positive, then* $\beta > \langle \mathbf{r}, \mathbf{A}\mathbf{p}\rangle / \langle \mathbf{p}, \mathbf{A}\mathbf{p}\rangle$; *if it is negative or zero, then any positive choice of* β *will do.*

Proof. Consider the CLF candidate

$$V(\mathbf{r}, \mathbf{p}) = \frac{1}{2}\langle \mathbf{r}, \mathbf{A}^{-1}\mathbf{r}\rangle + \frac{1}{2}\langle \mathbf{p}, \mathbf{A}\mathbf{p}\rangle. \tag{2.175}$$

Then

$$\begin{aligned}\dot{V} &= \langle \dot{\mathbf{r}}, \mathbf{A}^{-1}\mathbf{r}\rangle + \langle \dot{\mathbf{p}}, \mathbf{A}\mathbf{p}\rangle \\ &= \langle -\alpha\mathbf{A}\mathbf{p}, \mathbf{A}^{-1}\mathbf{r}\rangle + \langle \mathbf{r} - \beta\mathbf{p}, \mathbf{A}\mathbf{p}\rangle \\ &= -\alpha\langle \mathbf{p}, \mathbf{r}\rangle + \langle \mathbf{r}, \mathbf{A}\mathbf{p}\rangle - \beta\langle \mathbf{p}, \mathbf{A}\mathbf{p}\rangle,\end{aligned}$$

whence it follows that appropriate choices of α and β that make $\dot{V}$ negative semidefinite are as follows:

$$\text{if } \langle \mathbf{r}, \mathbf{p}\rangle \neq 0, \quad \alpha = \frac{\langle \mathbf{r}, \mathbf{A}\mathbf{p}\rangle}{\langle \mathbf{r}, \mathbf{p}\rangle}, \quad \beta > 0, \tag{2.176}$$

$$\text{if } \langle \mathbf{r}, \mathbf{p}\rangle = 0, \quad \beta \text{ such that } \langle \mathbf{r}, \mathbf{A}\mathbf{p}\rangle - \beta\langle \mathbf{p}, \mathbf{A}\mathbf{p}\rangle < 0. \tag{2.177}$$

Since β and $\langle \mathbf{p}, \mathbf{A}\mathbf{p}\rangle$ are positive, it follows that the choice of β in (2.177) depends on the sign of $\langle \mathbf{r}, \mathbf{A}\mathbf{p}\rangle$: If this inner product is positive, then $\beta > \langle \mathbf{r}, \mathbf{A}\mathbf{p}\rangle / \langle \mathbf{p}, \mathbf{A}\mathbf{p}\rangle$; if it is negative or zero, any positive choice of β will do. Since $\dot{V}$ is only semidefinite and α and β are functions of the state variables $\mathbf{r}$, $\mathbf{p}$, LaSalle's theorem (Theorem 1.27) can be applied. It states that the trajectories of the conjugate gradient flow (2.163) and (2.164) will approach the maximal invariant set $\mathcal{M}$ in the set

$$\mathcal{G} := \{(\mathbf{r}, \mathbf{p}) : \dot{V} = 0\}. \tag{2.178}$$

Invariance of $\mathcal{M} \subset \mathcal{G}$ means that any trajectory of the controlled system starting in $\mathcal{M}$ remains in $\mathcal{M}$ for all t.

Note that $\dot{V} = 0$ can occur only if (2.176) occurs and that this implies $\dot{V} = -\beta\langle \mathbf{p}, \mathbf{A}\mathbf{p}\rangle$, so that $\mathcal{G}$ can be alternatively characterized as $\{\mathbf{p} = \mathbf{0}\}$. From (2.163), $\mathbf{p} = \mathbf{0} \Rightarrow \dot{\mathbf{r}} = \mathbf{0}$, which, in turn, implies that $\mathbf{r}$ is constant. From (2.164), $\mathbf{p} = \mathbf{0}$ implies that $\dot{\mathbf{p}} = \mathbf{r}$. Since $\mathbf{r} = c$ (constant), this means that c must be zero (otherwise $\mathbf{p} = \mathbf{0}$ could not occur). Global asymptotic stability of the origin now follows from LaSalle's theorem (Theorem 1.27). □

Note that the Liapunov function could be chosen as

$$V(\mathbf{r}, \mathbf{p}) = \frac{1}{2}\langle \mathbf{r}, \mathbf{A}^{-1}\mathbf{r}\rangle + \frac{1}{2}\langle \mathbf{p}, \mathbf{Q}\mathbf{p}\rangle,$$

where $\mathbf{Q}$ is any positive definite matrix. In particular, the choice $\mathbf{Q} = \mathbf{I}$ results in the simple choice of parameters $\alpha = 1$, $\beta > 0$, demonstrating that the continuous version of the conjugate gradient method can even utilize constant parameters, as opposed to the discrete conjugate gradient method, where the "parameters" α and β must be chosen as functions of the state vectors $\mathbf{r}$ and $\mathbf{p}$. Notice, however, that in the continuous-time case, they are chosen either as constants or in feedback form, rather than being regarded as some arbitrary functions of time that must be chosen to stabilize (2.163) and (2.164). This makes it possible to use LaSalle's Theorem (Theorem 1.27) to obtain a global asymptotic stability result. The choice of α and β given in Theorem 2.11 corresponds to the choices made in the discrete conjugate gradient iteration. Note that a choice of initial conditions consistent with the discrete conjugate gradient iteration is $\mathbf{r}_0 = \mathbf{b} - \mathbf{A}\mathbf{x}_0$ and $\mathbf{p}_0 = \mathbf{r}_0$.

There is, of course, a close connection between these second-order dynamical systems for optimization and the second-order dynamical systems for zero finding proposed in (2.57). In fact, if, for example, the algorithm DC1 from Table 2.2 is applied to the problem of finding the zeros of the gradient ∇f (which is the minimization problem studied at the beginning of this section), then (2.57) for algorithm DC1 becomes

$$\ddot{\mathbf{x}} + \boldsymbol{\Gamma}\dot{\mathbf{x}} + [\nabla^2 f(\mathbf{x})]\nabla f(\mathbf{x}) = \mathbf{0}, \tag{2.179}$$

noting that the Hessian $\nabla^2 f(\mathbf{x})$ is a symmetric matrix. Comparing (2.168) and (2.179), it becomes clear that the algorithm DC1 is an instance of the HBF method; however, the former is derived from the CLF method, without recourse to mechanical analogies. Algorithms DC3 and DC4 are, to the best of our knowledge, new second-order dynamical systems for the zero finding/optimization task, and further research is needed to verify their effectiveness.

In closing, we call the reader's attention to two quotes. The first is a paragraph from Alber [Alb71] that is as relevant today (with some minor changes in the buzz words) as when it was written three decades ago:

> *The increasing interest in continuous-descent methods is due firstly to the fact that tools for the numerical solution of systems of ordinary differential equations are now well developed and can thus be used in conjunction with computers; secondly, continuous methods can be used on analog computers [neural networks]; thirdly, theorems concerning the convergence of these methods and theorems concerning the existence of solutions of equations and of minimum points of functionals are formulated under weaker assumptions than is the case for the analogous discrete processes.*

Similar justification for the consideration of continuous versions of well-known discrete-time algorithms can be found in [Chu88, Chu92].

The second quote is from Bertsekas's encyclopedic book [Ber99] on nonlinear programming:

> *Generally, there is a tendency to think that difficult problems should be addressed with sophisticated methods, such as Newton-like methods. This is often true, particularly for problems with nonsingular local minima that are poorly conditioned. However, it is important to realize that* often the reverse is true, *namely that for problems with "difficult" cost functions and singular local minima, it is best to use simple methods such as (perhaps diagonally scaled) steepest descent with simple stepsize rules such as a constant or diminishing stepsize. The reason is that methods that use sophisticated descent directions and stepsize rules often rely on assumptions that are likely to be violated in difficult problems. We also note that for difficult problems, it may be helpful to supplement the steepest descent method with features that allow it to deal better with multiple local minima and peculiarities of the cost function. An often useful modification is the heavy ball method....*

2.4 Notes and References

Continuous-time dynamical systems for zero finding

Continuous algorithms have been investigated in the Russian literature [Gav58, AHU58, Ryb65b, Ryb69b, Ryb69a, Alb71, Tsy71, KR76, Tan80] as well as the Western literature [Pyn56, BD76, Sma76, HS79] and the references therein. More recently, Chu [Chu88] developed a systematic approach to the continuous realization of several iterative processes in numerical linear algebra. A control approach to iterative methods is mentioned in [KLS89], but not developed as in this book.

Other terms for continuous algorithms that have been, or are, in fashion, are analog circuits, analog computers and more recently, neural networks.

Tsypkin [Tsy71] was one of the first to formulate optimization algorithms as control problems and to raise some of the questions studied in this book. An early discussion of the continuous- and discrete-time Newton methods can be found in [Sil80]. The discrete formulation of CLFs is from [AMNC97], which, in turn, is based on [Son89, Son98]. The idea of introducing a taxonomy of iterative methods in section 2.3 is mentioned in [KLS89], although, once again, the discussion in this reference is not put in terms of CLFs, and there is only a brief mention of control aspects of the problem. Variable structure Newton methods were derived by the present authors in [BK04a].

The continuous Newton method and its variants have been the subject of much investigation [Tan80, Neu99, DK80a, DK80b, DS75, Die93, RZ99, ZG02, HN05]. Branin's method, a much studied variant of the continuous Newton method, was originally proposed in [Dav53a, Dav53b] and, since then, has been studied in many papers: [ZG02] contains many references to this literature. Path following, continuation, and homotopy methods and their interrelationships and relationship to Branin's method are discussed in [ZG02, Neu99, Qua03].

The so-called gradient enhanced Newton method is introduced in [Gra03], where its connections with the Levenberg–Marquardt method are discussed. The latter method is a prototypical team method, which combines two (or more) algorithms, in an attempt to get a hybrid algorithm that has good features of both component algorithms. In Barán, Kaskurewicz, and Bhaya [BKB96] hybrid methods (resp., team algorithms) are put in a control framework and their stability analyzed in the context of asynchronous implementations.

CLF technique

As far as using Liapunov methods in the analysis of iterative methods is concerned, contributions have been made in both Russian [EZ75, VR77] and Western literature [Hur67, Ort73]. The generalized distance functions used in [Pol71, Pol76] can be regarded as Liapunov functions, as has been pointed out in [Ber83]. The Liapunov technique is extremely powerful and can be used to determine basins of convergence [Hur67], as well as to analyze the effects of roundoff errors [Ort73] and delays [KBŠ90].

Early use of a quadratic CLF to study bilinear systems occurs in [Qui80, RB83]. Recent descriptions of the CLF approach and the LOC approach can be found in [Son98, VG97], respectively.

Iterative methods for linear and nonlinear systems of equations

The classic reference for iterative methods for nonlinear equations is [OR70]; a recent survey, including both local and global methods, is [Mar94].

The interested reader is invited to compare the control approach to the conjugate gradient algorithm developed above with other didactic approaches, such as those in [SW95, She94], or an analysis from a z-transform signal processing perspective [CW00]. In our view, the control approach is natural and this is borne out by its simplicity.

Continuous-time systems for optimization

The discussion of the continuous version of the conjugate gradient algorithm and its connection to the well-known BPM method (much used in neural network training) is based on [BK04b], which also contains an analysis of the time-invariant continuous conjugate gradient algorithm.

The BPM algorithm has been much analyzed in the neural network literature, both theoretically and experimentally (see, e.g., [YCC95, YC97, KP99, PS94, HS95] and references therein). The paper [BK04b] shows that the analyses of Torii and Hagan [TH02] and Qian [Qia99] of the time-invariant BPM method are special cases of the conjugate gradient method.

Early papers on "continuous iterative methods" and "analog computing" (see, e.g., [Pol64, Ryb69b, Tsy71]) proposed the use of dynamical systems to compute the solution of various optimization problems. A detailed analysis of the HBF system is carried out in [AGR00], and further developments are reported in [AABR02].

Relationship between continuous and discrete dynamical systems

This is a large subject: Aspects of asymptotic and qualitative behaviors of continuous dynamical systems and their discrete counterparts are treated in the monograph [SH98] and, from a control viewpoint, in [Grü02].

The relationship between convergence rates of discrete and continuous dynamical systems is treated in [HN04].

Another important aspect of the relationship between a continuous dynamical system and its associated discrete dynamical system has to do with the discretization or integration method used. Different integration methods for continuous algorithms are discussed in [Bog71, IPZ79, DS80, Bre01].

Chapter 3
Optimal Control and Variable Structure Design of Iterative Methods

The problem of designing best algorithms of optimization is very similar to the problem of synthesis for discrete or continuous systems which realize these algorithms. Unfortunately, we cannot yet use the powerful apparatus of the theory of optimal systems for obtaining the best algorithms. This is due to the fact that in the modern theory of optimal systems it is assumed that the initial vector $\mathbf{x}_0$ *and the final vector* $\mathbf{x}^*$ *are known. In our case,* $\mathbf{x}^*$ *is unknown, and moreover, our goal is to find* $\mathbf{x}^*$*. Therefore, we are forced to consider only the "locally" best algorithms.*

—Ya. Z. Tsypkin [Tsy71, p. 36, sec. 2.18]

This chapter pursues the objective of "best" or optimal algorithms mentioned in the quote above and shows that there is a simple way to escape from the pessimism of Tsypkin's statement. The main tool used in this chapter is optimal control theory, and an additional objective is to show that this theory can be brought to bear on the problems of finding zeros and minimizing a given function. In particular, optimal control theory can be used to arrive at the concept of variable structure iterative methods, already met in the previous chapter, and these methods are explored further in this chapter. Another application, with a different flavor, of dynamic programming ideas, from discrete-time optimal control theory to unconstrained minimization problems, closes the chapter.

Given a function $\mathbf{f} : \mathbb{R}^n \to \mathbb{R}^n$, the zero finding problem is that of finding a vector $\mathbf{x}^* \in \mathbb{R}^n$ such that $\mathbf{f}(\mathbf{x}^*) = \mathbf{0}$. On the other hand, given a function $f : \mathbb{R}^n \to \mathbb{R}$, the problem of minimizing f is that of finding $\mathbf{x}^* \in \mathbb{R}^n$ such that $f(\mathbf{x}^*) \leq f(\mathbf{x})$ for all $\mathbf{x}$ in some neighborhood of $\mathbf{x}^*$ (local minimum), or in the whole of $\mathbb{R}^n$ (global minimum). Clearly, a zero finding problem for $\mathbf{f}$ can be turned into a minimization problem for $g : \mathbb{R}^n \to \mathbb{R}$, where, for example, $g(\mathbf{x})$ is chosen as $\mathbf{f}(\mathbf{x})^T\mathbf{f}(\mathbf{x})$. Conversely, a minimization problem for a differentiable function $f : \mathbb{R}^n \to \mathbb{R}$ can be approached by solving the problem of finding the zeros of the gradient $\nabla f : \mathbb{R}^n \to \mathbb{R}^n$.

In the zero finding problem as well as the minimization problem, the approach of section 2.1 is taken, starting with a continuous-time dynamical system. The novelty in this chapter is the introduction of an associated *cost* or *objective function*. With certain choices of cost function and system dynamics, in both zero finding and minimization problems, the optimal control turns out to be of the so-called bang-bang type, leading to an overall closed-loop system which has a switching or discontinuous element. Such dynamical systems, described by ODEs with discontinuous right-hand sides, constitute a special class of the so-called variable structure systems, which are also studied further in this chapter as well as the next, in the context of gradient dynamical systems. Optimal control theory provides a natural motivation for the introduction of variable structure control.

In the zero finding problem (2.1), a possible difficulty in the application of optimal control methods was pointed out by Tsypkin in the quote above. However, just a year after the publication of the quote, Branin [Bra72] and, subsequently, Chao and de Figueiredo [CdF73], realized that, if one changes variables from $\mathbf{x}$ to $\mathbf{f}(\mathbf{x})$, then in the $\mathbf{f}$-coordinates, this problem is taken care of, since now the known initial state $\mathbf{f}(\mathbf{x}_0)$ should be driven to the final state $\mathbf{f}(\mathbf{x}^*)$ which must be zero, in order that the zero finding problem be solved. With this elementary but crucial observation (which was, in fact, used in Chapter 2) in hand, it is possible to motivate the optimal control and variable structure approaches to the zero finding problem.

3.1 Optimally Controlled Zero Finding Methods

This section shows that iterative methods can be approached from the point of view of optimal control in at least two different ways, which are described in this section and the next.

The optimally controlled method presented in this section, in addition to being interesting in its own right, also provides another control perspective on the Newton–Raphson algorithm discussed in section 2.1.

Given $\mathbf{f} : \mathbb{R}^n \to \mathbb{R}^n$ such that $\mathbf{f} : \mathbf{x} \mapsto \mathbf{f}(\mathbf{x})$, define $\mathbf{y}(\mathbf{x}) = \mathbf{f}(\mathbf{x})$, as in (2.4). Assume also that $\mathbf{x}(\cdot)$ is a function of time t, i.e., that there exists a function $t \mapsto \mathbf{x}(t)$ that represents the evolution of the vector $\mathbf{x}$ as it starts from some initial value $\mathbf{x}_0$ and converges to a final value $\mathbf{x}_1 := \mathbf{x}(t_1)$ at some final time t_1, such that $\mathbf{f}(\mathbf{x}_1) = \mathbf{0}$.

A natural approach, in terms of continuous algorithms, is to take a dynamical system described by equations (2.3), (2.4) and introduce a criterion or objective function of the state vector that is to be optimized subject to the system dynamics (i.e., along its trajectories).

More specifically, given an initial point $\mathbf{x}_0$, the problem we wish to solve is that of finding an optimal trajectory starting from $\mathbf{x}_0$ and terminating in a zero $\mathbf{x}^*$ of a function $\mathbf{f} : \mathbb{R}^n \to \mathbb{R}^n$.

3.1.1 An optimal control–based Newton-type method

To start with, the change of coordinates discussed in the introduction to this chapter is applied; i.e., instead of working in $\mathbf{x}$-coordinates, consider a change of coordinates to $\mathbf{y} = \mathbf{f}(\mathbf{x})$. This means that an initial state $\mathbf{x}_0$ corresponds to an initial state $\mathbf{y}_0 = \mathbf{f}(\mathbf{x}_0)$ in the $\mathbf{y}$ coordinate system. Following Chao and Figueiredo [CdF73], consider the following dynamical system:

$$\dot{\mathbf{y}} = -\mathbf{y} + \mathbf{u}, \quad \mathbf{y}(0) = \mathbf{y}_0. \tag{3.1}$$

When no control is applied ($\mathbf{u} = \mathbf{0}$), (3.1) has an exponentially stable equilibrium at $\mathbf{y} = \mathbf{0}$. Assume that it is desired to transfer the initial state $\mathbf{y}_0$ to the final state $\mathbf{0}$ in the time interval

$[0, t_1]$. Then, for the zero finding problem, a possible *cost function* is the so-called minimum fuel (mf) consumption performance index, defined as follows:

$$J_{\text{mf}} = \frac{1}{2}\int_0^{t_1} \|\mathbf{u}\|^2 \mathrm{d}t. \tag{3.2}$$

The optimal control problem is to find a control input $\mathbf{u}$ such that the cost function J_{mf} is minimized subject to the boundary conditions

$$\mathbf{y}(\mathbf{x}_0) = \mathbf{y}_0, \qquad \mathbf{y}(\mathbf{x}(t_1)) = \mathbf{0}, \tag{3.3}$$

where t_1 is the specified final or terminal time. Since the boundary conditions are specified at the two end points of the optimization time interval, the optimal control problem just formulated is, in the language of differential equations, a two-point boundary value problem (TPBVP). Following the standard Hamiltonian approach [PBGM62, AF66] to the calculation of optimal control and the corresponding trajectories, one can write the Hamiltonian H for this problem:

$$H(\boldsymbol{\lambda}, \mathbf{y}, \mathbf{u}) = \frac{1}{2}\|\mathbf{u}\|^2 + \boldsymbol{\lambda}^T(-\mathbf{y} + \mathbf{u}), \tag{3.4}$$

where $\boldsymbol{\lambda}$ is the Lagrange multiplier vector. The corresponding equations for optimality are

$$\begin{aligned} \dot{\mathbf{y}} &= \partial H(\boldsymbol{\lambda}, \mathbf{y}, \mathbf{u})/\partial \boldsymbol{\lambda}, \\ \dot{\boldsymbol{\lambda}} &= -\partial H(\boldsymbol{\lambda}, \mathbf{y}, \mathbf{u})/\partial \mathbf{y}, \\ 0 &= \partial H(\boldsymbol{\lambda}, \mathbf{y}, \mathbf{u})/\partial \mathbf{u}. \end{aligned} \tag{3.5}$$

Thus, from (3.4),

$$\begin{aligned} \dot{\mathbf{y}} &= \partial H(\boldsymbol{\lambda}, \mathbf{y}, \mathbf{u})/\partial \boldsymbol{\lambda} = -\mathbf{y} + \mathbf{u}, \\ \dot{\boldsymbol{\lambda}} &= -\partial H(\boldsymbol{\lambda}, \mathbf{y}, \mathbf{u})/\partial \mathbf{y} = \boldsymbol{\lambda}, \\ \mathbf{u} + \boldsymbol{\lambda} &= 0. \end{aligned} \tag{3.6}$$

The optimal control input $\mathbf{u}$ and the corresponding trajectory $\mathbf{y}$ are determined by solving (3.6), subject to the boundary conditions (3.3), and turn out to be

$$\mathbf{u} = \left(\frac{2\exp(-t_1)\exp(t)}{\exp(-t_1) - \exp(t_1)}\right)\mathbf{y}_0 = -\frac{\exp[-(t_1 - t)]}{\sinh t_1}\mathbf{y}_0 \tag{3.7}$$

and

$$\mathbf{y} = \mathbf{y}_0 \sinh(t_1 - t)/\sinh t_1. \tag{3.8}$$

By the chain rule, the time derivative of the vector $\mathbf{y}$ can be written as follows:

$$\dot{\mathbf{y}} = \mathbf{D}_{\mathbf{f}}(\mathbf{x})\frac{\mathrm{d}\mathbf{x}}{\mathrm{d}t}, \tag{3.9}$$

where $\mathbf{D}_{\mathbf{f}}(\mathbf{x})$ denotes the Jacobian matrix (also denoted as $\partial \mathbf{f}/\partial \mathbf{x}$).

From (3.6), (3.7), and (3.9), assuming that the Jacobian matrix $\mathbf{D}_{\mathbf{f}}(\mathbf{x})$ is invertible, one can write

$$\dot{\mathbf{x}} = \mathbf{D}_{\mathbf{f}}(\mathbf{x})^{-1}\left(-\mathbf{y} - \mathbf{y}_0 \exp(-t_1 - t)/\sinh t_1\right). \tag{3.10}$$

To solve this differential equation in order to find the trajectories $\mathbf{x}(t)$, one can use the standard forward Euler method as follows. Divide the time interval $[0, t_1]$ into N equal intervals of width $\Delta t = h$ (i.e., h is the stepsize of the Euler method). Define the time instants

$$t_{k+1} = t_k + h, \quad k = 0, 1, 2, \dots, N-1. \tag{3.11}$$

Then the discretization of (3.10), which can be referred to as an *optimal iterative method* to calculate the zeros of a function $\mathbf{f}$, is as follows:

$$\mathbf{x}_{k+1} = \mathbf{x}_k - \alpha_k \mathbf{D_f}(\mathbf{x}_k)^{-1}\mathbf{f}(\mathbf{x}_k), \quad k = 0, 1, 2, \dots, N-1, \tag{3.12}$$

where

$$\alpha_k = h \coth(t_1 - t_k). \tag{3.13}$$

Recalling that $\mathbf{f}(\mathbf{x}(t_1)) = \mathbf{0}$, it follows that $\mathbf{x}(t_1) = \mathbf{x}^*$, i.e., is a zero of $\mathbf{f}(\cdot)$. Thus, if the discretization error of the Euler method is small, then $\mathbf{x}_N \approx \mathbf{x}^*$. Since the final time t_1 can be arbitrarily specified, the approximate solution $\mathbf{x}$ can be obtained in a finite number of steps (N).

The interesting feature of this optimal iterative method is that it leads to a variant of the discrete-time Newton method (2.69) in which the stepsize is time varying or, more precisely, optimally controlled. The fixed "gain" (or stepsize) α of (2.69) is replaced by the variable gain or stepsize $\alpha_k := h \coth(t_1 - t_k)$ in the optimal iterative method (3.12).

As remarked in section 2.2, observe that the use of a different discretization method, for example, the Runge–Kutta method, to solve (3.10) would lead to a different discrete iterative method. In this connection, see the notes and references to Chapter 2.

One obvious feature of the above method is that the choice of minimum fuel consumption cost function, although it would be well motivated in a control problem, is not particularly natural for the zero finding context. Similarly, the choice of the dynamical system (3.1) could also be questioned. Of course, a pragmatic view of these remarks is that the choices made above lead to an interesting variation of the Newton–Raphson algorithm, and this is sufficient justification. On the other hand, other cost functions or performance indices, such as the minimum time performance index $J_{\text{mt}} := \int_0^t 1\, dt$ or a norm of the final state $J_{\text{fs}} = \mathbf{x}(t_1)^T \mathbf{S} \mathbf{x}(t_1)$ for some positive definite matrix $\mathbf{S}$, as well as other underlying dynamical systems, could be introduced, and the next section does just this, in order to motivate another class of iterative methods, the so-called variable structure iterative methods.

3.2 Variable Structure Zero Finding Methods

In this section, choices of dynamical system and cost function different from those made in the previous section are shown to lead to variable structure iterative methods. To this end, using the change of coordinates $\mathbf{y} = \mathbf{f}(\mathbf{x})$ made in the previous section, consider the $\mathbf{y}$-coordinate analog of system (2.3), i.e.,

$$\dot{\mathbf{y}} = \mathbf{u}, \tag{3.14}$$

where $\mathbf{u}$ is a control input that is to be chosen adequately, in order to move $\mathbf{x}_0$ to a zero $\mathbf{x}^*$ of $\mathbf{f}(\cdot)$ in the time interval $[0, t_1]$, using bounded controls:

$$\mathbf{u} \in \mathcal{U} := \{\mathbf{u} : \|\mathbf{u}\|_\infty \leq 1\}. \tag{3.15}$$

Observe that, in contrast to the system (3.1) which is exponentially stable in $\mathbf{y}$-coordinates with $\mathbf{u} = \mathbf{0}$, the system (3.14) is only stable for zero input.

In order to minimize the final state $\mathbf{y}(t_1)$, the natural choice of cost function or performance index is

$$J_{\text{fs}} = \frac{1}{2}\mathbf{y}(t_1)^T\mathbf{y}(t_1), \tag{3.16}$$

which is a quadratic function that attains a minimum value of $\mathbf{0}$, if $\mathbf{y}(t_1) = \mathbf{0}$, as desired. The subscript fs serves as a reminder that the cost function requires minimization of the final state $\mathbf{y}(t_1)$ in the 2-norm.

The zero finding problem can now be formulated as the following optimal control problem:

$$\begin{array}{ll} \min & J_{\text{fs}} \\ \text{s.t.} & \dot{\mathbf{y}} = \mathbf{u}, \quad \mathbf{u} \in \mathcal{U}. \end{array} \tag{3.17}$$

This optimal control problem has a particularly simple form and can be solved directly. Observe that

$$\frac{\mathrm{d}[(1/2)\mathbf{y}^T\mathbf{y}]}{\mathrm{d}t} = \mathbf{y}^T\dot{\mathbf{y}}.$$

Thus

$$\int_0^{t_1} \mathbf{y}^T\dot{\mathbf{y}}\,\mathrm{d}t = \int_0^{t_1} \frac{\mathrm{d}[(1/2)\mathbf{y}^T\mathbf{y}]}{\mathrm{d}t}\,\mathrm{d}t = (1/2)[\mathbf{y}(t_1)^T\mathbf{y}(t_1) - \mathbf{y}(0)^T\mathbf{y}(0)]. \tag{3.18}$$

Since $\mathbf{y}(0)^T\mathbf{y}(0)$ is a constant value, this means that an equivalent way of writing (3.17) is as follows:

$$\min_{\mathbf{u}\in\mathcal{U}} \int_0^{t_1} \mathbf{y}^T\dot{\mathbf{y}}\,\mathrm{d}t = \min_{\mathbf{u}\in\mathcal{U}} \int_0^{t_1} \mathbf{y}^T\mathbf{u}\,\mathrm{d}t, \tag{3.19}$$

where the substitution $\dot{\mathbf{y}} = \mathbf{u}$ has been made under the integral sign. From this formulation, it is clear that the choice of $\mathbf{u} \in \mathcal{U}$ that minimizes the integral in (3.19) should be

$$\mathbf{u} = -\operatorname{sgn}\mathbf{y}, \tag{3.20}$$

leading to the closed-loop optimally controlled system

$$\dot{\mathbf{y}} = -\operatorname{sgn}\mathbf{y}. \tag{3.21}$$

Writing (3.21) in $\mathbf{x}$-coordinates using the relation $\dot{\mathbf{y}} = \mathbf{D_f}\dot{\mathbf{x}}$ gives

$$\dot{\mathbf{x}} = -\mathbf{D_f}^{-1}\operatorname{sgn}\mathbf{f}(\mathbf{x}), \tag{3.22}$$

and the Newton variable (NV) structure method (2.21) has been re-derived as an optimally controlled system.

From Theorem 1.38, any choice of diagonally stable matrix $\mathbf{P}$ in the system $\dot{\mathbf{w}} = \mathbf{P}\operatorname{sgn}(\mathbf{w})$ results in finite-time convergence, via a sliding mode, to the sliding equilibrium $\mathbf{w} = \mathbf{0}$. This means that, in (3.20), the control $\mathbf{u}$ can be chosen as

$$\mathbf{u} = \mathbf{P}\operatorname{sgn}\mathbf{y}, \tag{3.23}$$

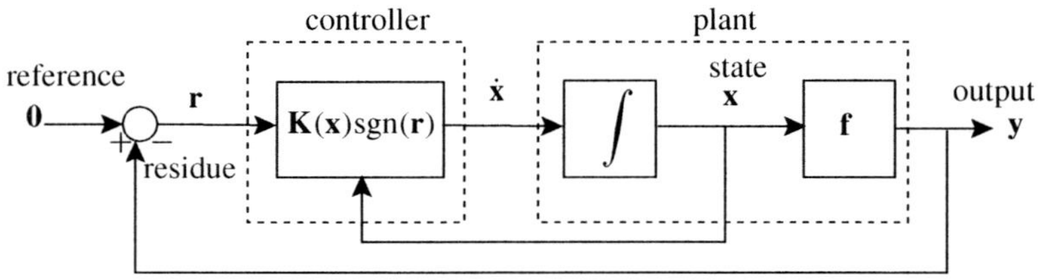

Figure 3.1. *Control system representation of the variable structure iterative method* (3.22). *Observe that this figure is a special case of Figure* 2.1A.

leading to a variable structure Newton method with gain $\mathbf{P}$:

$$\dot{\mathbf{x}} = -\mathbf{D}_\mathbf{f}^{-1}\mathbf{P}\,\mathrm{sgn}\,\mathbf{f}(\mathbf{x}). \tag{3.24}$$

Of course, with this choice, $\mathbf{u} \notin \mathcal{U}$. Observe that, although this can easily be remedied by choosing $\mathbf{u} = (1/\|\mathbf{P}\|_\infty)\mathbf{P}\,\mathrm{sgn}\,\mathbf{y}$, the new choice also leads to a stable closed-loop system. In fact, (3.24) has a faster rate of convergence than (3.22) and this is one motivation for the use of variable structure control, without the necessity of an associated cost function that is to be optimized.

The system is shown in Figure 3.1, and as before, in control terms, a regulation problem in which the output must become equal to the zero reference input must be solved. The novelty with discontinuous control is that this problem is solved in finite time instead of asymptotically.

Choosing the Persidskii-type Liapunov function

$$V(\mathbf{r}) := \|\mathbf{r}\|_1 = \sum_{i=1}^{n} \int_0^{r_i} \mathrm{sgn}\,\tau\, d\tau, \tag{3.25}$$

as in Theorem 1.38, and choosing

$$\mathbf{K}(\mathbf{x}) := \alpha\mathbf{D}_\mathbf{f}^{-1}(\mathbf{x}),\ \alpha > 0, \tag{3.26}$$

which corresponds to choosing $\mathbf{P} = \alpha\mathbf{I}$ in (3.24), leads to

$$\dot{V} = -\alpha\,\mathrm{sgn}(\mathbf{r})^T\mathrm{sgn}(\mathbf{r}) \le -\alpha, \quad \text{whenever } \mathbf{r} \ne \mathbf{0}. \tag{3.27}$$

Integrating this differential inequality leads to the estimate

$$V(t) = \|\mathbf{r}(t)\|_1 \le V(0) - \alpha t, \tag{3.28}$$

showing that the 1-norm of $\mathbf{r}$ can be expected to become zero in finite time less than or equal to $V(0)/\alpha$.

The developments in this section and the previous one can be summarized as follows. An optimal control problem is specified by giving a dynamical system, a cost function, and appropriate boundary conditions or constraints. The choices made are specified in Table 3.1, using the notation $\mathbf{y} = \mathbf{f}(\mathbf{x})$.

Table 3.1. *Choices of dynamical system, cost function, and boundary conditions/ constraints that lead to different optimal iterative methods for the zero finding problem.*

System	Cost function	Boundary conditions/ Constraints	Section
$\dot{\mathbf{y}} = -\mathbf{y} + \mathbf{u}$	$\frac{1}{2}\int_0^{t_1} \mathbf{u}^T \mathbf{u} dt$	$\mathbf{y}(0) = \mathbf{f}(\mathbf{x}_0)$ $\mathbf{y}(t_1) = \mathbf{f}(\mathbf{x}(t_1) = \mathbf{0}$	3.1
$\dot{\mathbf{y}} = \mathbf{u}$	$\frac{1}{2}\mathbf{y}(t_1)^T\mathbf{y}(t_1)$	$\mathbf{u} \in \mathcal{U} := \{\mathbf{u} : \|\mathbf{u}\|_\infty \leq 1\}$	3.2

The message that emerges from this table and the discussion in this section and the previous one is as follows. Each choice of a dynamical system, cost function, and boundary condition or constraints leads to a different iterative method, optimal for the choices made, for the zero finding problem. Of course, the proof of the pudding is in the eating, so whatever the choices are, the end result should be a robust, easily implementable algorithm. One way of ensuring this is to observe that the algorithm resulting from the optimal control method is a variant of some well-known algorithm (section 3.1). The other is to ensure that the resulting system is asymptotically stable about the point that is desired to be computed, which was the case with system (3.21) with respect to the point $\mathbf{y}(t_1) = \mathbf{0}$. Thus, two good choices appear above and the adventurous reader is invited to add other rows to Table 3.1, resulting in new optimal algorithms for the zero finding problem.

3.2.1 A variable structure Newton method to find zeros of a polynomial function

This subsection shows that the Kokotović–Šiljak method [KŠ64] of finding zeros of a polynomial with complex coefficients is actually a type of variable structure Newton method, introduced in Chapter 2; this is done by writing it as a discontinuous Persidskii system, with the corresponding diagonal-type CLF. The polynomial zero finding problem is an old and venerable one with a huge literature; the reader is referred to [McN93], which has references to more than 3000 articles and books on this topic.

The problem is to find zero(s) of a polynomial with complex coefficients:

$$P(z) = \sum_{k=0}^{n} a_k z^k, \quad a_k = b_k + ic_k, \ a_n \neq 0. \tag{3.29}$$

The Kokotović–Šiljak method starts by setting $z = \sigma + i\omega$ in (3.29) to get

$$P(z) = R(\sigma, \omega) + iI(\omega, \sigma), \tag{3.30}$$

where $R(\sigma, \omega)$ and $I(\omega, \sigma)$ are, respectively, the real and imaginary parts of the polynomial P. The idea is to use gradient descent to minimize the real-valued objective function

$$V(\sigma, \omega) := |R(\sigma, \omega)| + |I(\omega, \sigma)|. \tag{3.31}$$

More specifically, the method follows a trajectory of the gradient dynamical system

$$\begin{bmatrix} \dot{\sigma} \\ \dot{\omega} \end{bmatrix} = -h\nabla V(\sigma, \omega), \quad h > 0, \tag{3.32}$$

starting from an arbitrary initial point (σ_0, ω_0), and h is the control that is to be determined. Calculating the right-hand side of (3.32) for V in (3.31) gives

$$\begin{bmatrix} \dot{\sigma} \\ \dot{\omega} \end{bmatrix} = -h\nabla V = -h \begin{bmatrix} \partial R/\partial\sigma & \partial I/\partial\sigma \\ \partial R/\partial\omega & \partial I/\partial\omega \end{bmatrix} \begin{bmatrix} \operatorname{sgn}(R) \\ \operatorname{sgn}(I) \end{bmatrix}. \tag{3.33}$$

It is argued in [KŠ64, Šil69] that since $V = |R| + |I|$ has the following properties:

(i) V is nonnegative,

(ii) derivatives $\partial V/\partial\sigma$, $\partial V/\partial\omega$ exist,

(iii) zeros of V are located at the zeros of P,

(iv) zeros of V are the only minima of V,

(v) the time derivative dV/dt is always negative,

then all trajectories of (3.33) will have zeros of V as limit points when $t \to \infty$ in (3.33). This argument is certainly valid for gradient dynamical systems in which the right-hand side is C^1, as seen in Chapter 1. However, (3.33) represents a gradient dynamical system with a *discontinuous* right-hand side. Thus, for greater rigor, (3.33) is now analyzed using a result on stability of dynamical systems with discontinuous right-hand sides.

To do this, observe that since $P(z)$ is analytic, by the Cauchy–Riemann equations, $\partial R/\partial\sigma = \partial I/\partial\omega$ and $\partial R/\partial\omega = -\partial I/\partial\sigma$ and consequently the matrix

$$\mathbf{M} := \begin{bmatrix} \partial R/\partial\sigma & \partial I/\partial\sigma \\ \partial R/\partial\omega & \partial I/\partial\omega \end{bmatrix} = \begin{bmatrix} u & -v \\ v & u \end{bmatrix} \tag{3.34}$$

that appears on the right-hand side of (3.33) has the property

$$\mathbf{M}^T\mathbf{M} = \begin{bmatrix} u^2 + v^2 & 0 \\ 0 & u^2 + v^2 \end{bmatrix}, \tag{3.35}$$

which is a positive definite matrix, provided that

$$u^2 + v^2 = (\partial R/\partial\sigma)^2 + (\partial R/\partial\omega)^2 = (\partial I/\partial\sigma)^2 + (\partial I/\partial\omega)^2 > 0 \quad \text{for all } \sigma, \omega. \tag{3.36}$$

In brief, the Kokotović–Šiljak method consists of numerical (originally, analog) integration of (3.33) taking advantage of the following fact. Defining $z^k = X_k + iY_k$, observe that $R = \sum_{k=0}^{n} a_k X_k$ and $I = \sum_{k=0}^{n} a_k Y_k$, and furthermore that

$$\begin{aligned} X_k &= 2\sigma X_{k-1} - (\sigma^2 + \omega^2)X_{k-2}, \qquad & X_0 &= 1,\ X_1 = \sigma, \\ Y_k &= 2\sigma X_{k-1} - (\sigma^2 + \omega^2)X_{k-2}, \qquad & Y_0 &= 0,\ Y_1 = \omega, \end{aligned}$$

so that fast recursive computation of R, I, and their derivatives (and hence $\mathbf{M}$) is possible in terms of X_k and Y_k, which are called Šiljak polynomials in [Sto95].

Since the dynamical system (3.33) consists of a set of differential equations whose right-hand sides are linear combinations of (almost sigmoidal) nonlinearities, it could be regarded as a neural network that finds zeros of a polynomial.

The following theorem follows readily from the developments above.

Theorem 3.1. *Consider the polynomial $P(z)$ in* (3.30) *and the associated discontinuous gradient dynamical system* (3.33). *If assumption* (3.36) *holds and parameter h in* (3.33) *is chosen as*

$$h = k\left[(\partial R/\partial\sigma)^2 + (\partial R/\partial\omega)^2\right]^{-1}, \tag{3.37}$$

then every trajectory of (3.33) *converges to some zero of $P(z)$ through a sliding mode. In particular, a trajectory that starts from initial condition $R(0), I(0)$ attains a zero in finite time t_Z that is upper bounded as follows:*

$$t_Z \leq (|R(0)| + |I(0)|)k^{-1}. \tag{3.38}$$

Proof. By the chain rule,

$$\begin{bmatrix} \dot{R} \\ \dot{I} \end{bmatrix} = \begin{bmatrix} \partial R/\partial\sigma & \partial R/\partial\omega \\ \partial I/\partial\sigma & \partial I/\partial\omega \end{bmatrix} \begin{bmatrix} \dot{\sigma} \\ \dot{\omega} \end{bmatrix} = \mathbf{M}^T \begin{bmatrix} \dot{\sigma} \\ \dot{\omega} \end{bmatrix}. \tag{3.39}$$

Thus the dynamics of the Kokotović–Šiljak method (3.33) in R, I coordinates is

$$\begin{bmatrix} \dot{R} \\ \dot{I} \end{bmatrix} = -h\mathbf{M}^T\mathbf{M} \begin{bmatrix} \mathrm{sgn}(R) \\ \mathrm{sgn}(I) \end{bmatrix}, \tag{3.40}$$

or, equivalently,

$$\begin{bmatrix} \dot{R} \\ \dot{I} \end{bmatrix} = -h \begin{bmatrix} u^2 + v^2 & 0 \\ 0 & u^2 + v^2 \end{bmatrix} \begin{bmatrix} \mathrm{sgn}(R) \\ \mathrm{sgn}(I) \end{bmatrix}. \tag{3.41}$$

Observe that this is a Persidskii-type system with a discontinuous right-hand side of the type introduced in Chapter 1, where the following nonsmooth integral-of-nonlinearity diagonal-type CLF was proposed:

$$V(R, I) = \int_0^R \mathrm{sgn}(\tau)\mathrm{d}\tau + \int_0^I \mathrm{sgn}(\tau)\mathrm{d}\tau = |R| + |I|. \tag{3.42}$$

The time derivative of V_1 along the trajectories of (3.41) is given by

$$\dot{V} = [\mathrm{sgn}(R) \quad \mathrm{sgn}(I)] \begin{bmatrix} \dot{R} \\ \dot{I} \end{bmatrix} = -h[\mathrm{sgn}(R) \quad \mathrm{sgn}(I)]\mathbf{M}^T\mathbf{M} \begin{bmatrix} \mathrm{sgn}(R) \\ \mathrm{sgn}(I) \end{bmatrix} \tag{3.43}$$

or, from (3.35), as

$$\dot{V} = -h(u^2 + v^2)[\mathrm{sgn}(R)^2 + \mathrm{sgn}(I)^2]. \tag{3.44}$$

If assumption (3.36) holds, the choice of the control parameter h as in (3.37) yields

$$\dot{V}(t) = -k[\mathrm{sgn}(R)^2 + \mathrm{sgn}(I)^2] \leq -k. \tag{3.45}$$

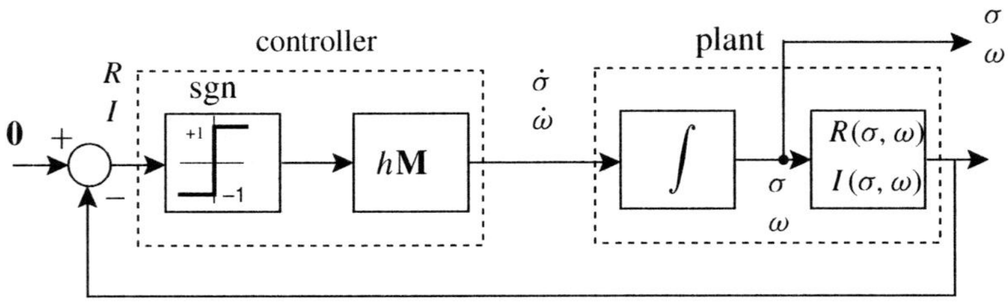

Figure 3.2. *Control system implementation of dynamical system* (3.33). *Observe that the controller is a special case of the controller represented in Figure* 2.12.

The differential inequality (3.45) furnishes the upper bound

$$t_Z \leq \frac{V(R(0), I(0))}{k} = \frac{|R(0)| + |I(0)|}{k}$$

for the time t_Z taken for the Liapunov function V_1 to go to zero along trajectories of (3.41), establishing the stated finite-time upper bound for the Kokotović–Šiljak method to find a zero of the polynomial P. □

In order to justify the remark that the Kokotović–Šiljak method is a type of variable structure (and consequently finite-time) Newton method, consider the block diagram representation in Figure 3.2. It shows that the Kokotović–Šiljak dynamical system (3.33) can be represented as a particular case of Figure 2.12, with an appropriately chosen plant and a switching controller. The gain of the controller is the scalar h, and the choice (3.37) essentially inverts the matrix $\mathbf{M}^T\mathbf{M} = (u^2 + v^2)\mathbf{I}$.

Observe that Theorem 3.1 only proves convergence of the trajectories to some zero of the polynomial. It is therefore of interest to see if the method can be extended or modified to take into account prior knowledge of a region in which zeros of the polynomial are known to exist. Alternatively, it may be of interest to look for zeros in a specified region. This leads to a natural extension that is motivated and further exploited in section 4.4.

Extension of the nonsmooth Kokotović–Šiljak method

The nonsmooth CLF analysis of the previous section allows extensions of the Kokotović–Šiljak method. For example, suppose that bounds on the zeros are known or desired: $\sigma < a$, $\omega < b$. Let the new variables R_1 and I_1 be defined accordingly:

$$R_1 := \sigma - a, \quad I_1 := \omega - b \;\Rightarrow \dot{\sigma} = \dot{R}_1, \quad \dot{\omega} = \dot{I}_1. \tag{3.46}$$

Thus it is desired to find, if it exists, a zero of the polynomial in the region $\mathcal{S}$ where $R_1 < 0$ and $I_1 < 0$, i.e.,

$$\mathcal{S} := \{(\sigma, \omega) : \sigma < a, \omega < b\}. \tag{3.47}$$

Foreshadowing the developments in Chapter 4, a new objective function is formed by adding terms that penalize violations of the given bounds to the original objective function (3.31) as follows:

$$V_2(\sigma, \omega) := h_1(|R| + |I|) + h_2(\max(0, R_1) + \max(0, I_1)). \tag{3.48}$$

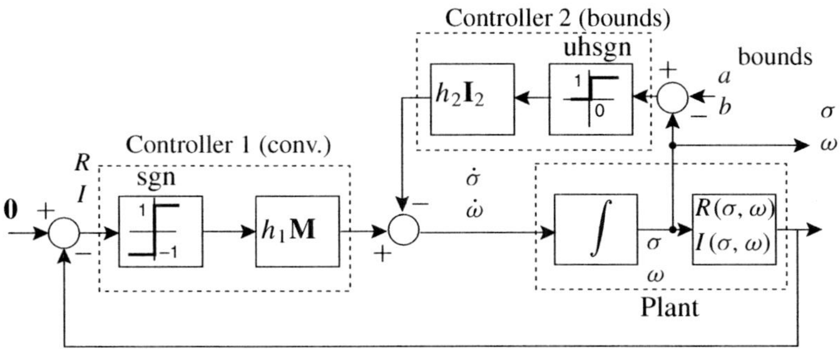

Figure 3.3. *Control system implementation of dynamical system* (3.50). *Observe that controller* 1 *is responsible for the convergence of the real and imaginary parts R and I to zero, while controller* 2 *is responsible for maintaining σ and ω below the known upper bounds. If lower bounds are known as well, a third controller is needed to implement these bounds.*

This new objective function is now minimized by gradient descent. Observe that the derivative of $\max(0, x)$ for $x \neq 0$ is the upper half signum function uhsgn$(\cdot)$ defined in Chapter 1. Thus, the new gradient system is

$$\begin{bmatrix} \dot{\sigma} \\ \dot{\omega} \end{bmatrix} = -\nabla V_2(\sigma, \omega) = -\begin{bmatrix} \partial V_2/\partial\sigma \\ \partial V_2/\partial\omega \end{bmatrix}. \tag{3.49}$$

Observing that $\partial R_1/\partial\sigma = \partial I_1/\partial\omega = 1$ and $\partial R_1/\partial\omega = \partial I_1/\partial\sigma = 0$, (3.49) can be written as

$$\begin{bmatrix} \dot{\sigma} \\ \dot{\omega} \end{bmatrix} = -h_1\mathbf{M}\begin{bmatrix} \text{sgn}(R) \\ \text{sgn}(I) \end{bmatrix} - h_2\mathbf{I}_2\begin{bmatrix} \text{uhsgn}(R_1) \\ \text{uhsgn}(I_1) \end{bmatrix}, \tag{3.50}$$

where $\mathbf{I}_2$ denotes a two-by-two identity matrix. A block diagram representation of (3.50) is shown in Figure 3.3.

We can now state a result concerning (3.50).

Lemma 3.2. *Consider the polynomial $P(z)$ in* (3.30). *For all initial conditions* (σ_0, ω_0) $\in \mathbb{R} \times \mathbb{R}$ *and all choices of $h_1 > 0$ and $h_2 > 0$, trajectories of* (3.50) *converge to the region $\mathcal{S}$.*

This lemma states that the reaching phase occurs with respect to the region $\mathcal{S}$ for all initial conditions and all choices of the gains h_1 and h_2. It does not, however, guarantee convergence to a zero of $P(z)$, since this depends on the position of the zeros in relation to the region $\mathcal{S}$ as well as the nature of the vector field defined by the right-hand side of (3.33).

Proof. From (3.50), (3.46), and (3.39), it follows that the augmented system that describes the dynamics of R_1, I_1, R, and I is

$$\begin{bmatrix} \dot{R}_1 \\ \dot{I}_1 \\ \dot{R} \\ \dot{I} \end{bmatrix} = \begin{bmatrix} -h_2\mathbf{I}_2 & -h_1\mathbf{M} \\ -h_2\mathbf{M}^T & -h_1\mathbf{M}^T\mathbf{M} \end{bmatrix} \begin{bmatrix} \text{uhsgn}(R_1) \\ \text{uhsgn}(I_1) \\ \text{sgn}(R) \\ \text{sgn}(I) \end{bmatrix}. \tag{3.51}$$

Factoring the first matrix on the right-hand side of (3.51) and defining

$$\mathbf{z} := (R_1, I_1, R, I)^T, \quad \mathbf{f}(\mathbf{z}) := (\text{uhsgn}(R_1), \text{uhsgn}(I_1), \text{sgn}(R), \text{sgn}(I)),$$

(3.51) can be written as

$$\dot{\mathbf{z}} = -\begin{bmatrix} \mathbf{I}_2 & \mathbf{M} \\ \mathbf{M}^T & \mathbf{M}^T\mathbf{M} \end{bmatrix} \begin{bmatrix} h_2\mathbf{I}_2 & 0 \\ 0 & h_1\mathbf{I}_2 \end{bmatrix} \mathbf{f}(\mathbf{z}), \tag{3.52}$$

which is recognizable as a Persidskii-type system, for which the corresponding nonsmooth diagonal-type CLF V_a is

$$V_a = h_2 \int_0^{R_1} \text{uhsgn}(\tau)d\tau + h_2 \int_0^{I_1} \text{uhsgn}(\tau)d\tau + h_1 \int_0^{R} \text{sgn}(\tau)d\tau + h_1 \int_0^{I} \text{sgn}(\tau)d\tau. \tag{3.53}$$

Observe that V_a is just V_2 written in a manner that makes calculation more obvious. Indeed, the time derivative of V_a along the trajectories of (3.52) can be written as

$$\dot{V}_a = \nabla V_a^T \dot{\mathbf{z}}, \tag{3.54}$$

where

$$\nabla V_a = \begin{bmatrix} h_2\text{uhsgn}(R_1) \\ h_2\text{uhsgn}(I_1) \\ h_1\text{sgn}(R) \\ h_1\text{sgn}(I) \end{bmatrix} = \begin{bmatrix} h_2\mathbf{I}_2 & 0 \\ 0 & h_1\mathbf{I}_2 \end{bmatrix} \mathbf{f}(\mathbf{z}). \tag{3.55}$$

Defining

$$\mathbf{H} = \begin{bmatrix} h_2\mathbf{I}_2 & 0 \\ 0 & h_1\mathbf{I}_2 \end{bmatrix} \quad \text{and} \quad \mathbf{W} = -\begin{bmatrix} \mathbf{I}_2 & \mathbf{M} \\ \mathbf{M}^T & \mathbf{M}^T\mathbf{M} \end{bmatrix}$$

and substituting (3.55) and (3.52) in (3.54) yields

$$\dot{V}_a = \mathbf{f}(\mathbf{z})^T\mathbf{H}^T\mathbf{W}\mathbf{H}\mathbf{f}(\mathbf{z}) \le 0, \tag{3.56}$$

since, under the assumption (3.36), the matrix $\mathbf{W}$ is clearly negative semidefinite. A little more work allows the conclusion that $\dot{V}_a$ is actually always negative. Observe that in order for $\dot{V}_a$ to be zero, the vector $\mathbf{Hf}(\mathbf{z})$ must belong to the nullspace of $\mathbf{W}$, and furthermore, in the reaching phase (i.e., outside the region $\mathcal{S}$), the function uhsgn assumes the value 1, so that

$$\mathbf{Hf}(\mathbf{z}) = [h_2 \ \ h_2 \ \ h_1\text{sgn}(R) \ \ h_1\text{sgn}(I)]^T. \tag{3.57}$$

The matrix $\mathbf{W}$ can be calculated from (3.34) and (3.35), and it is then easily seen that condition $\mathbf{WHf}(\mathbf{z}) = \mathbf{0}$ is equivalent to

$$\begin{bmatrix} u\,\text{sgn}(R) - v\,\text{sgn}(I) & 1 \\ v\,\text{sgn}(R) + u\,\text{sgn}(I) & 1 \end{bmatrix} \begin{bmatrix} h_1 \\ h_2 \end{bmatrix} = \begin{bmatrix} 0 \\ 0 \end{bmatrix}. \tag{3.58}$$

Defining $\mathbf{A}$ as the matrix on the left-hand side of (3.58), we observe that

$$\det \mathbf{A} = u\,\mathrm{sgn}(R) - v\,\mathrm{sgn}(I) - v\,\mathrm{sgn}(R) - u\,\mathrm{sgn}(I) \tag{3.59}$$

which, for the allowed values ± 1 of the sgn functions, implies that $\det \mathbf{A} = \pm 2u$ or $\pm 2v$. This means that, unless u, v become identically zero (which only occurs for special or trivial $P(z)$), the only solution of (3.58) is the trivial one $h_1 = h_2 = 0$. The overall conclusion is that, for $h_1, h_2 > 0$, $\dot{V}_a < 0$. □

If lower bounds are also known (i.e., we have box constraints), then a third controller is needed to implement these bounds. A similar use of variable structure controllers is made in the context of optimization problems in Chapter 4.

Existing implementations of the Kokotović–Šiljak method [Moo67, Sto95] use $V_2 = R^2 + I^2$ instead of (3.31). This leads to a smooth gradient dynamical system, for which a similar analysis can be made; however, finite-time convergence of the original Kokotović–Šiljak method [KŠ64] is lost. The smooth method also has quadratically convergent modifications for multiple zeros [Sto95].

Numerical examples of polynomial zero finding

This section shows examples of the trajectories obtained by numerical integration of the dynamical system (3.50) for a polynomial taken from [KŠ64, Šil69], in which zeros of the same polynomial were found using a smooth gradient dynamical system and several different initial conditions, instead of using bounds for the zeros, as proposed in Lemma 3.2.

It is proposed to find the zeros of the seventh degree polynomial

$$P(s) = s^7 + 0.3s^6 + 1.2s^5 + 0.09s^4 + 0.52s^3 + 0.08s^2 + 0.16s + 0.16, \tag{3.60}$$

which has the zeros

$$\begin{aligned} s_1 &= -0.5, & s_{2,3} &= 0.5 \pm i0.5,\\ s_{4,5} &= \pm i\sqrt{0.8}, & s_{6,7} &= -0.4 \pm i0.8. \end{aligned}$$

The dynamical system (3.50) corresponding to (3.60) was numerically integrated using the forward Euler method.

The σ-ω phase plane plot in Figure 3.4 shows the trajectories converging globally to the real zero of $P(s)$ located at $s_1 = -0.5 + i0.0$, with the bounds chosen as $a = 0$ and $b = 0.3$.

Figure 3.5 shows how appropriate choices of bounds can cause trajectories starting from the same initial condition to converge to different zeros of the polynomial. An initial condition is given from which all zeros (modulo conjugates) can be found by choosing appropriate bounds a and b in (3.47).

Numerical integration of (3.50) leads to discretization chatter, which may force the use of lower bounds on the gains h_2 and h_1, even though Lemma 3.2 allows for the use of any positive values of h_1, h_2. In practice, a large enough value of h_2 relative to h_1 is needed to ensure that the trajectories are attracted to the region $\mathcal{S}$ instead of a basin of attraction of a zero that is close to, but outside, the region $\mathcal{S}$.

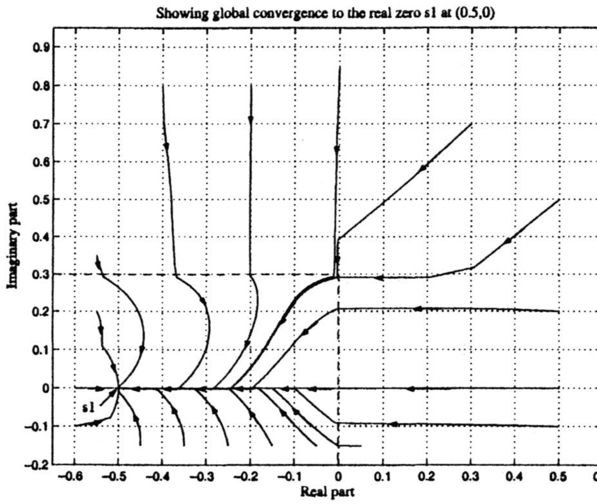

Figure 3.4. *Trajectories of the dynamical system* (3.50) *corresponding to the polynomial* (3.60), *showing global convergence to the real root* $s_1 = -0.5$, *with the bounds chosen as* $a = 0$ *and* $b = 0.3$ *and* $h_1 = 1, h_2 = 10$ *(see Figure* 3.3*). The region* $\mathcal{S}$ *determined by the bounds* a *and* b *is delimited by the dash-dotted line in the figure and contains only one zero* (s_1) *of* $P(z)$.

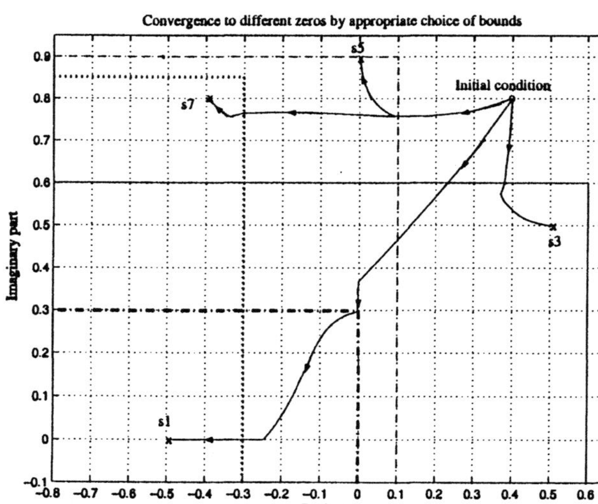

Figure 3.5. *Trajectories of the dynamical system* (3.50), *all starting from the initial condition* $(\sigma_0, \omega_0) = (0.4, 0.8)$, *converging to different zeros of* (3.60) *by appropriate choice of upper bounds* a *and* b: $(a, b) = (0, 0.3) \to s_1$, $(a, b) = (0.6, 0.6) \to s_3$, $(a, b) = (0.1, 0.9) \to s_5$, $(a, b) = (-0.3, 0.85) \to s_7$. *In all cases* $h_1 = 1, h_2 = 10$.

3.2.2 The spurt method

In the context of the variable structure iterative methods for zero finding, this subsection calls attention to the so-called spurt method, which was proposed to find zeros of the affine function $\mathbf{Ax} - \mathbf{b}$, i.e., a variable structure method to solve a linear system of equations $\mathbf{Ax} = \mathbf{b}$. Emelin, Krasnosel'skiI, and Panskih [EKP74] took their inspiration from the theory of variable structure systems [Utk92] to propose an iterative method for symmetric positive definite matrices $\mathbf{A}$ of the type (2.119), with the scalar parameter α_k switching between two values in accordance with a rule to be described below. As will be seen from the brief description below, the spurt iterative method is also related to the class of Krylov subspace methods discussed in sections 2.3.1 and 2.3.2. It can be seen as a Richardson method with a choice of control parameter α_k different from that used in (2.129). It is a method that is motivated by variable structure ideas rather than optimal control theory.

Equation (2.119) is rewritten as

$$\mathbf{x}_{k+1} = \mathbf{x}_k + \alpha_k \mathbf{r}_k, \tag{3.61}$$

where, as usual, $\mathbf{r}_k = \mathbf{b} - \mathbf{Ax}_k$, and α_k is a scalar parameter that is defined below.

First, a positive number γ is defined as follows:

$$\gamma > 0 \text{ such that } \mathbf{I} - \gamma\mathbf{A} \text{ positive semidefinite and } \rho(\mathbf{I} - \gamma\mathbf{A}) < 1. \tag{3.62}$$

A threshold parameter $\theta \geq 0$ and a nonnegative number $\delta \geq 0$ are also chosen. Then α_k switches between the numbers γ and δ in accordance with the following rules:

$$\begin{aligned} \alpha_k &= \gamma \text{ if either } k = 0 \text{ or } \alpha_{k-1} = \delta \text{ or } \frac{\langle \mathbf{r}_k, \mathbf{r}_k \rangle}{\langle \mathbf{r}_{k-1}, \mathbf{r}_{k-1} \rangle} < \theta, \\ \alpha_k &= \delta \text{ if } k > 0, \alpha_{k-1} = \gamma \text{ and } \frac{\langle \mathbf{r}_k, \mathbf{r}_k \rangle}{\langle \mathbf{r}_{k-1}, \mathbf{r}_{k-1} \rangle} \geq \theta. \end{aligned} \tag{3.63}$$

The iterative method defined by (3.61)–(3.63) is called the *spurt method* determined by the triple (γ, δ, θ) and is shown in the (by now familiar) control representation in Figure 3.6.

If an iteration is called a γ-iteration whenever $\alpha_k = \gamma$ and a δ-iteration whenever $\alpha_k = \delta$, then, from the construction of the iterative method, it follows that a δ-iteration is always followed by a γ-iteration and that the vector $\mathbf{x}_1$ is always computed by a γ-iteration. Computer experiments reported in [EKP74] indicate that (i) if the threshold parameter θ is chosen to be too large, then all vectors $\mathbf{x}_k$ are computed by γ-iterations and the spurt method reduces to an ordinary iterative method; (ii) if the threshold parameter θ is too small, then the spurt method diverges; (iii) there is a range of values for the threshold parameter θ for which the average frequency of appearance of δ-iterations does not depend on the value of θ. The experimental fact of "stabilization" of the frequency of appearance of δ-iterations was also explained theoretically in [EKP74] and the theorem obtained also permits an estimate of the speed of convergence of the spurt method, which, in turn, provides a guideline for the choice of the parameters δ and θ.

Let $(\lambda_i, \mathbf{v}_i)$, $i = 1, \ldots, n$, denote the eigenvalue–eigenvector pairs of the matrix $\mathbf{A}$. Let $k(\delta)$ be defined as the least index of the eigenvalues λ_i that are larger than $2/\delta - \lambda_1$, if

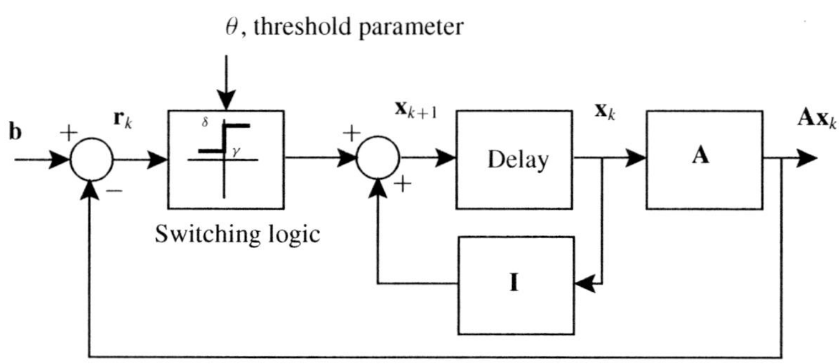

Figure 3.6. *Spurt method as standard plant with variable structure controller.*

such exist, and equal to $n+1$, if not. Let

$$\phi_i = \frac{\ln|1-\delta\lambda_i| - \ln|1-\delta\lambda_1|}{\ln|1-\gamma\lambda_1| - \ln|1-\gamma\lambda_i|}, \tag{3.64}$$

$$\phi_\infty := \max\{1, \phi_i : \langle \mathbf{x}_0, \mathbf{v}_i\rangle \neq \langle \mathbf{A}^{-1}\mathbf{b}, \mathbf{v}_i\rangle,\ i \geq k(\delta)\}. \tag{3.65}$$

Suppose that m steps of the computation have been performed using the spurt method and that, during these steps, γ-iterations have occurred $\Gamma(m)$ times and δ-iterations $\Delta(m)$ times, so that $\Gamma(m) + \Delta(m) = m$. In this case, the following theorem holds.

Theorem 3.3 [EKP74]. *Assume that the initial condition* $\mathbf{x}_0$ *is such that*

$$\langle \mathbf{x}_0, \mathbf{v}_i\rangle \neq \langle \mathbf{A}^{-1}\mathbf{b}, \mathbf{v}_i\rangle \tag{3.66}$$

and furthermore that the parameters δ *and* θ *satisfy the relation*

$$0 < 1 - \gamma(2/\delta - \lambda_1) < \theta < 1 - \gamma\lambda_1. \tag{3.67}$$

Then the following inequality holds:

$$\sup_m |\Gamma(m) - \phi_\infty \Delta(m)| < \infty. \tag{3.68}$$

Note that condition (3.66) holds generically (i.e., for almost all $\mathbf{x}_0$) so that it should not be considered restrictive. Let ε be an arbitrarily small positive number. Then from Theorem 3.3 it follows that, for all sufficiently large j, we have the inequality

$$\left| \frac{\Gamma(m+j) - \Gamma(m)}{\Delta(m+j) - \Delta(m)} - \phi_\infty \right| \leq \varepsilon, \qquad m = 0, 1, 2, \ldots;\ \ j \geq j(\varepsilon). \tag{3.69}$$

Thus the ratio $\Gamma(m)/\Delta(m) \to \phi_\infty$ as $m \to \infty$. The number ϕ_∞ is referred to as the *limiting porosity* of the iterations (3.61). Note that the limiting porosity computed for the same initial condition $\mathbf{x}_0$ and a fixed value of δ does not depend on the values of θ admissible under the conditions of Theorem 3.3.

In order to state a theorem on the rate of convergence, the following quantity needs to be defined:

$$\lambda_{\text{ef}} := [(1-\gamma\lambda_1)^{\phi_\infty}(1-\delta\lambda_1)]^{1/(1+\phi_\infty)}. \tag{3.70}$$

Theorem 3.4 [EKP74]. *Under the assumptions of Theorem 3.3, the spurt method converges to the exact solution of* $\mathbf{Ax} = \mathbf{b}$. *In addition, the following inequalities hold:*

$$c\lambda_{\text{ef}}^m \|\mathbf{x}_0 - \mathbf{A}^{-1}\mathbf{b}\| \le \|\mathbf{x}_m - \mathbf{A}^{-1}\mathbf{b}\| \le d\lambda_{\text{ef}}^m \|\mathbf{x}_0 - \mathbf{A}^{-1}\mathbf{b}\|, \tag{3.71}$$

where the positive numbers c and d depend on $\mathbf{x}_0$.

The inequalities (3.71) show that, under the assumptions of Theorem 3.3, the iterates $\mathbf{x}_m$ converge to $\mathbf{A}^{-1}\mathbf{b}$ with the speed of a geometric progression with ratio λ_{ef}. This ratio does not depend on the admissible values of θ in the interval from $1-\gamma(2/\delta-\lambda_1)$ to $1-\gamma\lambda_1$, since the choice of θ affects only the values of the numbers c and d. Computer experiments show that θ should be chosen as close as possible to the lower limit.

Other possibilities exist: The δ-iterations can be replaced by steepest descent iterations, and experimental results [EKP74] indicate much faster convergence than the spurt method, but, on the other hand, stability in the face of roundoff errors decreases appreciably. The reader is referred to [EKP74] for theoretical and practical details omitted here.

Other hybrid methods have been proposed in the literature; for example, in [BRZ94] a hybrid of the Lanczos and conjugate-gradient methods is proposed and analyzed, but, rather than switch between one method and another, it uses linear combinations of the iterates generated by each method.

Observe also that the Levenberg–Marquardt method can be considered as a variable structure method that switches between the steepest descent and Newton methods.

A general model of such methods, also referred to as *team algorithms*, in which each algorithm or "structure" is considered as a member of a team, was given in [BKB96] and a general convergence result derived, using the Liapunov function technique mentioned in [KBŠ90].

3.3 Optimal Control Approach to Unconstrained Optimization Problems

Most function minimization algorithms use line search along a descent direction to minimize the function in the chosen direction. This is clearly only locally optimal. In the numerical solution of the problem, we are interested in computing the trajectory that takes us optimally from the initial guess to the minimum point, rather than one that is locally optimal at every step. From this viewpoint, computation of the trajectory is clearly an optimal control problem. It is also obvious that, in order for the optimal control problem to be well defined, some constraints must be imposed on the direction vector, which is interpreted as the control. For if not, the straight line from the initial point to the minimum point is always an optimal trajectory, but one that cannot be constructed, since the minimum point is not known a priori. Another reason for constraints on the direction vector is that if there are none, impulsive control will allow the minimum point to be reached in an arbitrarily short time.

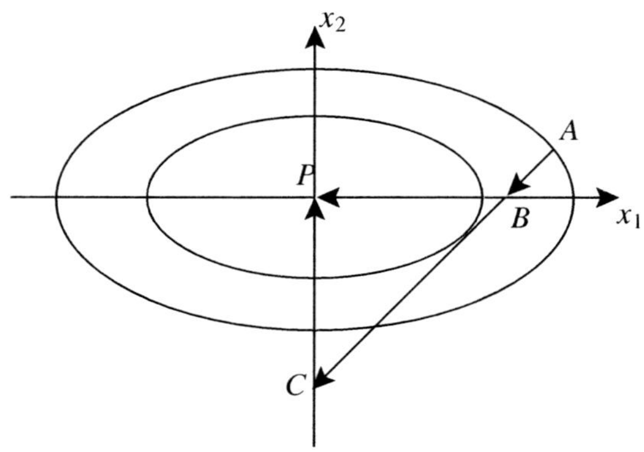

Figure 3.7. *Level curves of the quadratic function* $f(\mathbf{x}) = x_1^2 + 4x_2^2$*, with steepest descent directions at A, B, C and efficient trajectories ABP and ACP (following* [Goh97]*).*

As general motivation for an optimal control approach to unconstrained optimization problems, we consider a simple example from Goh [Goh97].

Consider the problem of minimizing the quadratic function

$$f(\mathbf{x}) = x_1^2 + 4x_2^2. \tag{3.72}$$

The standard iterative steepest descent algorithm for minimizing this function has the drawback that its convergence is slow near minima because the trajectory zigzags frequently as the computed point approaches the minimum point. It is not hard to see that this is caused by the fact that, in the standard algorithm, the steplength is chosen so that the function is minimized in the steepest descent direction at each iteration. As an illustration of this, consider the quadratic function (3.72) and let the starting point be A $= (2, 1)$ and let ABC be the line in the steepest descent direction through A. This line intersects the x_1-axis at the point B $= (1.5, 0)$ and the x_2-axis at the point C $= (0, -3)$. Note also that the lines BP and CP are in the steepest descent directions at points B and C, respectively. It is plain to see (Figure 3.7) that trajectories ABP and ACP provide efficient trajectories (two iterations) to the minimum point P $= (0, 0)$, even though $f(\mathbf{x})$ is not minimized by either point B or point C in the steepest descent direction through A. The lesson to be learned from this simple example is that, in the computation of a trajectory to a minimum, it may not be desirable to choose the steplength so as to minimize the function in a prescribed direction in each iteration. Goh [Goh97] refers to this as the "difference between long-term and short-term optimality." This example serves to motivate the introduction of a method that uses local information to compute trajectories that are long-term "optimal" (i.e., efficient).

Optimal control formulation of unconstrained minimization problem

Given an initial point $\mathbf{x}_0$, the problem we wish to solve is that of finding an optimal trajectory starting from $\mathbf{x}_0$ and terminating in the minimizer $\mathbf{x}^*$ of a function $f : \mathbb{R}^n \to \mathbb{R}$.

We seek a control function $\mathbf{u}(t)$ which generates a trajectory $\mathbf{x}(t)$ from $\mathbf{x}_0$ to $\mathbf{x}(T)$ such that $f(\mathbf{x}(T))$ is minimized. Let

$$\frac{\mathrm{d}\mathbf{x}}{\mathrm{d}t} = \mathbf{u}, \tag{3.73}$$

and let $\mathbf{g}(\mathbf{x})$ and $\mathbf{G}(\mathbf{x})$ be the gradient and Hessian of $f(\mathbf{x})$, respectively. Along a continuous and piecewise smooth trajectory, we have

$$\frac{\mathrm{d}f(\mathbf{x})}{\mathrm{d}t} = \mathbf{g}(\mathbf{x})^T \frac{\mathrm{d}\mathbf{x}}{\mathrm{d}t} = \mathbf{g}(\mathbf{x})^T \mathbf{u}. \tag{3.74}$$

Integrating (3.74) gives

$$f(\mathbf{x}(T)) = f(\mathbf{x}(0)) + \int_0^T \mathbf{g}(\mathbf{x}(t))^T \mathbf{u}(t)\mathrm{dt}. \tag{3.75}$$

Thus we are led to the problem of minimizing $f(\mathbf{x}(T))$ subject to the dynamics (3.73), which, in view of (3.75), can be written as the following optimal control problem:

$$\begin{aligned} &\min \left\{ f(\mathbf{x}(0)) + \textstyle\int_0^T \mathbf{g}(\mathbf{x}(t))^T \mathbf{u}(t)\mathrm{dt} \right\} \\ &\text{s.t. } \mathrm{d}\mathbf{x}/\mathrm{d}t = \mathbf{u}, \qquad \mathbf{x}(0) = \mathbf{x}_0. \end{aligned} \tag{3.76}$$

Since $f(\mathbf{x}(0))$ is a constant, the problem (3.76) is similar in form to (3.19), suggesting that the solution should be $\mathbf{u} = -\mathrm{sgn}\,\mathbf{g}$. However, notice that with such a solution, the resulting closed-loop system would be $\dot{\mathbf{x}} = -\mathrm{sgn}\,\mathbf{g}$, which cannot be guaranteed to be asymptotically stable around the value of $\mathbf{x}$ that minimizes f, for which $\mathbf{g}(\mathbf{x}) = \mathbf{0}$. Since the objective is to find a point $\mathbf{x}$ for which $\mathbf{g}(\mathbf{x}) = \mathbf{0}$, if the ith component g_i of $\mathbf{g}$ becomes zero, intuitively speaking, it would be efficient to choose a control in such a way that this component is maintained at zero, while attempting to reduce the other nonzero components, g_j, $j \neq i$, to zero, using controls of the type sgn g_j. This intuition is made precise in what follows.

Since the control enters both the system equations as well as the objective function linearly, it may be expected that a singular solution will be found [Goh66, BJ75]. It turns out that any optimal, normal, and totally singular control for this problem is indeterminate, so that, as might be expected intuitively, some or all components of the optimal control vector may be impulsive and therefore not usable in practice. A simple way to remedy this is to impose bounds on the control as follows:

$$-\beta_i \leq u_i(t) \leq \beta_i, \quad \beta_i > 0, \quad i = 1, 2, \dots, n. \tag{3.77}$$

When these bounds are imposed, it is known from the theory of optimal control that an optimal trajectory will consist of two types of segments. In one type, all components of the control attain extreme values ($-\beta_i$ or β_i), and in this case, the onomatopoeic terminology *bang-bang arc* is used. Controls at extreme values are also referred to as *saturated.* The other type of segment has some control values saturated (i.e., equal to $\pm\beta_i$) and others at intermediate values (i.e., in the interior of the interval $[-\beta_i, \beta_i]$) and is called a *bang-intermediate arc*. With these preliminaries, a conceptual algorithm can be formulated.

Conceptual iterative method based on an optimal control formulation

Consider the problem of minimizing a strictly convex function $\mathbf{f} : \mathbb{R}^n \to \mathbb{R}$. A standard iterative method to compute the minimum can be written as follows:

$$\mathbf{x}_{k+1} = \mathbf{x}_k + h_k \mathbf{u}_k, \tag{3.78}$$

where $\mathbf{x}_k$ is the estimate of the desired minimizing vector $\mathbf{x}^*$ at the kth iteration, h_k is a scalar stepsize and $\mathbf{u}_k$ is the so-called search direction in which the function is being minimized. Note that the usual notation for the search direction vector is $\mathbf{d}_k$, but here the notation $\mathbf{u}_k$ is chosen to alert the reader that it is going to be thought of as a control input to the discrete-time system (3.78), in contrast to the choice of the stepsize as control input in section 2.2. In fact, for simplicity, it will be assumed that the steplength h_k is chosen as a constant $h > 0$. The assumption (3.77) means that the control vector varies inside a hypercube $U_k := [-\beta_1, \beta_1] \times \cdots \times [-\beta_n, \beta_n]$. Since the system (3.78) is affine in the control, this means that the set of states that can be generated by these choices of control will also be a hypercube with its sides parallel to the axes. This set is usually referred to as a *reachability set* and denoted Z_k. In what follows, to further simplify matters, it will be assumed that $\beta_i = \beta$ for all i, so that $U_k = [-\beta, \beta]^n$.

Now consider the following conceptual algorithm to find the minimizer $\mathbf{x}^*$ of a strictly convex function $f(\cdot)$.

Algorithm 3.3.1 [Goh's conceptual algorithm]
Given:
Initial guess $\mathbf{x}_0$, initial search direction $\mathbf{u}_0$, constant stepsize $h > 0$
$k = 0$
while $(\nabla f(\mathbf{x}_k) \neq \mathbf{0})$
Compute the reachability set $Z_k := \{\mathbf{x} = \mathbf{x}_k + h\mathbf{u} : \mathbf{u} \in U_k\}$
$\mathbf{u}_k := \text{argmin}_{\mathbf{u} \in Z_k} f(\mathbf{x}_k + h\mathbf{u})$
$\mathbf{x}_{k+1} = \mathbf{x}_k + h\mathbf{u}_k$
$k \leftarrow k + 1$
end while

Example 3.5. In order to get an intuitive understanding of this algorithm, consider its application to the (trivial) problem of minimizing the function $f(\mathbf{x}) = x_1^2 + x_2^2$. The minimum is obviously $\mathbf{x}^* = \mathbf{0}$. The gradient vector is $\nabla f(\mathbf{x}) = [2x_1 \quad 2x_2]^T$. Let the stepsize $h = 1$ and the bound on each component of the search direction also be chosen as $b_i = 1$ for all i and the initial condition given as $\mathbf{x}_0 = [3.5 \quad 2.5]^T$.

The reachability set Z_0 is a square centered at $\mathbf{x}_0$ and with northeast (NE) and southwest (SW) vertices at $[2.5 \quad 1.5]^T$ and $[4.5 \quad 3.5]^T$, respectively. The function $f(\cdot)$ is minimized on Z_0 by choosing $\mathbf{u}_0 = [-1 \quad -1]^T$ and the minimum is attained at the SW corner of Z_0, so that $\mathbf{x}_1 = \mathbf{x}_0 + h\mathbf{u}_0 = [2.5 \quad 1.5]^T$. The reachability set Z_1 centered at the new iterate $\mathbf{x}_1$ has NE and SW vertices at $[3.5 \quad 2.5]^T$ and $[1.5 \quad 0.5]^T$, respectively, and, once again, the minimum of $f(\cdot)$ occurs at the SW corner for the choice $\mathbf{u}_1 = [-1 \quad -1]^T$. From the geometry of this problem, shown in Figure 3.8, it is clear that the minimizing points $\mathbf{x}_i$ occur at the points of tangency of the level sets and the reachability sets. The progress of the iteration is shown in Table 3.2.

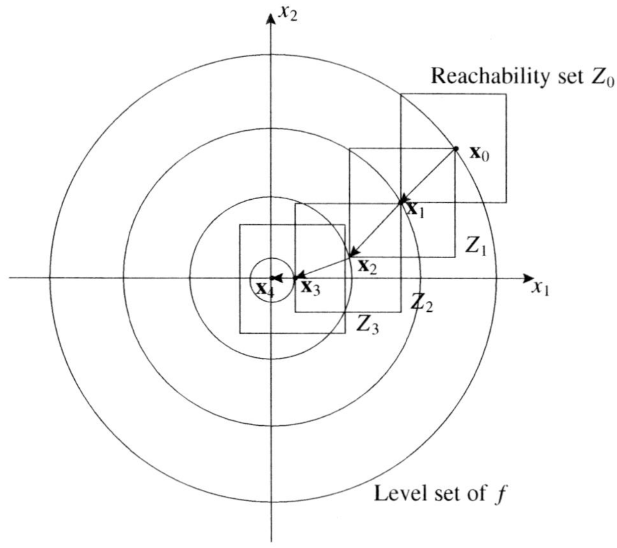

Figure 3.8. *The stepwise optimal trajectory from initial point* $\mathbf{x}_0$ *to minimum point* $\mathbf{x}_4$ *generated by Algorithm* 3.3.1 *for Example* 3.5. *Segments* $\mathbf{x}_0$–$\mathbf{x}_1$ *and* $\mathbf{x}_1$–$\mathbf{x}_2$ *are bang-bang arcs, while segment* $\mathbf{x}_2$–$\mathbf{x}_3$ *is bang-intermediate. The last segment,* $\mathbf{x}_3$–$\mathbf{x}_4$, *is a Newton iteration.*

Table 3.2. *Showing the iterates of Algorithm* 3.3.1 *for Example* 3.5.

k	$\mathbf{x}_k$	Z_k (SE)	Z_k (NW)	$\mathbf{u}_k$	$\nabla f(\mathbf{x}_k)$	$f(\mathbf{x}_k)$
0	$(3.5, 2.5)$	$(2.5, 1.5)$	$(4.5, 3.5)$	$(-1, -1)$	$(7, 5)$	18.5
1	$(2.5, 1.5)$	$(1.5, 0.5)$	$(3.5, 2.5)$	$(-1, -1)$	$(5, 3)$	8.5
2	$(1.5, 0.5)$	$(0.5, -0.5)$	$(2.5, 1.5)$	$(-1, -0.5)$	$(3, 1)$	2.5
3	$(0.5, 0)$	$(-0.5, -1)$	$(1.5, 1)$	$(-0.5, 0)$	$(1, 0)$	0.25
4	$(0, 0)$	$(-1, -1)$	$(1, 1)$	–	$(0, 0)$	0

Some observations on this example help to connect it to the earlier discussion on bang-bang and bang-intermediate arcs. The reachable set Z_k corresponding to the starting point $\mathbf{x}_k$ is a closed and bounded hypercube centered at $\mathbf{x}_k$, given the constraints (3.77). Hence the strictly convex function $f(\cdot)$ has a unique minimum point $\hat{\mathbf{x}}_{k+1}$ in the convex set Z_k. Three possibilities exist for this minimum point $\hat{\mathbf{x}}_k$: it can occur (i) at a corner of the hypercube Z_k, (ii) on a face of the hypercube Z_k, or (iii) in the interior of the hypercube Z_k. Since a convex function is being minimized on a hypercube, by the Kuhn–Tucker theorem, if case (i) or (ii) occurs, then the direction vector $\mathbf{u}_k$ is determined by the conditions

$$(\mathbf{u}_k)_i = -\beta \operatorname{sgn}(\nabla f(\mathbf{x}_{k+1}))_i \quad \text{if } (\nabla f(\mathbf{x}_{k+1}))_i \neq 0, \tag{3.79}$$

$$(\nabla f(\mathbf{x}_k + h\mathbf{u}_k))_i = 0 \quad \text{if } (\nabla f(\mathbf{x}_{k+1}))_i = 0, \tag{3.80}$$

where the ith component of a vector $\mathbf{v}$ is denoted as $(\mathbf{v})_i$.

In case (iii), the condition

$$\nabla f(\mathbf{x}_k + h\mathbf{u}_k) = \mathbf{0} \tag{3.81}$$

holds and, by convexity, the point $\mathbf{x}_k + h\mathbf{u}_k$ is also the global minimum of the function $f(\cdot)$.

Clearly, condition (3.79) can be called the generator of bang-bang directions. Observe that in (3.80), it is necessary to use a zero finding method, such as, for example, Newton's method, in order to find the value of $\mathbf{u}_k$ that makes the ith component of the gradient vector equal to zero, and this, in general, could be as hard as solving the original optimization problem. This is one reason why Algorithm 3.3.1 must be regarded as conceptual. Condition (3.80) can be called an intermediate direction generator (the term *nonlinear partial Newton method* is used in [Goh97]). Note that if condition (3.81) is satisfied, then the iterative process has terminated and the minimum point is $\mathbf{x}_k + h\mathbf{u}_k$. Finally, note that the parameter β which determines the size of the reachability set and the stepsize parameter h strongly influence the number of iterations to convergence.

The inverse Liapunov function problem

It is well known that it is usually very difficult to construct a suitable Liapunov function that establishes stability, either global or in a large region, of a given system of differential or difference equations. On the other hand, a simple but crucial observation is that the construction of an algorithm to compute the minimum of an unconstrained function can be interpreted as an inverse Liapunov function problem, which is defined as follows.

Definition 3.6. *Given a function*

$$V(\mathbf{x}) = f(\mathbf{x}) - f(\mathbf{x}^*),$$

the inverse Liapunov function problem *is that of constructing a system of differential equations*

$$\frac{d\mathbf{x}}{dt} = \mathbf{u}(\mathbf{x})$$

or a system of difference equations

$$\mathbf{x}_{k+1} = \mathbf{x}_k + h_k\mathbf{u}_k$$

such that $V(\mathbf{x})$ is a Liapunov function of a dynamical system with a (stable) equilibrium at $\mathbf{x}^*$.

Recalling the basic theory, a smooth Liapunov function $V(\mathbf{x})$ can be used to give a set of sufficient conditions for the stability of the equilibrium $\mathbf{x}^*$ of the system of differential equations:

$$\frac{d\mathbf{x}}{dt} = \mathbf{f}(\mathbf{x}), \quad \mathbf{x}(0) = \mathbf{x}_0. \tag{3.82}$$

The sufficient conditions to be satisfied by $V(\mathbf{x})$ for global asymptotic stability of the point $\mathbf{x}^*$ are as follows:

(i) $V(\mathbf{x}) > 0$, $\mathbf{x} \neq \mathbf{x}^*$, and $V(\mathbf{x}^*) = 0$;

(ii) $\mathrm{d}V(\mathbf{x})/\mathrm{d}t = [\nabla V(\mathbf{x})]^T \mathbf{f}(\mathbf{x}) < 0, \ \mathbf{x} \neq \mathbf{x}^*$;

(iii) $V(\mathbf{x})$ is radially unbounded.

Condition (iii) is true only if the level sets of $V(\mathbf{x})$ are nested, closed, and bounded, and, in addition, $V(\mathbf{x})$ must tend to infinity as the norm of $\mathbf{x}$ tends to infinity, i.e., is radially unbounded. Condition (i) implies that $\mathbf{x}^*$ is the only stationary point of the Liapunov function $V(\mathbf{x})$ and condition (ii) that $\mathbf{x}^*$ is the only equilibrium point of the dynamical system.

In the discrete-time case, i.e., for the system of difference equations

$$\mathbf{x}_{k+1} = \mathbf{f}(\mathbf{x}_k), \tag{3.83}$$

the equilibrium $\mathbf{x}^*$ is globally asymptotically stable if there exists a continuous function with properties (i) and (iii), while condition (ii) is replaced by

$$\Delta V(\mathbf{x}_k) := V(\mathbf{x}_{k+1}) - V(\mathbf{x}_k) < 0 \quad \text{for all } \mathbf{x}_k. \tag{3.84}$$

Asymptotic stability in a region occurs if the conditions above are met in a finite region and implies that all trajectories starting from a point inside this finite region converge to the point $\mathbf{x}^*$.

Example 3.7. The function

$$f(\mathbf{x}) = [x_1^2/(1 + x_1^2)] + x_2^2$$

clearly has a minimum at $\mathbf{x}^* = \mathbf{0}$. To find this minimum, suppose that the continuous-time steepest descent system

$$\frac{\mathrm{d}\mathbf{x}}{\mathrm{d}t} = \mathbf{u}(\mathbf{x}) = -\nabla f(\mathbf{x})$$

is used. Defining $V(\mathbf{x})$, as in Definition 3.6,

$$\frac{\mathrm{d}V(\mathbf{x})}{\mathrm{d}t} = \frac{\mathrm{d}f(\mathbf{x})}{\mathrm{d}t} = \nabla f(\mathbf{x})^T \mathbf{u} = -\nabla f(\mathbf{x})^T \nabla f(\mathbf{x}) < 0, \quad \mathbf{x} \neq \mathbf{x}^*,$$

so that $V(\mathbf{x})$ is a Liapunov function. Actually, it turns out that the level sets of $f(\mathbf{x})$ are nested, closed, and bounded for a positive constant $c < 1$, so that the steepest descent dynamical system is convergent in the region $f(\mathbf{x}) \leq c < 1$. On the other hand, for $f(\mathbf{x}) > 1$, the level sets of $f(\mathbf{x})$ are not closed and bounded, so no conclusion can be drawn about global convergence, if $V(\mathbf{x})$ is used as a Liapunov function.

Example 3.8. The Rosenbrock function

$$f(\mathbf{x}) = 100(x_1^2 - x_2)^2 + (1 - x_1)^2 \tag{3.85}$$

is a famous example of a nonconvex function used to test optimization algorithms, with a global minimum at $(1, 1)$. Despite its nonconvexity, all level sets are nested, closed, and bounded, and $f(\mathbf{x})$ is radially unbounded. Thus the function $V(\mathbf{x}) = f(\mathbf{x}) - f(1, 1)$ is a Liapunov function that proves global asymptotic stability of the equilibrium point $(1, 1)$ for the continuous-time steepest descent dynamical system.

With this background, a continuous-time method for optimization can be developed.

Continuous-time optimization method

As pointed out at the end of section 2.3.2, continuous-time methods have certain advantages over their discrete-time counterparts. They may be viewed as prototypes for the development of practical (discrete) iterative algorithms. In addition, in the optimization context, the specific advantage that they present, with respect to discrete methods, is that the choice of descent direction can be studied without the onus of having to choose an appropriate steplength. Of course, if a continuous-time method is globally convergent, then there will exist a choice of steplength that will make the corresponding discrete iterative algorithm convergent for that particular choice of descent direction.

To arrive at a prototypical continuous-time method, following Goh [Goh97], the idea is to use the conceptual method discussed above. The main result can be stated as follows.

Theorem 3.9 [Goh97]. *Given a strictly convex positive definite function $f : \mathbb{R}^n \to \mathbb{R}$ with minimum point at $\mathbf{x}^*$, consider the dynamical system*

$$\frac{\mathrm{d}\mathbf{x}}{\mathrm{d}t} = \mathbf{u}, \quad \mathbf{x}(0) = \mathbf{x}_0,$$

where $\mathbf{u}(\cdot)$ is to be chosen so that all its trajectories, starting from $\mathbf{x}_0$ in a given region, converge to the minimum point $\mathbf{x}^*$. This is achieved, with at most n switches in the control $\mathbf{u}(t)$ (descent direction), when it is chosen as follows, using the notation $\mathbf{g} := \nabla f(\mathbf{x})$ and $g_i := (\nabla f(\mathbf{x}))_i$:

$$\text{If } g_i \neq 0, \text{ then } u_i(t) = -\beta \operatorname{sgn}(g_i), \tag{3.86}$$

and if $g_i = 0$, then u_i is determined by the equations

$$\mathrm{d}g_i/\mathrm{d}t = (\partial g_i/\partial x_i)u_i + \sum_{j \neq i}(\partial g_i/\partial x_j)u_j = 0. \tag{3.87}$$

Proof. Consider the mapping

$$\mathbf{y} = \mathbf{g}(\mathbf{x}). \tag{3.88}$$

By the assumption of strict convexity of f, it follows that the Jacobian of $\mathbf{g}$, which is the Hessian of $\mathbf{f}$ and is denoted as $\mathbf{G}$, is positive definite and therefore nonsingular. Thus the mapping (3.88) between the $\mathbf{x}$-space and the $\mathbf{g}$-space is locally invertible, and furthermore, every principal minor of $\mathbf{G}$ is positive definite, which means that (3.86) and (3.87) can indeed be used to consistently determine $\mathbf{u}$.

Now consider the Liapunov function

$$W(\mathbf{g}) := \sum_i |g_i|. \tag{3.89}$$

Since $|x| = x \operatorname{sgn}(x)$, therefore, by the chain rule, $\mathrm{d}|x|/\mathrm{d}t = \dot{x} \operatorname{sgn}(x) + x\, \mathrm{d}\,[\operatorname{sgn}(x)]/\mathrm{d}t = \dot{x} \operatorname{sgn}(x)$. The second term is zero because sgn is a piecewise constant function, which has zero derivative everywhere, except possibly at $x = 0$. Thus

$$\frac{\mathrm{d}W}{\mathrm{d}t} = \sum_i \dot{g}_i \operatorname{sgn}(g_i).$$

Now $\dot{g}_i = (\partial g_i/\partial \mathbf{x})^T \dot{\mathbf{x}} = (\mathbf{G})_i^T \mathbf{u}$, where $(\mathbf{G})_i$ denotes the ith row of the Hessian $\mathbf{G}$ and $\mathbf{u}$ is given by (3.86) and (3.87). This means that $\mathbf{u}$ is a vector with element u_i such that $\dot{g}_i = 0$ when $g_i = 0$, and the remaining elements $u_j = -b\,\mathrm{sgn}(g_j)$, when $g_i \neq 0$. Overall this amounts to picking out a principal minor of the Hessian matrix, as follows:

$$\frac{\mathrm{d}W}{\mathrm{d}t} = -\beta \sum_i \sum_j \mathrm{sgn}(g_i)\mathbf{G}_{ij}\,\mathrm{sgn}(g_j).$$

This is always negative definite since $\mathbf{G}$ is positive definite, and the conclusion is that $\mathrm{d}W/\mathrm{d}t$ is negative definite in $\mathbf{g}$-space, so that all trajectories converge to the origin $\mathbf{g} = \mathbf{0}$. To complete the proof, observe that from (3.87) it follows that if g_i becomes equal to zero, then u_i is chosen so that it remains equal to zero thereafter. In $\mathbf{g}$-space, if $\mathbf{u}$ is determined by (3.86) and (3.87), then a trajectory remains on an axis plane once it intersects the latter. This means that there is at most one switch in each component of the control vector $\mathbf{u}(\mathbf{x}(\mathbf{g}))$. □

An interesting corollary of this theorem is that the convergence time is finite. Also, the assumption of strict convexity can be weakened [Goh97], essentially by using Liapunov stability theory, which only requires the level sets to be nested, closed, and bounded, *but not necessarily convex* (see Examples 3.7 and 3.8). With a function that is not necessarily convex, it may happen that $\partial g_i/\partial x_i = 0$, so that (3.87) cannot be used to determine u_i, as was the case in Theorem 3.9. It is then necessary to use a control u_i that moves the point $\mathbf{x}$ away from points where $\partial g_i/\partial x_i$ or other principal minors of the Hessian matrix $\mathbf{G}$ become small. Such so-called relaxed controls may violate the control constraints so as to allow movement along the surfaces $g_i(\mathbf{x}) = 0$ and permit extension of the continuous algorithm of Theorem 3.9 to the nonconvex case. Details will lead too far afield here but can be found in [Goh97]. Instead, some examples are given.

Example 3.10 [Goh97]. Let $f(\mathbf{x}) := \alpha x_1^2 + x_2^2$, where $\alpha = 1{,}000{,}000$. The curves $g_1 = 2{,}000{,}000 x_1 = 0$ and $g_2 = 2x_2 = 0$ determine the behavior of the trajectories that converge to the minimum point $(0, 0)$. The method described in Theorem 3.9, starting from initial point $\mathbf{x}_0 = (2, 1)$ and with constraints $-1 \leq u_i \leq 1, i = 1, 2$, generates the trajectories shown in Table 3.3. The trajectory reaches (and stops at) the minimum point when $t = 2$ and is composed of two piecewise linear segments. This can be thought of as the equivalent of two iterations in a corresponding discrete-time iterative algorithm. Since the function is strictly convex (for any α), all control values are admissible (i.e., satisfy the constraints) and the trajectory obtained is an optimal control trajectory. Furthermore, as can be seen, the well-known adverse effects of the imbalance in the scaling of the variables do not occur for this algorithm, in contrast to the continuous-time steepest descent and Newton methods, which, in this example, both display asymptotic convergence.

Example 3.11 [Goh97]. Returning to the nonconvex Rosenbrock function of Example 3.8 and choosing $\beta = 1$, the locus of points that satisfy $g_1 = 0$ and $g_2 = 0$ are calculated. In this case $g_1 = 0$ yields $200(x_1^2 - x_2)(2x_1) - 2(1 - x_1) = 0$, which describes two disjoint curves and a singular point at $(-0.13572, 0.06026)$, where $\partial g_1/\partial x_1 = 0$. Thus, in this case, Theorem 3.9 is not applicable without modification. Moreover, $g_2 = 0$ yields

Table 3.3. *Trajectories of the dynamical system described in Theorem 3.9, corresponding to $f(\mathbf{x}) := \alpha x_1^2 + x_2^2$.*

$u_1 = -1$	$u_2 = -1$	$x_1 = 2 - t$	$x_2 = 1 - t$	$0 \le t \le 1$
$u_1 = -1$	$u_2 = 0$	$x_1 = 1 - (t - 1)$	$x_2 = 0$	$1 \le t \le 2$

$-100(x_1^2 - x_2) = 0$, i.e., $x_1^2 = x_2$. Using (3.87) yields $2x_1\dot{x}_1 = \dot{x}_2$, i.e., $2x_1 u_1 = u_2$, whence, in order for the constraint on u_2 to be satisfied, it is necessary that $|x_1| \le 0.5$. Thus, in order to maintain $g_2 = 0$ or, equivalently, in order for the trajectories to follow the curve $x_1^2 = x_2$, it becomes necessary to use relaxed controls, leading to trajectories that converge efficiently but not optimally.

In closing, it should be mentioned that more research needs to be done in order to generate practical, efficient, and robust algorithms from the prototype methods presented in this section. Our hope is that an intrepid reader will take up this challenge successfully.

3.4 Differential Dynamic Programming Applied to Unconstrained Minimization Problems

This section follows Murray and Yakowitz [MY81] and describes how a class of *unconstrained nonlinear programming (NLP) problems* can be transcribed into a class of discrete-time *multistage optimal control problems (MOCPs)* and then solved by a technique known as *differential dynamic programming (DDP)* [JM70], motivated, as will be seen below, by the possibility of reducing computational effort.

Motivation for treating NLP problems as multistage optimal control problems

While it is well known that discrete-time optimal control problems can be treated as NLP problems and treated by the algorithms developed for the latter [CCP70], dynamic programming techniques have proven to be more successful for many problems. Roughly speaking, this can be attributed to the decomposition into stages which characterizes dynamic programming. This has the consequence that the amount of computation grows approximately linearly with the number n of decision times, whereas for other methods the growth rate is faster (e.g., n^3 for the Newton–Raphson method). This motivated Murray and Yakowitz [MY81] to investigate the reverse direction, namely, representing an NLP problem as a discrete-time multistage optimal control problem, and the motivation is now described in more detail.

A discrete-time optimal control problem is described first. Consider a discrete-time dynamical system

$$\mathbf{x}_{i+1} = \mathbf{f}_i(\mathbf{x}_i, \mathbf{u}_i), \quad i = 1, \ldots, n, \quad \mathbf{x}_1 = \overline{\mathbf{x}}_1, \quad \text{given.} \tag{3.90}$$

Consider a cost function $V(\cdot)$ defined as

$$V(\mathbf{u}) = \sum_{i=1}^{n} L_i(\mathbf{x}_i, \mathbf{u}_i) + L_{n+1}(\mathbf{x}_{n+1}), \tag{3.91}$$

where $\mathbf{u}$ is the control vector, defined as

$$\mathbf{u} := (\mathbf{u}_1, \mathbf{u}_2, \ldots, \mathbf{u}_n). \tag{3.92}$$

With no constraints on the controls, the discrete-time MOCP is to minimize (3.91) subject to the dynamical constraint (3.90) by suitable choice of the control sequence $\mathbf{u}_i$, $i = 1, \ldots, n$. The dimension of the state vectors $\mathbf{x}_i$ is the same for each *stage* i and is denoted by r, i.e., $\mathbf{x}_i \in \mathbb{R}^r$ for all i. The common dimension of the control is denoted by s, $\mathbf{u}_i \in \mathbb{R}^s, i = 1, \ldots, n$. In the dynamic programming literature, the terms *single-stage loss function* for L_i and *dynamical law* for $\mathbf{f}_i$ are often used.

Bellman, in his seminal book [Bel57], argued that the possibility of decomposing the optimal control into stages confers many advantages; the main one in the present context is now detailed. Observe that, for a control problem with n stages and s dimensional control vectors, there are ns unknowns to be determined. This means that the computational effort of dynamic programming methods grows linearly with the number of stages. On the other hand, when a control problem is solved by NLP methods, the growth of computational effort is usually faster. For example, the growth rate is $(ns)^3$ for the Newton–Raphson method and $(ns)^2$ for quasi-Newton methods.

In contrast, the DDP method, suitably applied in this context, leads to a method for which the computational effort grows as ns^3. Furthermore, there is a fairly large class of NLP problems in n variables that can be written as a discrete-time optimal control problem with n stages and with $s = 1$. This means that the computational effort for the DDP method, applied to this class of NLP problems, grows linearly in the total number of variables. In fact, it is known that DDP is close to a stagewise Newton method, so that it inherits a quadratic convergence property [Yak86], but, because of the growth rate of computational effort, the former becomes superior to the latter as the number of optimization variables increases.

Transcription of NLP problems into MOCPs

Many of the classical test problems or benchmarks such as those due to Rosenbrock, Himmelblau, Wood, Powell, and others, were specifically designed to have a particular difficult feature such as narrow, "banana-shaped" valleys, singular Hessian matrices at the extrema, and so on. In order to obtain multivariable functions (i.e., functions of n variables), a common strategy is to take identical test functions in each variable and sum them. Often, such test functions are polynomial functions of the individual optimization variables.

Before attempting a general description of classes of NLP problems that can be transcribed into MOCPs suitable for the application of the DDP method (or any dynamic programming method), several examples of successful transcriptions are given, writing the problem both in NLP and control notation. The general notational conventions are shown in Table 3.4.

Example 3.12 Minimization of extended Rosenbrock's function written as an MOCP. The extended Rosenbrock function, introduced in [Ore74], is a summation of Rosenbrock functions in the individual variables y_i:

$$V(\mathbf{y}) := \sum_{i=2}^{n} (100(y_i - y_{i-1}^2)^2 + (1 - y_{i-1})^2). \tag{3.93}$$

Table 3.4. *Notation used in NLP and optimal control problems.*

Name	Notation	Name	Notation
NLP variables	y_i	Control variables	u_i
NLP variable vector	$\mathbf{y} = (y_1, \ldots, y_n)$	Control vector	$\mathbf{u} = (u_1, \ldots, u_n)$
NLP objective function	$V(\mathbf{y})$	Cost function	$V(\mathbf{u})$ (see (3.91))

The standard identification of scalar control variables with scalar optimization variables, as in Table 3.4, is made:

$$u_i = y_i. \tag{3.94}$$

The dynamical systems are defined as

$$x_{i+1} = u_i, \quad i = 1, \ldots, n-1, \tag{3.95}$$

while the single-stage loss functions are defined as

$$\begin{aligned} L_1 &= 0, \\ L_i(x_i, u_i) &= 100(u_i - x_i^2)^2 + (1 - x_i)^2, \quad i = 2, \ldots, n. \end{aligned} \tag{3.96}$$

Note that the MOCP defined by (3.95) and (3.96) have both state vector $\mathbf{x}_i$ and control vector $\mathbf{u}_i$ as scalars, i.e., $r = s = 1$.

Example 3.13 (Minimization of Oren's power function [Ore74] written as an MOCP). Oren's power function is defined as follows:

$$V(\mathbf{y}) = \left(\sum_{i=1}^{n} i y_i^2 \right)^2. \tag{3.97}$$

One simple possibility for transcription to an MOCP is as follows. The system dynamics is given by

$$x_{i+1} = x_i + i u_i^2, \quad i = 1, \ldots, n-1, \quad x_1 = 0, \tag{3.98}$$

whereas the single-stage loss functions are defined as

$$\begin{aligned} L_i &= 0, \quad i = 2, \ldots, n, \\ L_{n+1}(x_{n+1}) &= x_{n+1}^2, \end{aligned} \tag{3.99}$$

As in the case of Rosenbrock's function, for this example, once again, $r = s = 1$.

Example 3.14 (Minimization of Wood's function written as an MOCP). Wood's function is defined as

$$\begin{aligned} V(\mathbf{y}) := {} & 100(y_2 - y_1^2)^2 + (1 - y_1)^2 + 90(y_4 - y_3^2)^2 + (1 - y_3)^2 \\ & + 10.1[(y_2 - 1)^2 + (y_4 - 1)^2] + 19.8(y_2 - 1)(y_4 - 1). \end{aligned} \tag{3.100}$$

In this case, the variable assignment $u_i = y_i$ made in Table 3.4 is not followed, an interchange of variables being necessary to permit transcription as an MOCP.

Table 3.5. *Definition of two-dimensional state vectors, dynamical laws, and single-stage loss functions used in the transcription of Powell's function* (3.102) *to an MOCP of the form* (3.90), (3.91).

$\mathbf{x}_1$	$\mathbf{x}_2$	$\mathbf{x}_3$	$\mathbf{x}_4$
$(x_{1,1}, x_{1,2})$	$(x_{2,1}, x_{2,2})$	$(x_{3,1}, x_{3,2})$	$(x_{4,1}, x_{4,2})$
$\mathbf{f}_0(\mathbf{x}_0, u_0)$	$\mathbf{f}_1(\mathbf{x}_1, u_1)$	$\mathbf{f}_2(\mathbf{x}_2, u_2)$	$\mathbf{f}_3(\mathbf{x}_3, u_3)$
$(0, 0)$	(u_1, u_1)	$(u_2, x_{2,2})$	$(u_3, x_{3,2})$
$L_1(\mathbf{x}_1, u_1)$	$L_2(\mathbf{x}_2, u_2)$	$L_3(\mathbf{x}_3, u_3)$	$L_4(\mathbf{x}_4, u_4)$
0	$(x_{2,1} + 10u_2)^2$	$(x_{3,1} - 2u_3)^4$	$5(x_{4,1} - u_4)^2 + 10(x_{4,2} - u_4)^4$

Accordingly, let $u_1 = y_1$, $u_2 = y_2$, $u_3 = y_4$, and $u_4 = y_3$. Choosing the dynamics as in (3.95) for $i = 1, 2, 3$, the single-stage loss functions are defined as follows:

$$\begin{aligned} L_1 &= (1 - u_1)^2, \\ L_2(x_2, u_2) &= 100(x_2^2 - u_2)^2 + 10.1(u_2 - 1)^2, \\ L_3(x_3, u_3) &= 19.8(x_3 - 1)(u_3 - 1) + 10.1(u_3 - 1)^2, \\ L_4(x_4, u_4) &= 90(u_4^2 - x_4)^2 + (1 - u_4)^2. \end{aligned} \tag{3.101}$$

Example 3.15 Minimization of Powell's function written as an MOCP. Powell's function is defined as follows:

$$V(\mathbf{y}) = (y_1 + 10y_2)^2 + (y_2 - 2y_3)^4 + 5(y_3 - y_4)^2 + 10(y_1 - y_4)^4. \tag{3.102}$$

In this example, renaming of variables as in the previous example does not work. Observe that all terms contain two variables. If these variables had consecutive indices (i.e., y_i and y_{i-1}), then the approach used in the Rosenbrock and Wood examples could be used. However, the last term contains the variables y_1 as well as y_4. This means that, if the controls are identified with the optimization variables (i.e., $u_i = y_i$), then it is necessary to have a component of the state variable "remember" the value of a control variable . Specifically, it is necessary for a second-stage state vector component to carry the value of the first stage-control (u_1) to the last stage. The standard way of doing this is to augment the state vector of each stage to dimension 2; in this case, $\mathbf{x}_i \in \mathbb{R}^2$ for all i, so that $r = 2$. The reader can easily check that the definitions of the dynamical systems and loss functions in Table 3.5 transcribe Powell's problem to an MOCP.

Example 3.16 (Minimization of Fletcher and Powell's helical valley function written as an MOCP). Fletcher and Powell's helical valley function is defined as

$$V(\mathbf{y}) = 100(y_3 - 10\theta(y_1, y_2))^2 + 100(r(y_1, y_2) - 1)^2 + y_3^2, \tag{3.103}$$

where

$$2\pi\theta(y_1, y_2) := \begin{cases} \tan^{-1}(y_1/y_2) & \text{for } y_1 > 0, \\ \pi + \tan^{-1}(y_1, y_2) & \text{for } y_1 < 0, \end{cases} \tag{3.104}$$

and

$$r(y_1, y_2) = (y_1^2 + y_2^2)^{1/2}. \tag{3.105}$$

Table 3.6. *Definition of two-dimensional state vectors, dynamical laws, and single-stage loss functions used in the transcription of Fletcher and Powell's function* (3.103) *to an MOCP of the form* (3.90), (3.91).

$\mathbf{x}_1$	$\mathbf{x}_2$	$\mathbf{x}_3$
$(x_{1,1}, x_{1,2})$	$(x_{2,1}, x_{2,2})$	$(x_{3,1}, x_{3,2})$
$\mathbf{f}_0(\mathbf{x}_0, u_0)$	$\mathbf{f}_1(\mathbf{x}_1, u_1)$	$\mathbf{f}_2(\mathbf{x}_2, u_2)$
$(0, 0)$	$(u_1, 0)$	$(x_{2,1}, u_2)$
$L_1(\mathbf{x}_1, u_1)$	$L_2(\mathbf{x}_2, u_2)$	$L_1(\mathbf{x}_3, u_3)$
0	$100(r(x_{2,1}, u_2) - 1)^2$	$100(u_3 - 10\theta(x_{3,1}, x_{3,2}))^2 + u_3^2$

In the new coordinates defined in Table 3.6, (3.104) and (3.105) become, respectively,

$$r(x_{2,1}, u_2) = (x_{2,1}^2 + u_2^2)^{1/2} \tag{3.106}$$

and

$$2\pi\theta(x_{3,1}, x_{3,2}) := \begin{cases} \tan^{-1}(x_{3,1}/x_{3,2}) & \text{for } x_{3,1} > 0, \\ \pi + \tan^{-1}(x_{3,1}/x_{3,2}) & \text{for } x_{3,1} < 0. \end{cases} \tag{3.107}$$

As will be seen in the outline of the DDP method in what follows, once the NLP problem is transcribed to a DDP problem, it is necessary to specify nominal control inputs to get the DDP algorithm started in the initial iteration; these values are called *starting values* [MY81] and can, in principle, be specified arbitrarily.

General transcription strategy of NLP problem into MOCP

From the preceding examples, a general picture of a strategy for transcription emerges and is briefly outlined below.

1. Can the objective function $V(\mathbf{y})$ be decomposed into terms containing expressions in the individual variables (y_i)? If yes, do so. Otherwise, go to step 2.

2. Can the objective function $V(\mathbf{y})$ be decomposed into terms containing expressions in y_i, y_{i-1}? If yes, do these terms occur in order of increasing i in the individual terms?

3. Does the objective function $V(\mathbf{y})$ contain a summation term? If so, this indicates the presence of a discrete integrator and the dynamics will have a term of the type $x_{i+1} = x_i + g_i(u_i)$.

4. Choose y_i as u_j in suitable order ($u_j = y_i$; usually, but not always, $j = i$).

5. Attempt to write single-stage loss functions L_i that are functions of x_i, u_i alone, choosing the simplest possible dynamics compatible with the items above (usually $x_{i+1} = u_i$).

6. Otherwise, use state integrators to allow for "delays" (memory, as in Wood's example; in this case, the state can no longer be chosen as scalar).

Outline of DDP method

Once the transcription of an NLP problem to a discrete-time MOCP of the type (3.90), (3.91) has been carried out, as indicated above, the MOCP can be solved by any dynamic programming method. One of these methods, DDP, uses quadratic approximations of the functions involved. An abbreviated description of DDP is now given, the reader being referred to [JM70] for more detail.

The DDP approach starts from the choice of a nominal control sequence $\bar{\mathbf{u}} = (\bar{u}_1, \ldots, \bar{u}_n)$, which determines the nominal state trajectory $\bar{\mathbf{x}} = (\bar{\mathbf{x}}_1, \ldots, \bar{\mathbf{x}}_{n+1})$ from (3.90). The DDP approach then performs a succession of backward and forward computations or runs. On the backward run, in which the index i is decremented from n to 1, a feedback control law that determines $\mathbf{u}_i$ as a linear function of the state $\mathbf{x}_i$ is determined, as will be explained in detail in what follows. On the forward run that follows, with the index i running from 1 to n, the control law is used to obtain an improved value of the control variable, and the dynamics (3.90) used to calculate the next state, at every stage. This clearly defines an iterative method, although an iteration index is not required, since only the nominal values $(\bar{\mathbf{x}}_i, \bar{\mathbf{u}}_i)$ are carried from one iteration to the next.

On the backward run, the DDP method carries out the second-order Taylor expansion, about $(\bar{\mathbf{x}}_i, \bar{\mathbf{u}}_i)$, of the cost function

$$V_i(\mathbf{x}_i, \mathbf{u}_i) = L_i(\mathbf{x}_i, \mathbf{u}_i) + Q^*_{i+1}(f_i(\mathbf{x}_i, \mathbf{u}_i)), \quad V_n(\mathbf{x}_n, \mathbf{u}_n) = L_n(\mathbf{x}_n, \mathbf{u}_n), \tag{3.108}$$

where Q^*_{i+1} is the approximate optimal cost function, to be defined as a function of x_{i+1} in what follows. This yields the following expansion:

$$Q_i(\mathbf{x}_i, \mathbf{u}_i) = \frac{1}{2}\mathbf{x}_i^T \mathbf{D}_i \mathbf{x}_i + \mathbf{x}_i^T \mathbf{E}_i \mathbf{u}_i + \frac{1}{2}\mathbf{u}_i^T \mathbf{F}_i \mathbf{u}_i + \mathbf{g}_i^T \mathbf{x}_i + \mathbf{h}_i^T \mathbf{u}_i, \tag{3.109}$$

where the matrices $\mathbf{D}_i$, $\mathbf{E}_i$, and $\mathbf{F}_i$ are the second derivative matrices of $V_i(\mathbf{x}_i, \mathbf{u}_i)$ evaluated at $(\bar{\mathbf{x}}_i, \bar{\mathbf{u}}_i)$. The gradients of $Q_i(\mathbf{x}_i, \mathbf{u}_i)$ and $V_i(\mathbf{x}_i, \mathbf{u}_i)$ with respect to $\mathbf{x}_i$ and evaluated at $(\bar{\mathbf{x}}_i, \bar{\mathbf{u}}_i)$ have to be equal and this yields a condition that determines $\mathbf{g}_i$. The analogous condition involving $\mathbf{u}_i$ determines $\mathbf{h}_i$, completing the definition of the coefficients in (3.109).

For any state $\mathbf{x}_i$, the value $\mathbf{u}_i = \mathbf{u}_i(\mathbf{x}_i)$ which minimizes $Q_i(\mathbf{x}_i, \mathbf{u}_i)$ can be found, if it exists, by obtaining the control that makes the gradient of Q_i with respect to $\mathbf{u}_i$ equal to zero. The gradient of Q_i is easily calculated and the minimizing control has the form

$$\mathbf{u}_i(\mathbf{x}_i) = \mathbf{w}_i + \mathbf{Y}_i \mathbf{x}_i, \tag{3.110}$$

where

$$\mathbf{w}_i = -\mathbf{F}_i^{-1}\mathbf{h}_i \tag{3.111}$$

and

$$\mathbf{Y}_i = -\mathbf{F}_i^{-1}\mathbf{E}_i^T \tag{3.112}$$

if $\mathbf{F}_i^{-1}$ exists. Substituting (3.110) into (3.109) gives the approximate optimal cost function

$$Q_i^*(\mathbf{x}_i) = \frac{1}{2}\mathbf{x}_i^T \mathbf{A}_i \mathbf{x}_i + \mathbf{b}_i^T \mathbf{x}_i, \tag{3.113}$$

which means that $\mathbf{A}_i = \mathbf{D}_i - \mathbf{E}_i\mathbf{F}_i^{-1}\mathbf{E}_i^T$ and $\mathbf{b}_i = \mathbf{g}_i - \mathbf{E}_i\mathbf{F}_i^{-1}\mathbf{h}_i$. Note that constants have been ignored in the quadratic approximation of $V_i(\mathbf{x}_i, \mathbf{u}_i)$, since they have no effect on the control law.

After the backward run, the forward run uses the state dynamics (3.90) to generate the new states and controls, followed by the computation of the cost function (3.91). If improvement (i.e., reduction in cost) occurs, then the new controls become the nominal controls for the next DDP iteration. If, on the other hand, the new control law does not yield sufficient improvement over the nominal control law, then steplength reduction must be applied. Practical details and results of numerical experimentation can be found in [MY81], on which this section is based.

Finally, we call the reader's attention to the fact that there is a clear connection between the above description of the DDP method and the descriptions of the shooting method and iterative learning control (ILC), described in sections 5.2 and 5.2.2, respectively.

Computational requirements and convergence of the DDP method

It is worth noting that the arithmetic operations required for the calculations of the intermediate quantities $(\mathbf{A}_i, \mathbf{b}_i, \ldots, \mathbf{w}_i, \mathbf{Y}_i)$ are the same for every i. Thus for fixed state and control dimensions (r and s, respectively) and for fixed computational requirements for the evaluation of the required derivatives of $\mathbf{f}_i$ and L_i, the computational requirement of DDP grows linearly with the number of stages n.

In comparing DDP to Newton's method applied to the function $V(\cdot)$, it can be verified that the Hessian of $V(\cdot)$ need not have any zero components, and so there is no simplification in the Newton method solution of an optimization problem because it happens to be an MOCP.

The DDP method also has the good feature of global and quadratic convergence under mild regularity assumptions [Mur78], and good computational results are reported in [MY81], wherein it is affirmed that the DDP approach to optimization problems shows that, even for large-scale problems, the rapid convergence rate of a second-order method can be preserved, while maintaining relatively modest computational requirements.

3.5 Notes and References

Optimal control and optimally controlled methods

Section 3.1 follows [CdF73], which was published just two years after the Tsypkin quote mentioned at the beginning of this chapter.

The treatment in subsection 3.3 follows [Goh97], while the treatment of the spurt method in subsection 3.2.2 is based wholly on [EKP74].

Variable structure zero finding methods

Variable structure Newton methods were first proposed by the present authors in [BK04a].

Branin [Bra72] proposed a Newton-type algorithm with switching that bears some resemblance to (3.24) with $\mathbf{P} = \mathbf{I}$. In the notation of this book, it is written as

$$\dot{\mathbf{x}} = -\text{sgn}(\det \mathbf{D_f})\mathbf{D_f}^{-1}(\mathbf{x})\mathbf{f}(\mathbf{x}). \tag{3.114}$$

Later Hirsch and Smale [HS79] gave a detailed mathematical analysis of Branin's method and some variations of it based on what they called the *Newton vector*

$$\mathbf{n}(\mathbf{x}) := -\text{sgn}(\det \mathbf{D_f})\mathbf{D_f}^{-1}(\mathbf{x})\mathbf{f}(\mathbf{x})$$

as well as the *Newton transformation of length* ρ defined by $\mathbf{T}_\rho := \mathbf{x} + \alpha\mathbf{n}(\mathbf{x})$, $\alpha > 0$, $\|\alpha\mathbf{n}(\mathbf{x})\|_2 = \rho$. Since the basic iteration of their method is $\mathbf{x}_{k+1} = \mathbf{T}_\rho(\mathbf{x}_k)$, it can be written as

$$\mathbf{x}_{k+1} = \mathbf{x}_k - \alpha\, \text{sgn}(\det \mathbf{D_f})\mathbf{D_f}^{-1}(\mathbf{x}_k)\mathbf{f}(\mathbf{x}_k),$$

so that it is just the forward Euler discretization of Branin's method (3.114) with stepsize α chosen so that the norm of the correction term is ρ.

Branin's method has been studied intensively by many other authors: See [ZG02] and references therein.

MOCPs and DDP

MOCPs are defined and studied in the classic treatises [Bel57, BD64] as well as in [BH69]. The standard reference for DDP is [JM70].

Chapter 4

Neural-Gradient Dynamical Systems for Linear and Quadratic Programming Problems

At first glance, Liapunov's works were not related to optimization. However, this is not exactly the case. Liapunov developed stability theory for ordinary differential equations; in its simplest form it states that a solution $\mathbf{x}(t)$ *of the equation* $\dot{\mathbf{x}} = \mathbf{f}(\mathbf{x})$ *is stable if there exists a function* $V(\mathbf{x})$ *(the Liapunov function) such that* $\langle \nabla V(\mathbf{x}), \mathbf{f}(\mathbf{x}) \rangle < 0$*. We can take a reverse point of view: the differential equation above is a continuous-time method for minimizing* $V(\mathbf{x})$*. Thus the method provides a systematic tool for validation of the convergence of numerical methods of optimization.*

—B. T. Polyak [Pol02]

This chapter develops the point of view advocated in the quote above, adding the ingredients of control parameters and control Liapunov functions. From the perspective of Chapter 2, continuous algorithms developed in this chapter for finding extrema of functions are given in the form of ODEs whose equilibria are identical with the extrema being sought. In fact, linear and quadratic programming problems are solved using the class of ODEs called gradient dynamical systems (GDSs). These optimization problems are transformed into unconstrained optimization problems using exact penalty functions, which are then solved using a gradient descent method, i.e., a GDS. The key idea is to adjust the penalty parameters, interpreted as control parameters, in the resulting GDS, using a control Liapunov function (CLF) approach. Since the system being studied is a GDS, the energy or potential function whose gradient defines the right-hand side (of the GDS) is a natural Liapunov function. The fact that these GDFSs can be represented and implemented as a class of recurrent neural networks explains the neologism *neural-gradient dynamical systems* that occurs in the chapter title, essentially to call the reader's attention to the equivalence between a GDS approach to optimization and the recurrent neural network approach. A novelty that arises in the approach presented in this chapter is that nonsmooth CLFs have to be used since exact penalty functions are nonsmooth (i.e., have finitely many points of nondifferentiability). This technicality is handled using a generalized Persidskii-type result, presented in Chapter 1, that allows the treatment of a class of differential equations with discontinuous right-hand sides.

This GDS approach outlined above is quite general and therefore applicable in a wide variety of situations, such as neural networks for optimization problems, and solutions for nonsquare linear systems, for example, in least squares, least norm, and least absolute deviation senses. The latter applications lead, in turn, to a novel GDS approach to the linear and quadratic programming problems that arise in different types of support vector machines, as well as in the so-called K-winners-take-all problem. Connections to the continuous- and discrete-time iterative methods of Chapter 2 are also established.

There is a large literature on GDS methods for optimization. Liao and coworkers, in a paper [LQQ04] that surveys this area, write: "Dynamical (or ODE) system and neural network approaches for optimization have coexisted for two decades (and) ... share many common features and structures" and go on to lay out a general framework for what they term neurodynamical optimization. We have chosen to be more specific and concentrate on GDSs, which we term neural-gradient systems, in order to link the two approaches in the reader's mind and also because algorithms based on GDSs, as opposed to general dynamical systems, possess many good properties, such as efficient and computationally inexpensive implementation, the possibility of parallel and asynchronous implementations [BT89], and robustness to perturbations and errors [BT00]. Neural-gradient dynamical systems are studied in this chapter with these facts in mind.

Given a general constrained optimization problem

$$\begin{aligned} \min_{\mathbf{x}\in\mathbb{R}^n} \quad & f(\mathbf{x}), \\ \text{s.t.} \quad & \mathbf{g}(\mathbf{x}) \leq 0, \quad \mathbf{h}(\mathbf{x}) = 0, \end{aligned} \tag{4.1}$$

where the functions $f : \mathbb{R}^n \to \mathbb{R}$, $\mathbf{g} : \mathbb{R}^n \to \mathbb{R}^m$, and $\mathbf{h} : \mathbb{R}^n \to \mathbb{R}^p$, in the framework of [LQQ04], a neurodynamical optimization system is characterized by the following features:

(i) An energy function, also called a merit function and denoted $V(\mathbf{x})$, which is bounded below, and a dynamical system or ODE, denoted N, are to be specified.

(ii) The set of equilibria of the ODE must coincide with the set of (constrained) minima of problem (4.1).

(iii) The dynamical system must be asymptotically stable at any (isolated) solution of problem (4.1).

(iv) The time derivative of the energy function $V(\mathbf{x}(t))$ along the trajectories of the dynamical system N must be nonpositive for all time and zero if and only if $\mathbf{x}(t)$ is an equilibrium solution of the dynamical system N (i.e., $\mathrm{d}\mathbf{x}/\mathrm{d}t = \mathbf{0}$).

All the neural-gradient dynamical systems introduced in this chapter have these four features and thus may be considered neurodynamical optimizers in the sense of Liao and coworkers [LQQ04].

4.1 GDSs, Neural Networks, and Iterative Methods

This section establishes the connections between the topics indicated in the title of this section, within the general perspective given above. We start with a general discussion of GDSs and then give specific examples: Hopfield neural networks and a class of linear iterative methods for solving linear systems.

Building GDSs with control parameters

While it is well known that Hopfield neural networks are GDSs, in the reverse direction it is not equally well appreciated that, with suitable assumptions, a GDS can be interpreted as a neural network. In this chapter, we explore this connection in a systematic manner, as well as exploit the general framework of feedback dynamical systems, developed in the earlier chapters, to design this new class of dynamical systems with control parameters. More specifically, we will generalize by allowing constrained optimization problems, which are transformed into unconstrained penalty problems using an exact penalty function approach. The penalty parameters are considered as control inputs and the unconstrained problem is solved using a standard gradient descent method. This leads to GDSs, with the special feature that, due to the use of exact nonsmooth penalty functions, the resulting dynamical system has a discontinuous right-hand side, which introduces the need for some technicalities in the analysis. On the other hand, the penalty parameters or controls can be chosen using the control Liapunov function approach introduced in Chapter 2. A pictorial overview of the process described in this paragraph is shown in Figure 4.1.

GDSs for unconstrained optimization

The simplest class of unconstrained optimization problem is that of finding $\mathbf{x} \in \mathbb{R}^n$ that minimizes the real scalar function $E : \mathbb{R}^n \to \mathbb{R} : \mathbf{x} \mapsto E(\mathbf{x})$.

A point $\mathbf{x}^* \in \mathbb{R}^n$ is a *global minimizer* for $E(\mathbf{x})$ if and only if

$$E(\mathbf{x}^*) \leq E(\mathbf{x}) \quad \text{for all } \mathbf{x} \in \mathbb{R}^n$$

and a *local minimizer* for $E(\mathbf{x})$ if

$$E(\mathbf{x}^*) \leq E(\mathbf{x}) \quad \text{for all } \mathbf{x} \text{ in a ball } B(\mathbf{x}^*, \epsilon).$$

Assuming here, for simplicity, that the first and second derivatives of $E(\mathbf{x})$ with respect to $\mathbf{x}$ exist, then necessary and sufficient conditions for the existence of a local minimizer are

$$\nabla E(\mathbf{x}^*) = 0 \text{ and } \nabla^2 E(\mathbf{x}^*) > 0.$$

It is quite natural to design procedures that seek a minimum of a given function based on its gradient [Pol63, Ryb65b, Tsy71, Ber99]. More specifically, the class of methods known as steepest descent methods uses the gradient direction as the one in which the largest decrease (or descent) in the function value is achieved (see Chapter 2).

The idea of moving along the direction of the gradient of a function leads naturally to a dynamical system of the following form:

$$\frac{\mathrm{d}\mathbf{x}(t)}{\mathrm{d}t} = -\mathbf{M}(\mathbf{x}, t)\nabla E(\mathbf{x}), \tag{4.2}$$

where $\mathbf{M}(\mathbf{x}, t)$ is, in general, a positive definite matrix called the *learning matrix* in the neural network literature, which will later be identified as being the controller gain matrix. Integrating the differential equation (4.2) for a given arbitrary initial condition $\mathbf{x}(0) = \mathbf{x}_0$ corresponds to following a trajectory that leads to a vector $\mathbf{x}^*$ that minimizes $E(\mathbf{x})$. More

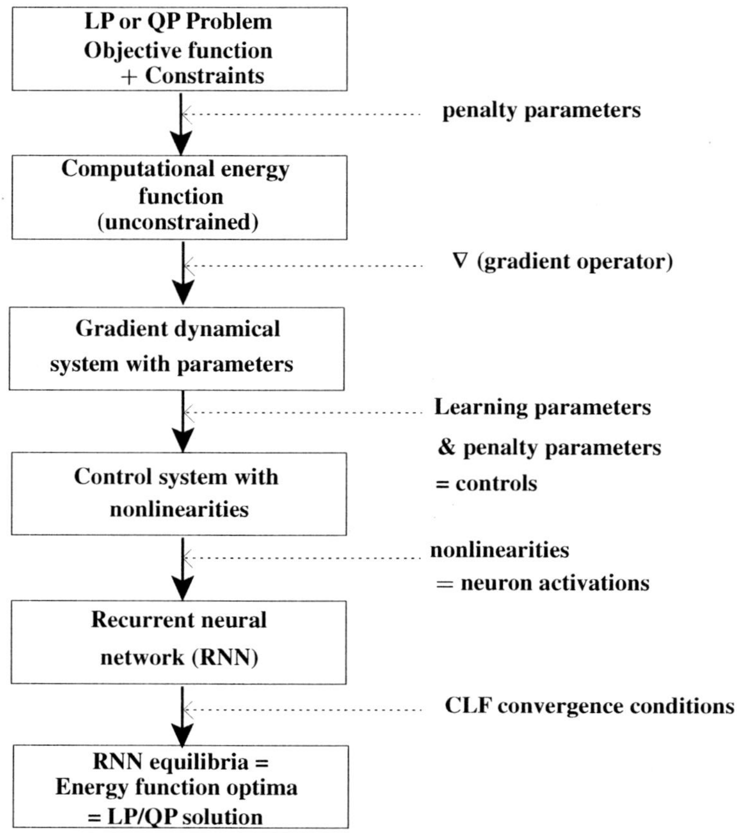

Figure 4.1. *This figure shows the progression from a constrained optimization problem to a neural network (i.e., GDS), through the introduction of controls (i.e., penalty parameters) and an associated energy function.*

precisely, provided that some appropriate conditions (given below) are met, the solution $\mathbf{x}(t)$ along time of the system (4.2) will be such that

$$\lim_{t\to\infty} \mathbf{x}(t) = \mathbf{x}^*. \tag{4.3}$$

In order to ensure that the above limit is attained, and consequently that a desired solution $\mathbf{x}^*$ to the minimization problem is obtained, we recall another connection that is to associate to function $E(\mathbf{x})$ the status, or the interpretation, of an energy function. In the present context, $E(\mathbf{x})$ is also frequently called a *computational energy function*. The notation $E(\mathbf{x})$ is chosen to help in recalling this connection.

Liapunov stability of GDSs

The natural Liapunov function associated to a GDS is now discussed in the present context.

Consider an unconstrained optimization problem and, to aid intuition, rename the objective function that is to be minimized as an *energy function* $E(\mathbf{x})$, where $\mathbf{x}$ is the vector of variables to be chosen such that $E(\cdot)$ attains a minimum. Thinking of $\mathbf{x}$ as a state vector, consider the GDS that corresponds to steepest descent or descent along the negative of the gradient of $E(\cdot)$, where the matrix $\mathbf{M}(\mathbf{x}, t)$, as in (4.2), can be thought of as a "gain" matrix that is positive definite for all $\mathbf{x}$ and t and whose role will become clear in what follows. Then the time derivative of E decreases along the trajectories of (4.2) because, for all $\mathbf{x}$ such that $\nabla E(\mathbf{x}) \neq \mathbf{0}$,

$$\frac{\mathrm{d}E}{\mathrm{d}t} = \nabla E(\mathbf{x})^T \dot{\mathbf{x}} = -\nabla E(\mathbf{x})^T \mathbf{M}(\mathbf{x}, t) \nabla E(\mathbf{x}) < 0. \tag{4.4}$$

Under the assumption that the matrix $\mathbf{M}(\mathbf{x}, t)$ is positive definite for all $\mathbf{x}$ and t, equilibria of (4.2) are points at which $\nabla E = \mathbf{0}$, i.e., are local minima of $E(\cdot)$. By Liapunov theory, the conclusion is that all isolated equilibria of (4.2), which are isolated local minima of $E(\,)$, are locally asymptotically stable, as seen in Chapter 1.

One of the simplest choices for $\mathbf{M}(\mathbf{x}, t)$ is clearly $\mathbf{M}(\mathbf{x}, t) = \mu\mathbf{I}$, where $\mu > 0$ is a scalar, $\mathbf{I}$ is the $n \times n$ identity matrix, and for this choice, system (4.2) becomes

$$\frac{\mathrm{d}\mathbf{x}}{\mathrm{d}t} = -\mu \nabla E(\mathbf{x}), \qquad \mathbf{x}(0) = \mathbf{x}_0. \tag{4.5}$$

Clearly, choice of the parameter μ affects the rate at which trajectories converge to an equilibrium. In general, the choice of the positive definite matrix $\mathbf{M}(\mathbf{x}, t)$ in (4.2) is guided by the desired rate of convergence to an equilibrium. In fact, the choice of $\mathbf{M}(\mathbf{x}, t)$, as a function of $\mathbf{x}$ (the state of the system) and time t, amounts to the use of time-varying state feedback control as seen in Chapter 1.

GDSs that solve linear systems of equations

Another perspective on the connection between the iterative methods studied as dynamical systems with control in Chapter 2 and the neural networks studied in this chapter is obtained by recalling that a zero finding problem can be recast as an optimization problem.

Given a positive definite matrix $\mathbf{A} \in \mathbb{R}^{n\times n}$ and a vector $\mathbf{b} \in \mathbb{R}^n$, consider the following quadratic energy function:

$$E(\mathbf{x}) = \frac{1}{2}\mathbf{x}^T \mathbf{A}\mathbf{x} - \mathbf{b}^T \mathbf{x}. \tag{4.6}$$

The corresponding GDS that minimizes $E(\cdot)$ is

$$\dot{\mathbf{x}} = -\mathbf{M}\nabla E(\mathbf{x}) = -\mathbf{M}(\mathbf{A}\mathbf{x} - \mathbf{b}). \tag{4.7}$$

Since trajectories of this GDS converge to the unique equilibrium, which is the solution $\mathbf{x}^*$ of the linear system

$$\mathbf{A}\mathbf{x} = \mathbf{b}, \tag{4.8}$$

this means that (4.7) can be said to "implement" a continuous algorithm to solve the linear system with positive definite coefficient matrix (4.8). Observe that the GDS (4.7) is clearly of the general feedback controller form (2.7) discussed in Chapter 2.

Since the gradient of $E(\mathbf{x})$ is $\nabla E(\mathbf{x}) = \mathbf{Ax} - \mathbf{b}$ and the Hessian matrix $\nabla^2 E(\mathbf{x}) = \mathbf{A}$ (which is positive definite), $\mathbf{x}^* = \mathbf{A}^{-1}\mathbf{b}$ is thus the unique global minimizer of the energy function $E(\mathbf{x})$.

The function

$$V(\mathbf{x}) = E(\mathbf{x}) - E(\mathbf{x}^*) \tag{4.9}$$

is a Liapunov function since, assuming that $\mathbf{M}$ is positive definite, clearly $\dot{V}(\mathbf{x}) = -(\mathbf{Ax} - \mathbf{b})^T\mathbf{M}(\mathbf{Ax} - \mathbf{b})$ is negative definite. The matrix $\mathbf{M}$ is interpreted as a learning matrix or, equivalently, as a feedback gain matrix. Comparing (4.7) with (2.109), it is clear that matrix $\mathbf{M}$ in (4.7) corresponds to the feedback gain matrix $\mathbf{K}$ in (2.109). In section 5.2.2, this feedback gain or learning matrix is also identified, in the appropriate context, as a preconditioner.

Minimizing the energy function $E(\mathbf{x})$ in (4.6) corresponds to a 2-norm minimization problem. If, instead, the 1-norm is chosen to formulate the minimization problem, a least absolute deviation solution is found and the resulting dynamical system can be represented as a neural network with discontinuous activation functions. For rectangular matrices, minimizing the square of the 2-norm of the residue leads to the normal equations and the solution in the least squares sense. These topics are explored further in section 4.2.

Neural networks

The term *artificial neural network* (ANN) refers to a large class of dynamical systems. In the taxonomy or "functional classification scheme" of [MW92], from a theoretical point of view, ANNs are divided into two classes: *feedforward* and *feedback* networks; the latter class is also referred to as *recurrent*. The feature common to both classes and, indeed, to all ANNs is that they are used because of their "learning" capability. From the point of view adopted in this book, this means that ANNs are dynamical systems that depend on some control parameters, also known as weights or gains, and therefore they can be described from the viewpoint of control systems [HSŻG92]. Once this is done, the general framework developed in Chapter 2 is applicable to ANN models. When a suitable adjustment of these parameters is achieved, the ANN is said to have learned some functional relationship between a class of inputs and a class of outputs.

Three broad classes of applications of ANNs can be identified.

In the first, a set of inputs and corresponding desired outputs is given, and the problem is to adjust the parameters in such a way that the system fits this data in the sense that if any of the given inputs is presented to it, the corresponding desired output is in fact generated. This class of problem goes by many names: Mathematicians call it a *function approximation problem* and the parameters are adjusted according to time-honored criteria such as least squared error, etc.; for statisticians, it is a *regression problem* and, once again, there is an array of techniques available; finally, in the ANN community, this is referred to as a *learning problem* and, on the completion of the process, the ANN is said to have *learned* a functional relationship between inputs and outputs.

In the second class, it is desired to design a dynamical system whose equilibria correspond to the solution set of an optimization problem. This used to be referred to as *analog computation*, since the idea is to start from some arbitrary initial condition and follow the trajectories of the dynamical system that converge to its equilibria, which are the optimum points that it is required to find.

For the third class, loosely related to the second, the objective is to design a dynamical system with multiple stable equilibria. In this case, if for each initial condition, the corresponding trajectory converges to one of the stable equilibria, then the network is referred to as an *associative memory* or *content-addressed memory*. The reason behind this picturesque terminology is that an initial condition in the basin of attraction of one of the stable equilibria may be regarded as a corrupted or noisy version of this equilibrium; when the ANN is "presented" with this input, it "associates" it with or "remembers" the corresponding "uncorrupted" version, which is the attractor of the basin (to which nearby trajectories converge). In this chapter, we will focus mainly on the first two classes of problems.

In the ANN community, a distinction is often made between the parameter adjustment processes in the first and second classes of problems. For instance, if a certain type of dynamical system, referred to as a *feedforward* ANN or *multilayer perceptron*, is used, then a popular technique for adjustment of the parameters (weights) is referred to as *backpropagation* and usually there is no explicit mention of feedback, so that control aspects of the weight adjustment process are not effectively taken advantage of. In the second class of problems, the parameters to be chosen are referred to as *gains* and are usually chosen once and for all in the design process, so that the desired convergence occurs. However, in this case, the ANN community refers to the feedback aspect by using the name *recurrent ANNs*. We will take the viewpoint that both classes of problems can profitably be viewed as feedback control systems to which the design and analysis procedures of Chapter 2, with appropriate adjustments, are applicable.

This chapter builds on the perspective of Chapter 2 by treating the learning and penalty parameters of an ANN as controls. Thus, this chapter extends the "iterative method equals dynamical system with control inputs" point of view to recurrent ANNs that solve certain classes of computational problems, such as linear and quadratic programming as well as pattern classification.

As a concrete example of the connection between GDSs and neural networks, we will now briefly discuss the well-known class of Hopfield–Tank neural networks from this perspective.

Hopfield–Tank neural networks written as GDSs

The basic mathematical model of a Hopfield neural network (see, e.g., [HT86, CU93]) in normalized form can be written as

$$\mathbf{C}\frac{d\mathbf{u}}{dt} = -\mathbf{K}\mathbf{u} + \mathbf{W}\boldsymbol{\phi}(\mathbf{u}) + \boldsymbol{\theta}, \tag{4.10}$$

where $\mathbf{u} = (u_1, u_2, \ldots, u_n)$, $\mathbf{C} = \text{diag}\,(\tau_1, \tau_2, \ldots, \tau_n)$, $\mathbf{K} = \text{diag}\,(\alpha_1, \alpha_2, \ldots, \alpha_n)$, $\boldsymbol{\phi}(\mathbf{u}) = (\phi_1(u_1), \phi_2(u_2), \ldots, \phi_n(u_n))$, $\boldsymbol{\theta} = (\theta_1, \theta_2, \ldots, \theta_n)$, and, finally, $\mathbf{W} = (w_{ij})$ is the symmetric interconnection matrix, where $w_{ji} = r_j G_{ji}$; $\theta_j = r_j I_j$; $\tau_j = r_j C_j > 0$; $\alpha_j = r_j/R_j > 0$, $\phi_i(\cdot)$ are the nonlinear activation functions that are assumed to be differentiable and monotonically increasing; r_j are the so-called scaling resistances, and G_{ji} are the conductances. The notation in the above equation originated from the circuit implementation of artificial neural networks where, in addition, C_j, R_j correspond, respectively, to capacitors and resistors and I_j to constant input currents.

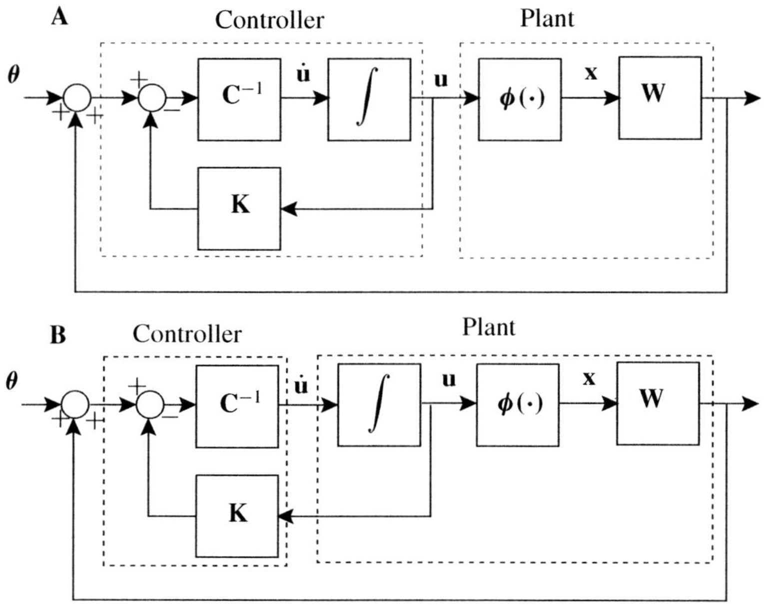

Figure 4.2. *Dynamical feedback system representations of the Hopfield–Tank network* (4.10). *In part* A, *the controller is dynamic and the plant static, whereas in part* B, *the controller is static, with state feedback, while the plant is dynamic.*

The recurrent Hopfield–Tank neural net can be viewed from a feedback control perspective in different ways and two possible block diagram representations are depicted in Figure 4.2, in which the interconnection or weight matrix has been shown as part of the plant. Of course, if $\mathbf{W}$ is being thought of as a feedback gain, it could be moved to the feedback loop.

If the Hopfield–Tank neural network is to be used as an associative memory, then it must have multiple equilibria, each one corresponding to a state that is "remembered" as well as a local minimum of an appropriate energy function. In fact, the gradient of this energy function defines a GDS that is referred to as a Hopfield–Tank associative memory neural network.

On the other hand, if the Hopfield–Tank neural network is to be used to carry out a global optimization task, then the energy function, which corresponds to the objective function of the optimization problem to be solved, should admit a unique global minimum.

These statements are now made precise.

Hopfield–Tank network as associative memory

Defining

$$\mathbf{x} = \boldsymbol{\phi}(\mathbf{u}), \tag{4.11}$$

$\dot{\mathbf{x}} = \mathbf{D}_{\phi}\dot{\mathbf{u}}$, and thus (4.10) can be written as

$$\dot{\mathbf{x}} = \mathbf{D}_{\phi}\mathbf{C}^{-1}(-\mathbf{K}\mathbf{u} + \mathbf{W}\mathbf{x} + \boldsymbol{\theta}). \tag{4.12}$$

Since $\boldsymbol{\phi}(\mathbf{u})$ is a diagonal function, with components $\phi_i(u_i)$ continuously differentiable and monotonically increasing, bounded and belonging to the first and third quadrants, two conclusions follow: $\mathbf{D}_{\boldsymbol{\phi}}$ is a positive diagonal matrix and $\boldsymbol{\phi}$ is invertible [OR70], so that

$$\mathbf{u} = \boldsymbol{\phi}^{-1}(\mathbf{x}), \tag{4.13}$$

and (4.12) can be written in $\mathbf{x}$-coordinates as

$$\dot{\mathbf{x}} = \mathbf{D}_{\boldsymbol{\phi}}\mathbf{C}^{-1}(-\mathbf{K}\boldsymbol{\phi}^{-1}(\mathbf{x}) + \mathbf{W}\mathbf{x} + \boldsymbol{\theta}). \tag{4.14}$$

Now, defining $\mathbf{k}$ as the vector containing the diagonal elements of the diagonal matrix $\mathbf{K}$, i.e.,

$$\mathbf{k} = (\alpha_1, \ldots, \alpha_n), \tag{4.15}$$

and the function $\boldsymbol{\phi}_I : \mathbb{R}^n \to \mathbb{R}^n$ as

$$\boldsymbol{\phi}_I(\mathbf{x}) := \left(\int_0^{x_1} \phi_1^{-1}(\tau)\mathrm{d}\tau, \ldots, \int_0^{x_n} \phi_n^{-1}(\tau)\mathrm{d}\tau \right), \tag{4.16}$$

an energy function $E : \mathbb{R}^n \to \mathbb{R}$ can be defined as follows:

$$E(\mathbf{x}) := -\frac{1}{2}\mathbf{x}^T\mathbf{W}\mathbf{x} - \boldsymbol{\theta}^T\mathbf{x} + \mathbf{k}^T\boldsymbol{\phi}_I(\mathbf{x}). \tag{4.17}$$

Note that this energy function is not necessarily positive definite; it is, however, continuously differentiable. Calculating the gradient of $E(\mathbf{x})$ gives

$$\nabla E(\mathbf{x}) = -\mathbf{W}\mathbf{x} - \boldsymbol{\theta} + \mathbf{K}\boldsymbol{\phi}^{-1}(\mathbf{x}). \tag{4.18}$$

From (4.14) and (4.18), it follows that we can write

$$\dot{\mathbf{x}} = -\mathbf{D}_{\boldsymbol{\phi}}\mathbf{C}^{-1}\nabla E(\mathbf{x}), \tag{4.19}$$

showing, as claimed above, that the Hopfield–Tank network is a GDS, recalling that the matrix $\mathbf{D}_{\boldsymbol{\phi}}\mathbf{C}^{-1}$ is positive diagonal. This means that results on gradient systems (see Chapter 1) can be used. In particular, it can be concluded that all trajectories of the Hopfield–Tank network tend to equilibria, which are extrema of the energy function, and furthermore, that all isolated local minima of the energy function $E(\cdot)$ are locally asymptotically stable. Thus, the Hopfield–Tank network functions as an associative memory that "remembers" its equilibrium states, in the sense that if it is presented with (i.e., initialized with) a state that is a slightly "corrupted" version of an equilibrium state, then it converges to this uncorrupted state.

Hopfield–Tank net as global optimizer

Under different hypotheses and a different choice of energy function it is possible to find conditions for convergence of the trajectories of (4.10) to a unique equilibrium state.

First, (4.10) is rewritten as

$$\frac{\mathrm{d}\mathbf{u}}{\mathrm{d}t} = -\mathbf{L}\mathbf{u} + \mathbf{T}\boldsymbol{\phi}(\mathbf{u}) + \boldsymbol{\eta}, \tag{4.20}$$

where $\mathbf{L} := \mathbf{C}^{-1}\mathbf{K}$ is a diagonal matrix, $\mathbf{T} := \mathbf{C}^{-1}\mathbf{W}$ is the diagonally scaled interconnection matrix, and $\boldsymbol{\eta} := \mathbf{C}^{-1}\boldsymbol{\theta}$ is a constant vector. Suppose that it is desired to analyze the stability of an equilibrium point $\mathbf{u}^*$. For Liapunov analysis, it is necessary to shift the equilibrium to the origin, using the coordinate change $\mathbf{z} = \mathbf{u} - \mathbf{u}^*$. In $\mathbf{z}$-coordinates, (4.20) becomes

$$\dot{\mathbf{z}} = -\mathbf{L}\mathbf{z} + \mathbf{T}\boldsymbol{\psi}(\mathbf{z}), \tag{4.21}$$

where $\boldsymbol{\psi}(\mathbf{z}) := (\psi_1(z_1), \ldots, \psi_n(z_n))$ and $\psi_i(z_i) := \phi_i(z_i + u_i^*) - \phi_i(u_i^*)$. It is easy to see that, under the smoothness, monotonicity, boundedness, and first-quadrant–third-quadrant assumptions on the ϕ_i, the functions ψ_i inherit these properties, and in fact, the function $\boldsymbol{\psi}$ has bounded components; i.e., for all i, $|\psi_i(z_i)| \leq b_i|z_i|$, so that, defining $\mathbf{B} = \text{diag}\,(b_1, \ldots, b_n)$, the following vector inequality holds (componentwise):

$$\mathbf{B}^{-1}\boldsymbol{\psi}(\mathbf{z}) \leq \mathbf{z}. \tag{4.22}$$

In order to define a computational energy function, we introduce the following notation: $\mathbf{p}$ is a vector with all components positive, $\mathbf{P}$ is a diagonal matrix with components of the diagonal equal to that of the vector $\mathbf{p}$, and

$$\boldsymbol{\psi}_S(\mathbf{z}) := \left(\int_0^{z_1} \psi_1(\tau)\mathrm{d}\tau, \ldots, \int_0^{z_n} \psi_n(\tau)\mathrm{d}\tau\right). \tag{4.23}$$

A computational energy function is then defined as follows:

$$\overline{E}(\mathbf{z}) = 2\mathbf{p}^T\boldsymbol{\psi}_S(\mathbf{z}). \tag{4.24}$$

From (4.23) and (4.24), it is clear that the computational energy function is of the Persidskii type, discussed in Chapter 1.

Calculating the time derivative of $\overline{E}(\cdot)$ along the trajectories of (4.21) gives

$$\dot{\overline{E}}(\mathbf{z}) = -2\boldsymbol{\psi}(\mathbf{z})^T\mathbf{P}\mathbf{L}\mathbf{z} + 2\boldsymbol{\psi}(\mathbf{z})^T\mathbf{P}\mathbf{T}\boldsymbol{\psi}(\mathbf{z}). \tag{4.25}$$

From (4.22), noticing that $\mathbf{PL}$ is a positive diagonal matrix, the following componentwise majorization is immediate:

$$\boldsymbol{\psi}(\mathbf{z})^T\mathbf{P}\mathbf{L}\mathbf{z} \geq \boldsymbol{\psi}(\mathbf{z})^T\mathbf{P}\mathbf{L}\mathbf{B}^{-1}\boldsymbol{\psi}(\mathbf{z}). \tag{4.26}$$

Thus, defining the positive diagonal matrix $\overline{\mathbf{L}} := \mathbf{L}\mathbf{B}^{-1}$, the following holds:

$$\begin{aligned}\dot{\overline{E}}(\mathbf{z}) &\leq -2\boldsymbol{\psi}(\mathbf{z})^T\mathbf{P}\overline{\mathbf{L}}\boldsymbol{\psi}(\mathbf{z}) + 2\boldsymbol{\psi}(\mathbf{z})^T\mathbf{P}\mathbf{T}\boldsymbol{\psi}(\mathbf{z}) \\ &= \boldsymbol{\psi}(\mathbf{z})^T[(\mathbf{T} - \overline{\mathbf{L}})^T\mathbf{P}^T + \mathbf{P}(\mathbf{T} - \overline{\mathbf{L}})]\boldsymbol{\psi}(\mathbf{z}) \\ &< 0,\end{aligned} \tag{4.27}$$

provided that

$$(\mathbf{T} - \overline{\mathbf{L}})^T\mathbf{P}^T + \mathbf{P}(\mathbf{T} - \overline{\mathbf{L}}) < 0. \tag{4.28}$$

An interconnection matrix $\mathbf{T}$ is said to be *additively diagonally stable* if there exist positive diagonal matrices $\mathbf{P}$ and $\overline{\mathbf{L}}$ such that (4.28) holds [KB00]. It turns out that, if the interconnection matrix $\mathbf{T}$ is additively diagonally stable, then it is also true that (4.20) has a unique equilibrium, and so it has just been proved, by the use of the energy function (4.24), which in this case is also a Liapunov function, that this unique equilibrium is globally asymptotically stable. Finally, in view of the definition of the computational energy function (4.24), it is easy to verify that (4.21) can be written as

$$\dot{\mathbf{z}} = -\mathbf{L}\mathbf{z} + \mathbf{T}\mathbf{P}^{-1}\nabla \overline{E}(\mathbf{z}). \tag{4.29}$$

The right-hand side of (4.29) contains two terms: a dissipative linear term $\mathbf{Lz}$ and a gradient term. The additive diagonal stability condition, which relates the matrix $\mathbf{TP}^{-1}$ that multiplies the gradient of the energy function and the matrix $\mathbf{L}$ of the linear term, ensures that the energy function works as a Liapunov function for the system. Note that, in this case, unlike the associative memory case, symmetry of the interconnection matrix is not required.

For more details on global stability of neural networks and on the matrix classes involved in this stability analysis, see [KB00].

Discrete-time neural networks

There are many different methods to derive discrete-time versions of the differential equations that describe the Hopfield model above [TG86, BKK96]. For example, the simple forward Euler method applied to (4.10), after normalization of stepsize and time constants, yields the following difference equation:

$$\mathbf{u}_{k+1} = (\mathbf{I} - \mathbf{K})\mathbf{u}_k + \mathbf{W}\boldsymbol{\phi}(\mathbf{u}_k) + \boldsymbol{\theta}. \tag{4.30}$$

Equation (4.30) describes a dynamical system of the form (1.77) that is also known as a discrete-time recurrent artificial neural network.

Another simple discrete-time version of the Hopfield model [CU93] is arrived at by choosing $\mathbf{K} = \mathbf{I}$ and using the change of variables (4.11), i.e., $\mathbf{x}_k = \boldsymbol{\phi}(\mathbf{u}_k)$, which yields, in the $\mathbf{x}$-variables,

$$\mathbf{x}_{k+1} = \boldsymbol{\phi}(\mathbf{W}\mathbf{x}_k + \boldsymbol{\theta}). \tag{4.31}$$

In digital implementations of (4.31) above, the nonlinear activation function $\phi_j(\cdot)$ commonly used is the signum function (1.83) which, in addition to being easily implemented, allows useful theoretical manipulations. This leads to the form

$$\mathbf{x}_{k+1} = \mathrm{sgn}(\mathbf{W}\mathbf{x}_k + \boldsymbol{\theta}). \tag{4.32}$$

Observe that the discrete-time model (4.32) is in fact is a nonlinear recurrence of the form (2.111), with the following substitutions: $\mathbf{I} - \mathbf{KA} = \mathbf{W}$ and $\mathbf{Kb} = \boldsymbol{\theta}$. Notice also that the signum function is applied after the summation (in (4.32)) or addition of state variables. There are other situations in which the nonlinearity occurs before the addition (i.e., $\mathbf{x}_{k+1} = \mathbf{W}\mathrm{sgn}(\mathbf{x}_k) + \boldsymbol{\theta}$; see [BKK96, KB00]).

The neural network expressed by system (4.32) is represented in block diagram form in Figure 4.3 and is, in fact, a discrete-time version of Figure 2.12, where the controller has unity gain.

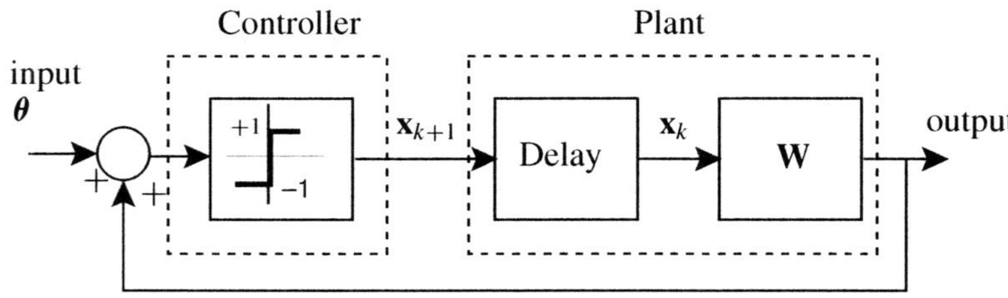

Figure 4.3. *The discrete-time Hopfield network* (4.32) *represented as a feedback control system.*

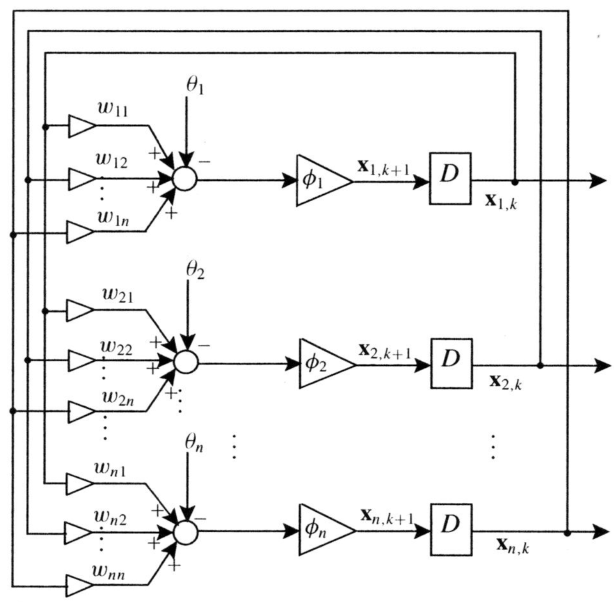

Figure 4.4. *The discrete-time Hopfield network* (4.31) *represented as an artificial neural network. The blocks marked "D" represent delays of one time unit.*

Iterative methods as discrete-time neural networks

The standard Jacobi and Gauss–Seidel iterative methods to solve the linear system $\mathbf{Ax} = \mathbf{b}$, as explained in section 2.3, can be viewed as linear versions of the neural network (4.32). The implementation of such neural networks using standard circuit components such as operational amplifiers, resistors, capacitors, etc., is discussed in [CU93].

Similarly, the well-known SOR method is known to improve the convergence rate of Jacobi or Gauss–Seidel-type iterative methods and is chosen here to give an illustration of the connections described above. This method, as pointed out in section 2.3, can be written as the iteration

$$\mathbf{x}_{k+1} = (\mathbf{I} - \mathbf{KA})\mathbf{x}_k + \mathbf{Kb}, \qquad \mathbf{K} = (\alpha^{-1}\mathbf{D} - \mathbf{E})^{-1}, \tag{4.33}$$

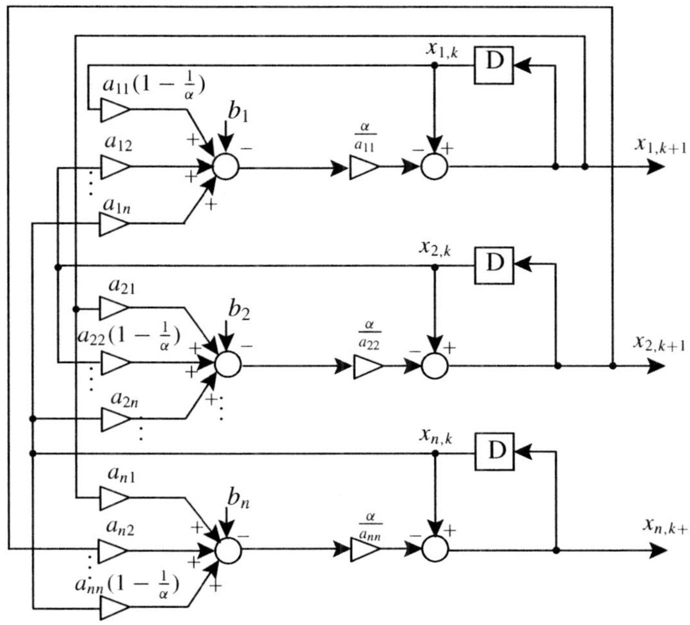

Figure 4.5. *Neural network realization of the SOR iterative method* (4.33).

which in turn describes the recurrent network represented in Figure 4.5 and where α is the relaxation parameter, which is normally chosen to maximize the rate of convergence [Var00, You71].

Once a general iterative algorithm has been put in the form (2.111) or a variant thereof, then the standard tools of control and systems theory can be used to carry out local and global analyses of its convergence behavior, and this is also true for the problem of synthesis of neural networks with desired properties.

In order to establish further connections between optimization procedures, in the spirit of the discussion above, and some other standard iterative methods, as well as the corresponding neural network implementations, consider the gradient system (4.7). Discretizing this ODE using the forward Euler method yields the discrete-time recurrent equation

$$\mathbf{x}_{k+1} = \mathbf{x}_k - \mathbf{M}_k(\mathbf{A}\mathbf{x}_k - \mathbf{b}), \quad \mathbf{x}_0 \text{ given}. \tag{4.34}$$

Note that this equation is exactly the same as (2.109), which represents a general iterative method to solve the linear equation $\mathbf{A}\mathbf{x} = \mathbf{b}$.

If, for example, the learning matrix $\mathbf{M}_k$ is chosen as the diagonal matrix $\mathbf{M}_k = \mu_k \mathbf{I}$, where

$$\mu_k = \frac{\langle \mathbf{r}_k, \mathbf{r}_k \rangle}{\langle \mathbf{r}_k, \mathbf{A}\mathbf{r}_k \rangle} \tag{4.35}$$

and where $\mathbf{r}_k$ is the residue at the kth iteration, then the iterative scheme becomes the well-known *Richardson iterative method* [You71] which was derived in section 2.3.1 using a control Liapunov function (see (2.128)).

If, on the other hand, the learning matrix is chosen as a constant diagonal matrix $\mathbf{M} = \text{diag}\ (\mu_1, \ldots, \mu_n)$, specifically with $\mu_i = a_{ii}^{-1} \neq 0$, then the resulting iterative scheme corresponds to the classical *Jacobi method*. The matrix $\mathbf{M}$ is also known as the *Jacobi preconditioner* (for more on preconditioning from a control perspective, see section 5.3). Going further along these lines, each choice of a learning matrix corresponds to an iterative method. In particular, for the choice $\mathbf{M} = (\mathbf{D} - \mathbf{E})^{-1}$, (4.34) becomes the *Gauss–Seidel method*, whereas for $\mathbf{M} = (\alpha^{-1}\mathbf{D} - \mathbf{E})^{-1}$, (4.34) becomes the *SOR method*, with relaxation parameter α, as already mentioned above. In other words, we recover the classification or taxonomy discussed in section 2.3.2, this time from a "learning matrix/neural network" perspective.

Notice that, in section 2.3, these learning matrices, interpreted as feedback control gain matrices, were found using control Liapunov functions, in order to ensure convergence of the general linear iterative method of the form (2.109).

The discussion above shows why artificial neural networks are suited to solve optimization problems and how they can be recast in terms of iterative methods and also in the basic control system structures discussed in Chapter 2. This point of view will be exploited further in what follows.

4.2 GDSs that Solve Linear Systems of Equations

Using a GDS with tunable parameters, which are viewed as control gains, this section establishes the basic control approach to the solution of a linear system of equations,

$$\mathbf{A}\mathbf{x} = \mathbf{b}, \tag{4.36}$$

where $\mathbf{A} \in \mathbb{R}^{m\times n}$, $\mathbf{b} \in \mathbb{R}^m$, $\mathbf{x} \in \mathbb{R}^n$. First, a unified view of the various possibilities that arise in the solution of linear systems is given using the concept of loss functions.

For the linear system (4.36), let the residue be defined here as

$$\mathbf{r}(\mathbf{x}) = (r_1(\mathbf{x}), \ldots, r_m(\mathbf{x})) := \mathbf{b} - \mathbf{A}\mathbf{x}. \tag{4.37}$$

Consider the energy or cost function

$$E_\rho(\mathbf{x}) = \sum_{i=1}^{m} \rho([\mathbf{b} - \mathbf{A}\mathbf{x}]_i) = \sum_{i=1}^{m} \rho(r_i(\mathbf{x})), \tag{4.38}$$

where $\rho(r_i)$ is a convex *loss function* and its derivative $\mathrm{d}\rho/\mathrm{d}r_i$ is known as the corresponding *influence function*.

The associated GDS is

$$\dot{\mathbf{x}} = -\mathbf{M}\nabla_{\mathbf{x}} E_\rho(\mathbf{x}). \tag{4.39}$$

For the choice of E_ρ in (4.38),

$$\nabla_{\mathbf{x}} E_\rho(\mathbf{x}) = -\mathbf{A}^T \nabla_{\mathbf{r}} \rho(\mathbf{r}). \tag{4.40}$$

Introducing the notation

$$\Psi(\mathbf{r}) := \nabla_{\mathbf{r}} \rho(\mathbf{r}), \tag{4.41}$$

Table 4.1. *Choices of objective or energy functions that lead to different types of solutions to the linear system* $\mathbf{Ax} = \mathbf{b}$*, when* $\mathbf{A}$ *has full rank (row or column). Note the presence of a constraint in the second row, corresponding to the least norm solution. The abbreviation LAD stands for least absolute deviation and the* r_i*'s in the last row refer to the components of the residue defined here as* $\mathbf{r} := \mathbf{b} - \mathbf{Ax}$.

Conditions on $\mathbf{A}$	Energy function	Solution type
$m \times n$, $m > n$, rank$(\mathbf{A}) = n$	$\|\mathbf{b} - \mathbf{Ax}\|_2^2$	Least squares
$m \times n$, $m < n$, rank$(\mathbf{A}) = m$	$\frac{1}{2}\|\mathbf{x}\|_2^2$ s.t. $\mathbf{Ax} = \mathbf{b}$	Least norm
$m \times n$, $m > n$, rank$(\mathbf{A}) = n$	$\frac{1}{2}\|\mathbf{W}^{1/2}(\mathbf{b} - \mathbf{Ax})\|_2^2$	Weighted least squares
$m \times n$, $m < n$, rank$(\mathbf{A}) = m$	$\sum_{i=1}^{m} \rho(r_i(\mathbf{x}))$	LAD, for $\rho = \|\cdot\|$

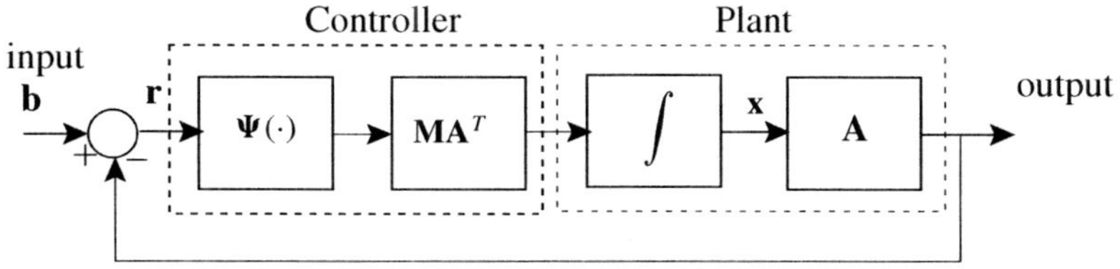

Figure 4.6. *Control system representation of GDS* (4.42). *Observe that the controller is a special case of the general controller* $\boldsymbol{\phi}(\mathbf{x}, \mathbf{r})$ *in Figure* 2.1A.

(4.39) becomes

$$\dot{\mathbf{x}} = \mathbf{MA}^T \Psi(\mathbf{r}). \tag{4.42}$$

Under different assumptions on the matrix $\mathbf{A}$, different choices of the cost function E, of the loss function ρ, and of the learning matrix $\mathbf{M}$ lead to different GDSs (i.e., neural networks), the trajectories of which converge to the solutions of the linear systems. These GDSs are continuous realizations of well-known methods such as the least squares, weighted least squares, and least absolute deviation methods. For the reader's convenience, these assumptions and choices are shown in Table 4.1, while the control system representation of the GDS (4.39) is shown in Figure 4.6. Observe that the first and third rows in Table 4.1 correspond to particular choices of the loss function $\rho(\cdot)$ shown in the fourth row. Many other choices of loss functions are discussed in [CA02]. In all cases, the GDS resulting from the steepest descent method applied to the corresponding energy function leads to a solution type corresponding to the choice of loss function.

Solution of a linear system using a GDS

This section considers, in greater detail, the problem of finding a solution, in the L_1-norm or least absolute deviation (LAD) sense (fourth row of Table 4.1), to a system of algebraic linear equations (4.36).

We consider the underdetermined case in which $\mathbf{A} \in \mathbb{R}^{m \times n}$ has full row rank, $m \leq n$, $\mathbf{x} \in \mathbb{R}^n$, and $\mathbf{b} \in \mathbb{R}^m$, as well as the overdetermined case in which $\mathbf{A} \in \mathbb{R}^{m \times n}$ has full column rank, $m \geq n$, $\mathbf{x} \in \mathbb{R}^n$, and $\mathbf{b} \in \mathbb{R}^m$.

The LAD or L_1 approach is to minimize the 1-norm of the residue, or, in other words, to choose the energy function as

$$E(\mathbf{x}) := \|\mathbf{r}(\mathbf{x})\|_1. \tag{4.43}$$

In other words, it is desired to solve the following unconstrained optimization problem:

$$\min_{\mathbf{x} \in \mathbb{R}^n} E(\mathbf{x}). \tag{4.44}$$

A solution of the optimization problem (4.44) is regarded as a solution of the system of linear equations (4.36) in the L_1 sense and is also commonly known as a *LAD solution*, specifically in the case when $\mathbf{A}$ has full column rank ($m > n$).

LAD solutions have many attractive properties [BS83, CA02] such as insensitivity to outliers (bad data) and sparsity (small number of nonzero components), and have therefore been the subject of much research in optimization as well as in neural networks (see, e.g., [War84, CU92, WCXC00, CA02] and the references therein). The purpose of this section is to point out that LAD solutions are easily found by GDSs with discontinuous right-hand sides.

Let $r_i : \mathbb{R}^n \to \mathbb{R}$ be the components of vector $\mathbf{r}$. The set $\Delta := \{\mathbf{x} : \mathbf{r}(\mathbf{x}) = \mathbf{0}\}$ is defined as

$$\Delta = \bigcap_{i=1}^{m} \Delta_i; \ \ \Delta_i := \{\mathbf{x} : r_i(\mathbf{x}) = 0\}. \tag{4.45}$$

The minimum of the energy function E in (4.43) is $\mathbf{r} = \mathbf{0}$; consequently, a solution of problem (4.44) is a vector $\mathbf{x}^* \in \mathbb{R}^n$ such that $\mathbf{x}^* \in \Delta$. Notice that E is convex in $\mathbf{r}$, thus its unique minimizer is the zero vector $\mathbf{r}^* = \mathbf{0}$ and it is nondifferentiable at $\mathbf{r} = \mathbf{0}$.

The optimization problem (4.44) is solved by gradient descent on the energy function (4.43); in other words, by following the trajectories of the GDS

$$\dot{\mathbf{x}} = -\mathbf{M}\nabla E(\mathbf{x}) = -\mathbf{M}\mathbf{A}^T \operatorname{sgn}(\mathbf{r}), \tag{4.46}$$

where $\mathbf{M} := \operatorname{diag}(\mu_1, \ldots, \mu_n)$, $\mu_i > 0$, is a positive diagonal matrix, and a neural network representation of this GDS is given in Figure 4.7.

Notice that the function $E(\mathbf{x})$ in (4.43) is nondifferentiable at $\Delta_i = 0$, so that the right-hand side of the associated GDS (4.46) is discontinuous. The solutions of (4.46) are considered in the sense of Filippov, and the set Δ is referred to as a *discontinuity set*. If the trajectories of (4.46) are confined to Δ, this motion is said to be a *sliding motion* or, equivalently, the system is said to be in *sliding mode*. Sliding may be *partial*, in the sense that the trajectories are confined to the intersection of manifolds $\Delta_i = 0$ for some subset of indices i. Further details about sliding modes can be found in [Utk92, Żak03].

Convergence analysis when matrix A has full row rank

This section analyzes only the simpler case of (4.36) in which $\mathbf{A}$ has full row rank. Convergence analysis is performed using a Persidskii form of the gradient system (4.46) in conjunction with the corresponding candidate diagonal-type Liapunov function. The Per-

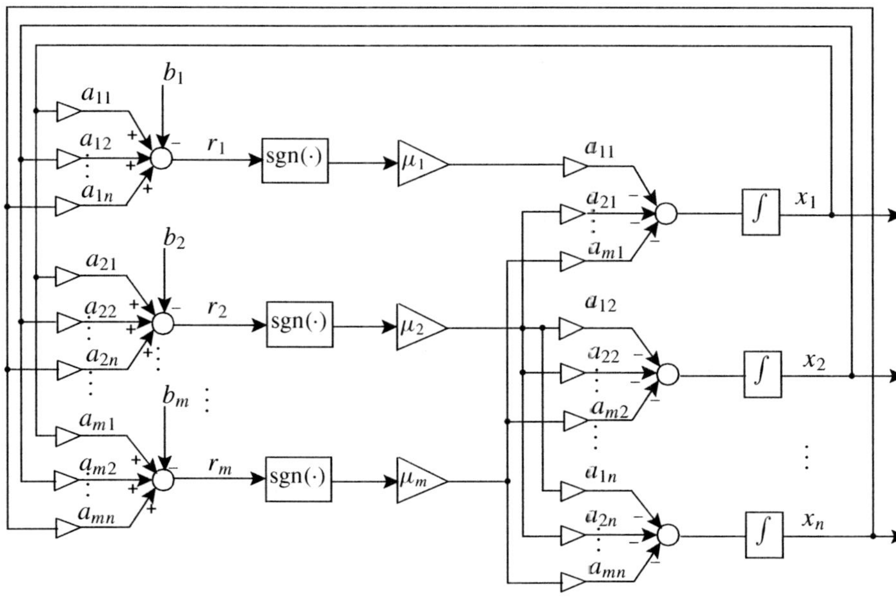

Figure 4.7. *A neural network representation of gradient system* (4.46).

sidskii form of (4.46) is obtained by premultiplying (4.46) by the matrix $\mathbf{A}$. Observe that since $\dot{\mathbf{r}} = \mathbf{A}\dot{\mathbf{x}}$, from (4.46) we get

$$\dot{\mathbf{r}} = -\mathbf{A}\ \mathbf{M}\ \mathbf{A}^T \mathrm{sgn}(\mathbf{r}). \tag{4.47}$$

The Persidskii system (4.47) is equivalent to the original gradient system (4.46). Defining $\sqrt{\mathbf{M}} := \mathrm{diag}\ (\sqrt{\mu_1}, \ldots, \sqrt{\mu_n})$, notice that the right-hand side of (4.47) can be written as $-\mathbf{A}\ \sqrt{\mathbf{M}}\sqrt{\mathbf{M}}\ \mathbf{A}^T \mathrm{sgn}(\mathbf{r})$, and using this observation, we can prove the following proposition.

Proposition 4.1. *The Persidskii system* (4.47) *is equivalent to the original gradient system* (4.46), *in the sense that* $\dot{\mathbf{r}} \equiv \mathbf{0}$ *if and only if* $\dot{\mathbf{x}} \equiv \mathbf{0}$.

Proof. If $\dot{\mathbf{x}} \equiv \mathbf{0}$, it is immediate that $\dot{\mathbf{r}} \equiv \mathbf{0}$. On the other hand, if $\dot{\mathbf{r}} \equiv \mathbf{0}$, then the vector $\sqrt{\mathbf{M}}\nabla E = \sqrt{\mathbf{M}}\mathbf{A}^T \mathrm{sgn}(\mathbf{r})$ belongs to the null space $\mathcal{N}(\mathbf{A}\sqrt{\mathbf{M}})$ of matrix $\mathbf{A}\sqrt{\mathbf{M}}$; however, $\sqrt{\mathbf{M}}\nabla E$ is a vector from the row space $\mathcal{R}(\sqrt{\mathbf{M}}\mathbf{A}^T)$ of $\mathbf{A}\ \sqrt{\mathbf{M}}$. Since $\mathcal{N}(\mathbf{A}\sqrt{\mathbf{M}}) \perp \mathcal{R}(\sqrt{\mathbf{M}}\,\mathbf{A}^T)$, the only possible solution for $\dot{\mathbf{r}} \equiv \mathbf{0}$ is $\sqrt{\mathbf{M}}\,\nabla E \equiv \mathbf{0}$, and consequently $\dot{\mathbf{x}} \equiv \mathbf{0}$. □

Proposition 4.1 is necessary since it ensures that the convergence results derived for the Persidskii system (4.47) also hold for the original gradient system (4.46).

Theorem 4.2. *The trajectories of system* (4.46) *converge, from any initial conditions, to the solution set of the system of linear equations* (4.36) *in finite time and remain in this set thereafter. Moreover, the convergence time* t_f *satisfies the bound* $t_f \le V(\mathbf{r}_0)/\lambda_{\min}(\mathbf{AMA}^T)$, *where* $\mathbf{r}_0 := \mathbf{r}(\mathbf{x}_0)$.

Proof. Since system (4.47) has a discontinuous right-hand side, we choose the nonsmooth candidate diagonal-type Liapunov function introduced in Chapter 1:

$$V(\mathbf{r}) = \sum_{i=1}^{m} \int_0^{r_i} \operatorname{sgn}(\tau)\, d\tau. \tag{4.48}$$

Observe that (i) $V(\mathbf{r}) > 0$ for $\mathbf{r} \neq \mathbf{0}$; (ii) $V(\mathbf{r}) = 0$ if and only if $\mathbf{r} = \mathbf{0}$; and (iii) $V(\mathbf{r})$ is radially unbounded. Furthermore, $V(\mathbf{r}(t))$ can be interpreted as a measure of the distance of the point $\mathbf{x}(t)$ on a trajectory to the set Δ. The time derivative of V along the trajectories of (4.47) is given by $\dot{V} = \nabla V^T \dot{\mathbf{r}}$, i.e.,

$$\dot{V}(\mathbf{r}) = -\operatorname{sgn}^T(\mathbf{r})\, \mathbf{A}\ \mathbf{M}\ \mathbf{A}^T \operatorname{sgn}(\mathbf{r}). \tag{4.49}$$

Notice that since $\mathbf{A}$ has full row rank and $\mathbf{M}$ is positive definite, then $\mathbf{A}\ \mathbf{M}\ \mathbf{A}^T$ is also positive definite. Consequently, $\dot{V} \equiv 0$ if and only if $\operatorname{sgn}(\mathbf{r}) \equiv \mathbf{0}$ implying $\dot{\mathbf{r}} \equiv \mathbf{0}$ and, from Proposition 4.1, $\dot{\mathbf{x}} \equiv \mathbf{0}$.

Consider system (4.47), the time derivative (4.49) of (4.48), and the partition of the set Δ, defined in (4.45). Considering the solutions of (4.47) in the sense of Filippov, two situations occur: first, when the trajectories have not reached any Δ_i, and second, when the trajectories have already reached some set Δ_i. The aim is to show that in both situations there exists a scalar $\varepsilon > 0$, such that $\dot{V} \leq -\varepsilon$, and thus to observe that Δ is an invariant set.

(i) $\mathbf{x} \notin \Delta_i$ for every i. In this case the trajectories are not confined to the surface of discontinuity and the solutions of (4.47) are considered in the usual sense. Since $\mathbf{A}$ has full row rank and $\mathbf{M}$ is a positive diagonal matrix, then the matrix $\mathbf{AMA}^T$ is positive definite, and using the Rayleigh principle and the fact that $\|\operatorname{sgn}(\mathbf{r})\|_2 = m^2 \geq 1$ for $r_i \neq 0$, we can write

$$\dot{V}(\mathbf{r}) \leq -\lambda_{\min}(\mathbf{AMA}^T)m^2 \leq -\lambda_{\min}(\mathbf{AMA}^T), \tag{4.50}$$

where $\lambda_{\min}(\mathbf{AA}^T) > 0$ is the smallest eigenvalue of $\mathbf{AA}^T$.

(ii) $\mathbf{x} \in \Delta_i$ for some indices i and almost all t in an interval $\mathcal{I}$. In this case the trajectories are confined to the sets Δ_i, except for at least one trajectory, resulting in a sliding motion in the sets Δ_i. Thus, the vectors $\mathbf{e}$ that describe this motion are subgradients of E at $\mathbf{x}$, i.e., $\dot{\mathbf{r}} = -\mathbf{AMe}$, $\mathbf{e} \in \partial E(\mathbf{x})$, where $\mathbf{e} = \mathbf{A}^T\mathbf{s}$ and $\mathbf{s} = (s_1, \ldots, s_n)^T$, with $s_i \in [-1, 1]$, for every i [Cla83]. Since there exists at least one index i such that $\mathbf{x} \notin \Delta_i$, then $\|\mathbf{s}\|_2^2 \geq 1$, and using (4.49) and the Rayleigh principle, we obtain the inequality (4.50).

From items (i) and (ii), it follows that the trajectories of (4.46) converge to the set Δ in finite time: Observe that (4.50) implies $V(t) \leq V(t_0) - \lambda_{\min}(\mathbf{AMA}^T)\, t$, so that the time t_f for r_i to reach zero does not exceed $V_0/\lambda_{\min}(\mathbf{AMA}^T)$.

Thus the trajectories of (4.47) reach the set Δ and remain in this set. From Proposition 4.1, this result also holds for the original gradient system (4.46), concluding the proof. □

Illustrative examples of the use of the GDS (4.46) in the solution of both over- and underdetermined linear systems $\mathbf{Ax} = \mathbf{b}$ are now given.

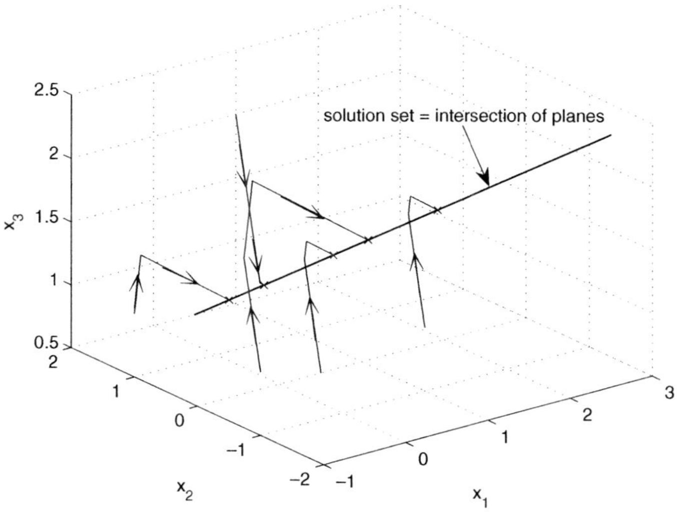

Figure 4.8. *Phase space plot of the trajectories of gradient system* (4.46) *for the underdetermined system described in Example* 4.3.

Example 4.3. Let

$$\mathbf{A} = \begin{bmatrix} 1 & 2 & 3 \\ 0 & 1 & 2 \end{bmatrix}, \quad \mathbf{b} = (6, 3), \quad \text{and} \quad \mathbf{x} = (x_1, x_2, x_3).$$

It is easy to check that the general solution of this system is $(x_1, x_2, x_3) = \mathbf{x}_p + \mathbf{x}_h = (0, 3, 0) + x_3(1, -2, 1) = (x_3, 3 - 2x_3, x_3)$.

Figure 4.8 shows the trajectories of the GDS (4.46) for different initial conditions. Observe that, since the solution of an underdetermined system is not unique, each initial condition results in a solution on the line defined by the intersection of the planes $x_1 + 2x_2 + 3x_3 = 6$ and $x_2 + 2x_3 = 3$, which, of course, is the sum of the particular solution $\mathbf{x}_p$ and a different homogeneous solution $\mathbf{x}_h$. Time plots of x_1, x_2, x_3 are shown for the initial condition $(0, 0, 0)$ in Figure 4.9, in which the finite-time convergence to the solution set is clearly visible.

Example 4.4. This simple example is from [War84]. The data in the table below are supposed to represent the function $z = y$.

y	1	2	3	4	5	6	7	8	9
z	0.75	2	3	4.25	4.75	6.5	7.25	0	8.88

In order to do an L_1 fit of a straight line to these data points, assume the relationship to be $z = a + by$. This results in a linear system $\mathbf{Ax} = \mathbf{b}$, where $\mathbf{A} \in \mathbb{R}^{9\times 2}$, $\mathbf{x} = (x_1, x_2) := (a, b)$, $\mathbf{b} \in \mathbb{R}^9$. The first column of $\mathbf{A}$ has all elements equal to 1, and the second column has elements equal to the first row (y values) in the above table, while $\mathbf{b}$ has elements equal to the second row (z values) in the above table.

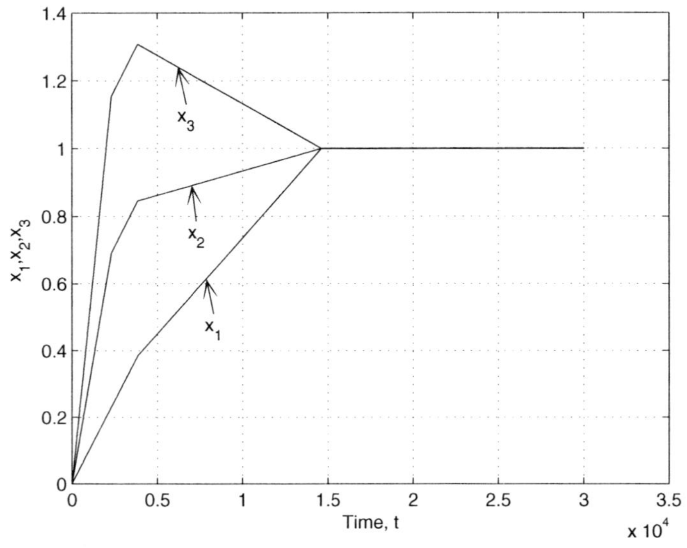

Figure 4.9. *Time plots of the trajectories of gradient system* (4.46) *for the underdetermined system described in Example* 4.3 *for the arbitrary initial condition* (0, 0, 0), *the solution being* $\mathbf{x} = (1, 1, 1)$.

Figure 4.10 shows the trajectories of the GDS (4.46) in the x_1–x_2 phase plane, converging globally to $(a, b) = (0.034, 0.983)$, from different initial conditions. This is a reasonably close fit to the "real" parameter values $(a, b) = (0, 1)$, despite the presence of an "outlier" (the eighth data point (8, 0) is clearly a bad point). For comparison, the L_2 or least squares fit to these data, obtained from the normal equations, is $(a, b) = (1.0475, 0.6212)$, which is clearly adversely affected by the bad point.

4.3 GDSs that Solve Convex Programming Problems

As mentioned in the introduction to this chapter, the gradient-based approach to optimization problems has a long history. However, it has been approached from a control viewpoint more recently, with major contributions coming from the work of Utkin, Żak, and coworkers, detailed in their books [Utk92, Żak03]. We will follow the exposition of these two authors, adding one ingredient, an integral-of-nonlinearity Persidskii-type Liapunov function, that simplifies the analysis, is a natural extension of the results derived in the earlier sections, and allows us to put this problem into the control framework proposed in this book.

We begin by outlining Utkin's general approach to convex programming problems using gradient descent on a penalized objective function. In subsequent sections, this general approach is specialized to some linear and quadratic programming problems using our simplified approach.

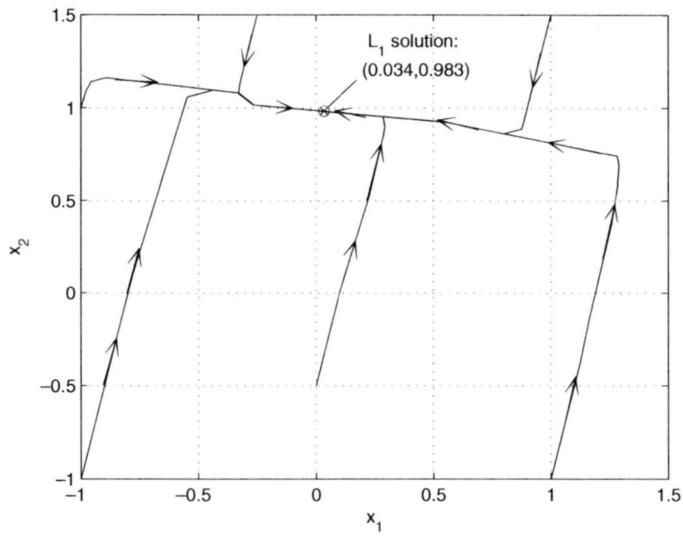

Figure 4.10. *Phase plane plot of the trajectories of gradient system* (4.46) *for the overdetermined system described in Example* 4.4.

Consider the convex programming problem

$$\begin{aligned} &\min f(\mathbf{x}) \\ &\text{s.t. } h_i(\mathbf{x}) = 0, \quad i = 1, \ldots, m, \\ &\qquad h_i(\mathbf{x}) \le 0, \quad i = m+1, \ldots, m+l, \end{aligned} \tag{4.51}$$

where $\mathbf{x} \in \mathbb{R}^n$, the function $f : \mathbb{R}^n \to \mathbb{R}$, as well as the functions h_i are continuously differentiable, $h_i(\mathbf{x}), i = 1, \ldots, m$ are linear, and $h_i(\mathbf{x}), i = m+1, \ldots, m+l$ are convex.

Applying the penalty function approach leads to the unconstrained minimization problem

$$\min E(\mathbf{x}),$$

where the energy function $E(\cdot)$ is defined as

$$E(\mathbf{x}) := f(\mathbf{x}) + \mathbf{h}(\mathbf{x})^T \mathbf{u}, \tag{4.52}$$

where

$$\mathbf{h}(\mathbf{x}) := (h_1(\mathbf{x}), \ldots, h_{m+l}(\mathbf{x})), \quad \mathbf{u} = (u_1, \ldots, u_{m+l}), \tag{4.53}$$

and the functions u_i are defined as follows:

$$u_i := \lambda_i \operatorname{sgn}(h_i(\mathbf{x})), \ \lambda_i > 0, \ \ i = 1, \ldots, m, \tag{4.54}$$

$$u_i := \lambda_i \operatorname{uhsgn}(h_i(\mathbf{x})), \ \lambda_i > 0, \ \ i = m+1, \ldots, m+l. \tag{4.55}$$

The choice of notation $\mathbf{u}$ is intended to alert the reader that the second term in (4.52) is being thought of as the control term, with $\mathbf{u}$ being the control input. From this viewpoint, the penalty parameters λ_i are control gains and design of a penalty function is equivalent to choice of control gains, as will be seen in the sections that follow.

The second term in (4.52) is called a *penalty function*, but we will also identify it as a CLF, one that will be denominated a *reaching phase CLF* and denoted as

$$V_{rp}(\mathbf{x}) := \mathbf{h}(\mathbf{x})^T\mathbf{u}, \tag{4.56}$$

which clearly satisfies the requirement

$$V_{rp}(\mathbf{x}) \begin{cases} = 0 & \text{inside the feasible region,} \\ > 0 & \text{outside the feasible region.} \end{cases} \tag{4.57}$$

For such a penalty function V_{rp}, there exists a positive number λ_0 such that for all $\lambda_i \geq \lambda_0$, the minimum of the function $E(\mathbf{x})$ (with no constraints) is equal to the constrained minimum of $f(\mathbf{x})$ in (4.51), reducing the constrained problem to an unconstrained one. This fact is shown in [Zan67], [Lue84, p. 389], using the terminology penalty function.

Note that, under the hypotheses on the functions h_i, $E(\mathbf{x})$ is a convex function. It remains to specify an estimate of the parameter λ_0 and a minimization procedure for the piecewise smooth function $E(\mathbf{x})$.

The transpose of the Jacobian matrix of the function $\mathbf{h}$ is denoted as the matrix $\mathbf{G}$ for notational simplicity in what follows

$$\mathbf{G} := \mathbf{D}_\mathbf{h}(\mathbf{x})^T = [\nabla h_1 \ \cdots \ \nabla h_{m+l}] \in \mathbb{R}^{n\times(m+l)}. \tag{4.58}$$

In order to determine the minimum of $E(\mathbf{x})$, the steepest descent algorithm is written formally for all points where the gradient of $E(\mathbf{x})$ exists, i.e., off the surfaces $h_i(\mathbf{x}) = 0$:

$$\dot{\mathbf{x}} = -\nabla E(\mathbf{x}) = -\nabla f - \mathbf{G}\mathbf{u}. \tag{4.59}$$

Note that the right-hand side of (4.59) will have discontinuities on the surfaces $h_i(\mathbf{x}) = 0$, which are therefore referred to as *discontinuity surfaces*. Off the discontinuity surfaces, the steepest descent trajectory is governed by (4.59) and consequently $E(\mathbf{x})$ is locally decreasing. If the trajectory does not intersect the discontinuity surfaces or if the intersection set of the latter has zero measure, then, by convexity of E, the trajectory would end at some stationary point that is the minimum of E, solving the problem under investigation. Points at which the right-hand side of (4.59) vanishes are referred to as *stationary points*.

As is well known from the theory of differential equations with discontinuous right-hand sides, as is the case of (4.59), sliding modes may occur when the trajectories lie on the intersection of discontinuity surfaces. In particular, if for some surface $h_i(\mathbf{x}) = 0$, the conditions

$$\lim_{h_i \to 0} h_i(\mathbf{x}) < 0 \quad \text{and} \quad \lim_{h_i \to 0-} h_i(\mathbf{x}) > 0 \tag{4.60}$$

hold, then a sliding mode occurs on this surface. This additional possibility (compared to a smooth gradient descent procedure) introduces some additional technicalities in the proof of convergence of the piecewise smooth steepest descent gradient algorithm to the extremum. Specifically, it is necessary to establish and verify conditions for the existence of sliding modes, analyze the sliding mode reduced-order dynamical system, specify the gradient procedure for a piecewise smooth function, give the conditions for the existence of an exact penalty function, and, finally, give conditions for the convergence of the gradient procedure. This route is followed in [Utk92], to which we refer the interested reader; however, these convergence conditions are difficult to verify since they depend on the knowledge of the extremum which we want to compute. Thus, as is the usual practice under such circumstances, stronger and verifiable sufficient conditions are now derived.

Choice of control gains using a CLF

The reaching phase CLF, denoted $V_{rp}(\mathbf{x})$, is now used to make the appropriate choices of the control gains. First a preliminary result is shown.

Lemma 4.5 [Utk92]. *If the feasible region is nonempty, then the vector* **Gu** *is always nonzero in the interior of the infeasible region.*

Proof. Each component u_i of the vector **u** can assume values of $\pm\lambda_i$ outside the discontinuity surfaces, or the value $u_{\text{eq},i}$, if the ith sliding mode has occurred. Thus, with a suitable partitioning of the vector **u**, we may write

$$\mathbf{G}\mathbf{u} = \mathbf{G}'\mathbf{u}'_{\text{eq}} + \mathbf{G}''\mathbf{u}''.$$

If this vector were to become zero in the infeasible region, then it is not difficult (see Property 3, sec. 3, Chap. 15 in [Utk92]) to show that the convex function V_{rp} would attain its minimum at this point. This, however, contradicts the property of positivity (4.57) in the exterior of the feasible region. $\square$

Observe that the norm of the vector **Gu** depends on the gradients $\nabla h_i(\mathbf{x})$ as well as the control gains λ_i (penalty function coefficients). Define

$$\lambda_0 := \min_i \lambda_i. \tag{4.61}$$

Assume that the following lower bound estimate holds:

$$\|\mathbf{G}\mathbf{u}\| \geq \lambda_0 g_0, \tag{4.62}$$

where $g_0 > 0$ is assumed to be known. Assume furthermore that

$$\|\nabla f\| \leq f_0, \tag{4.63}$$

where the nonnegative number f_0 is known. Then, the following theorem can be stated.

Theorem 4.6 [Utk92]. *With the notation established in the preceding paragraph, suppose that the control gains λ_i (coefficients of the penalty function) are chosen such that*

$$\lambda_0 = \min_i \lambda_i > \frac{f_0}{g_0}. \tag{4.64}$$

Then, the function $V_{rp}(\mathbf{x})$ defined in (4.56) *is a CLF with respect to the feasible region, and trajectories of* (4.59) *enter the feasible region in finite time.*

Proof. To see that the nonnegative function $V_{rp}(\mathbf{x})$ is a CLF and that (4.64) defines appropriate choices of the control gains, calculate the time derivative of $V_{rp}(\mathbf{x})$ defined in (4.56):

$$\dot{V}_{rp} = \frac{\mathrm{d}V_{rp}}{\mathrm{d}t} = \mathbf{u}^T\mathbf{G}^T(-\nabla f - \mathbf{Gu}) + \sum_{i=1}^{m+l} h_i(\mathbf{x})\frac{\mathrm{d}u_i}{\mathrm{d}t}. \tag{4.65}$$

If a function $h_i(\mathbf{x})$ differs from zero, then u_i = constant and $\mathrm{d}u_i/\mathrm{d}t = 0$ from the definition of the penalty functions. If a sliding mode occurs at the intersection of discontinuity surfaces, values of $\mathrm{d}u_i/\mathrm{d}t$ that are equal to $\mathrm{d}u_{\mathrm{eq}.i}/\mathrm{d}t$ will, in general, be different from zero; however, the corresponding function $h_i(\mathbf{x})$ will be equal to zero. In other words, all the products $h_i(\mathbf{x})\mathrm{d}u_i/\mathrm{d}t$ in the second term of (4.65) are zero and we can ignore the second term. From estimates (4.62) and (4.63), we arrive at the following differential inequality for the left-hand side of (4.65):

$$\dot{V}_{rp} = \frac{\mathrm{d}V_{rp}}{\mathrm{d}t} \leq \|\mathbf{Gu}\|\,(f_0 - \|\mathbf{Gu}\|) \leq -\lambda_0 g_0 f_0\left(\frac{g_0}{f_0}\lambda_0 - 1\right) < 0. \tag{4.66}$$

Since $V_{rp}(\mathbf{x}) \geq 0$, the finite-time property follows from (4.66). □

After this finite time has elapsed (i.e., V_{rp} has become zero), the trajectory enters the feasible region where $E(\mathbf{x}) = f(\mathbf{x})$ and the gradient procedure (4.59) converges, as desired, to the minimum of $f(\cdot)$.

This approach is very general but has the disadvantage of requiring estimates of bounds for the various gradients (f_0 and g_0). The next section specializes to a simpler class of convex programming problems that includes linear and quadratic programming problems. Using the same Liapunov function ($V_{rp}(\mathbf{x})$), but recognizing that it is a Persidskii-type Liapunov function, allows us to obtain simpler convergence conditions.

4.3.1 Stability analysis of a class of discontinuous GDSs

The objective of this subsection is to outline another approach, closely related to the one described above, for a special class of convex programming problems, obtained by considering only linear inequality constraints in (4.51). This approach was developed by Żak and coworkers (see [Żak03]) and is presented here with simplifications resulting from the use of the generalized Persidskii result (Theorem 1.38).

Consider the constrained optimization problem

$$\begin{array}{l} \min f(\mathbf{x}) \\ \text{s.t. } \mathbf{Ax} \leq \mathbf{b}, \end{array} \tag{4.67}$$

where $\mathbf{x} \in \mathbb{R}^n$, $f : \mathbb{R}^n \to \mathbb{R}$ is a C^1 convex function, $\mathbf{A} \in \mathbb{R}^{m\times n}$, $\mathbf{b} \in \mathbb{R}^m$.

In order to describe the main technical results, some notation and assumptions are needed. Let the feasible set be denoted as

$$\Omega := \{\mathbf{x} \in \mathbb{R}^n : \mathbf{Ax} \leq \mathbf{b}\}, \tag{4.68}$$

and the set of *local minimizers* of the optimization problem (4.67) be denoted as Γ.

Geometrically, the feasible region can be described as the intersection of m closed half-spaces, i.e., a convex polytope. Denoting the ith row of $\mathbf{A}$ as $\mathbf{a}_i^T$ and the ith component of the vector $\mathbf{b}$ as b_i, the hyperplanes associated with the m half-spaces are denoted H_i and described as

$$H_i = \{\mathbf{x} : \mathbf{a}_i^T \mathbf{x} = b_i\}, \quad i = 1, \ldots, m.$$

The union of all the H_i's is denoted $H := \cup_{i=1}^m H_i$, and finally, the active and violated constraint index sets are defined as follows:

$$\begin{aligned} I(\mathbf{x}) &:= \{i : \mathbf{a}_i^T \mathbf{x} = b_i\} \quad \text{(active constraint indices)}, \\ J(\mathbf{x}) &:= \{i : \mathbf{a}_i^T \mathbf{x} > b_i\} \quad \text{(violated constraint indices)}. \end{aligned}$$

We make the following assumptions:

(i) The objective function is convex on $\mathbb{R}^n$ and has continuous partial derivatives on $\mathbb{R}^n$, i.e., it is C^1.

(ii) The feasible set, polyhedron Ω, is nonempty and compact.

(iii) Each point in Ω is a regular point of the constraint; i.e., for any $\mathbf{x} \in \Omega$, the vectors $\mathbf{a}_i$, $i \in I(\mathbf{x})$, are linearly independent.

(iv) There are no redundant constraints; i.e., the matrix $\mathbf{A}$ has full row rank.

An important consequence of these assumptions is that the optimization problem (4.67) then belongs to the particular class of convex programming problem that is the minimization of a convex function over a polyhedron Ω. For such problems, any local minimizer is a global minimizer, and furthermore, the Karush–Kuhn–Tucker (KKT) conditions for optimality are both necessary and sufficient [Lue84, NS96, NW99].

Lemma 4.7 (KKT conditions). *Suppose that $\mathbf{x}^* \in \Omega$. Then $\mathbf{x}^*$ is a local (and hence global) minimizer of* (4.67) *if and only if there exist constants $\mu_1, \mu_2, \ldots, \mu_m$ such that*

$$\nabla f(\mathbf{x}^*) + \sum_{i \in I(\mathbf{x}^*)} \mu_i \mathbf{a}_i = \mathbf{0}.$$

In order to arrive at an energy function for this problem, the penalty function method used in the previous section is used, with a small modification in the term involving the objective function. More specifically, the objective function term is switched on only in the interior of the feasible region, and thus the energy function is defined as

$$E(\mathbf{x}, k_1, k_2) = k_1 f(\mathbf{x}) + k_2 \sum_{i=1}^{m} \max\{0, \mathbf{a}_i^T \mathbf{x} - \mathbf{b}_i\}, \tag{4.69}$$

where

$$k_1 = \begin{cases} 0 \text{ if } \mathbf{x} \notin \Omega, \\ 1 \text{ if } \mathbf{x} \in \Omega. \end{cases} \tag{4.70}$$

This means that the GDS $\dot{\mathbf{x}} = -\nabla E(\cdot)$ that minimizes $E(\cdot)$ in (4.69) can be written as follows:

$$\dot{\mathbf{x}} = -k_1 \nabla f(\mathbf{x}) - k_2 \mathbf{A}^T \operatorname{uhsgn}(\mathbf{A}\mathbf{x} - \mathbf{b}). \tag{4.71}$$

Note that the choice of k_1 made in (4.70) means that the first term on the right-hand side of (4.71) is switched off (i.e., becomes zero) outside the feasible set and is switched on inside the feasible set Ω. Similarly, the second term is switched off inside Ω, totally switched on outside Ω, and partially switched on at the boundary of Ω. The notation k_1, k_2 for the penalty parameters has been used to call attention to the fact that they will be thought of as feedback controls and chosen via a CLF.

The GDS (4.71) can also be written as

$$\dot{\mathbf{x}} = \begin{cases} -\nabla f(\mathbf{x}) \text{ if } \mathbf{x} \in \Omega, \\ -k_2 \sum_{i \in J(\mathbf{x})} \mathbf{a}_i \text{ if } \mathbf{x} \notin \Omega, \end{cases} \qquad \mathbf{x}(0) = \mathbf{x}_0. \tag{4.72}$$

It needs to be demonstrated that every trajectory of the GDS (4.71), for all initial states, converges to the solution set of the optimization problem (4.67). This occurs in the reaching and convergence phases, as mentioned in section 1.5. In the context of this section, these phases can be described as follows:

1. All trajectories that start from an infeasible initial state $\mathbf{x}_0 \notin \Omega$ reach the feasible set Ω in finite time and remain in it thereafter. This part of the trajectory is referred to as the *reaching phase*.

2. Trajectories inside the feasible set converge asymptotically to the solution set, and this part of the trajectory is referred to as the *convergence phase*.

The analysis of the GDS (4.71) in these two phases is now described briefly. Using the generalized Persidskii theorem, analysis of the reaching phase is simple. The main ideas of convergence phase proofs, which are more intricate, are only outlined.

Reaching phase analysis of a discontinuous GDS

Here we define $\mathbf{r} := \mathbf{A}\mathbf{x} - \mathbf{b}$, so that $\dot{\mathbf{r}} = \mathbf{A}\dot{\mathbf{x}}$ and, from (4.71), it follows that

$$\dot{\mathbf{r}} = -k_2 \mathbf{A}\mathbf{A}^T \text{uhsgn}(\mathbf{r}). \tag{4.73}$$

Convergence of the trajectories to the feasible set $\mathbf{r} \leq \mathbf{0}$ is shown using a Persidskii-type Liapunov function:

$$V_{rp}(\mathbf{r}) := \sum_{i=1}^{m} \max\{\mathbf{a}_i^T \mathbf{x} - b_i, 0\} = \sum_{i=1}^{m} \int_0^{r_i} \text{uhsgn}(\tau) d\tau. \tag{4.74}$$

In fact, it will be shown that $\mathbf{r} \to \mathbf{0}$ for any initial condition. Evaluating the time derivative of the Liapunov function (4.74) along the trajectories of (4.73) gives

$$\dot{V}_{rp} = -k_2 \text{uhsgn}(\mathbf{r})^T \mathbf{A}\mathbf{A}^T \text{uhsgn}(\mathbf{r}). \tag{4.75}$$

From the full row rank assumption, it follows that $\mathbf{A}\mathbf{A}^T$ is a positive definite matrix. Thus, from (4.75) and Rayleigh's principle, it follows that

$$\dot{V}_{rp} \leq -k_2 \lambda_{\min}(\mathbf{A}\mathbf{A}^T) \|\text{uhsgn}(\mathbf{r})\|_2^2 \leq -k_2 \lambda_{\min}(\mathbf{A}\mathbf{A}^T), \tag{4.76}$$

since, in the reaching phase, $\|\text{uhsgn}(\mathbf{r})\|_2^2 \geq 1$. Integration of the differential inequality (4.76) gives

$$V_{rp}(t) \leq V(0) - k_2 \lambda_{\min}(\mathbf{A}\mathbf{A}^T)t, \tag{4.77}$$

from which the finite-time estimate for $V(\cdot)$ to become zero follows:

$$t_z \leq V_{rp}(0)/k_2 \lambda_{\min}(\mathbf{A}\mathbf{A}^T). \tag{4.78}$$

Since $V_{rp}(\cdot)$ in (4.74) can clearly be interpreted as a measure of the distance in $\mathbf{r}$-coordinates to the feasible set Ω, this means that $\mathbf{r} \to \mathbf{0}$ in finite time, which is less than or equal to t_z. Note that the parameter k_2 can be used to adjust the finite-time estimate: A larger value of k_2 speeds up the reaching phase.

The observant reader will notice that this proof is isomorphic to that of Theorem 4.2 and in fact, in view of this, has been presented in less detail.

Convergence phase analysis of a discontinuous GDS

When the trajectories of (4.71) are confined to the feasible set, a Liapunov function can be written as

$$V_{cp}(\mathbf{x}) := f(\mathbf{x}) - f^*, \tag{4.79}$$

where f^* is the optimal value of f for the optimization problem (4.67). Note that $V_{cp}(\mathbf{x}) = 0$ if $\mathbf{x} \in \Gamma$, and $V_{cp}(\mathbf{x}) > 0$ if $\mathbf{x} \in \Omega \backslash \Gamma$.

The essential steps of the proof are to show that

1. the Liapunov function V_{cp} has negative time derivative for all $\mathbf{x} \notin \Gamma$ and, moreover, its derivative becomes zero only on the set of minimizers Γ;

2. the Liapunov function, thought of as the distance between $\mathbf{x}(t)$ and the set of minimizers Γ, satisfies the property $\lim_{t\to\infty} V_{cp}(\mathbf{x}(t)) = 0$.

The first step is easy to show when the trajectory is confined to the interior of the feasible set. When the trajectory is confined to the boundary of the feasible set, a little more work and the use of the KKT condition allow the proof of the first step. The second step is similar: Its proof is also divided into two parts, one part corresponding to trajectories in a certain neighborhood of the origin, intersected with the interior of the feasible set, and the other to trajectories in the same neighborhood intersected with the boundary of the feasible set. For all the details of this analysis, the reader is referred to [Żak03, p. 348–355].

Putting the reaching and convergence phase results together, the following theorem is proved.

Theorem 4.8. *Every trajectory of* (4.71), *from all initial conditions* $\mathbf{x}(0)$, *converges to a solution of problem* (4.67).

The proofs outlined above have several features that are worthy of note. First, conditions that possibly involve the controls (i.e., penalty parameters) ensure that the trajectories enter the feasible set, when starting from infeasible initial conditions. Furthermore, these reaching phase conditions guarantee that once trajectories enter the feasible set, they remain

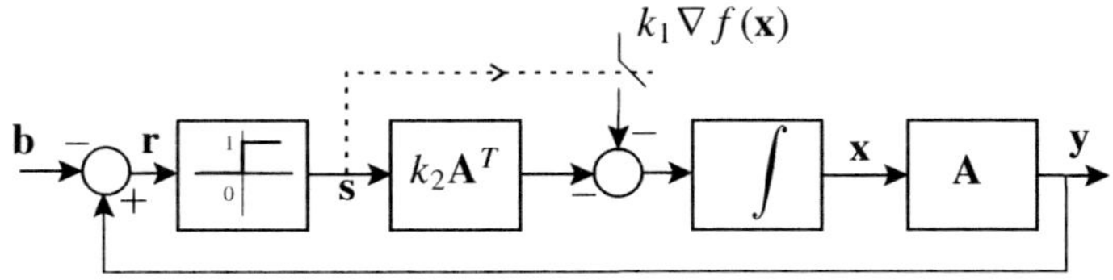

Figure 4.11. *Representation of* (4.71) *as a control system. The dotted line in the figure indicates that the switch on the input* $k_1 \nabla f(\mathbf{x})$ *is turned on only when the output* $\mathbf{s}$ *of the first block is the zero vector* $\mathbf{0} \in \mathbb{R}^m$*, i.e., when* $\mathbf{x}$ *is within the feasible region* Ω.

inside it. Second, in the interior of the feasible set, the GDS is a smooth GDS, which means that its trajectories tend to a zero of the gradient field. Some additional technicalities arise when trajectories hit boundaries of the feasible set, but it can be shown that, when this happens, trajectories tend to a point where the KKT conditions are satisfied, thus finding a solution to the constrained convex optimization problem.

In the remainder of this chapter, the technicalities mentioned in this section will be partially omitted and only the reaching phase conditions, which differ from application to application, will be given, since, once these conditions are satisfied, convergence to the desired solution is ensured.

Finally, in order to emphasize one of the important features of this approach, it should be observed that switching off the objective function gradient outside the feasible set has the advantage of simplifying the reaching phase dynamics (4.73). On the other hand, from the point of view of implementation, this switching requires additional logic that checks the feasibility of the current point (see Figure 4.11). The next section shows that, in the case of linear programming, this additional switching logic can be avoided, at the cost of an additional analysis and a condition to ensure reaching phase convergence.

4.4 GDSs that Solve Linear Programming Problems

This section applies the general GDS approach to convex programming problems, developed in the previous section, to the particular case of linear programming problems, while the next section treats the case of quadratic programming in a similar way. There is one difference, with respect to the gradient dynamical system (4.72), in which the gradient of the objective function is switched off outside the feasible region. In this section, as well as in the remainder of this chapter, this switching will not be used; i.e., the gradient of the penalized objective function is taken "as is," without any additional switching. This adds an additional term to the time derivative of the reaching phase Liapunov function that needs to be accounted for. On the other hand, the implementation as a control system does not require the additional switching logic used in (4.72).

In order to introduce this additional analysis in as simple a manner as possible, a couple of one-dimensional examples are given below.

Example 4.9. Consider the following linear programming problem in a single variable, with a single inequality constraint:

Table 4.2. *Solutions of a linear programming problem* (4.80) *for different possible combinations of signs of the parameters a, b, c. For a problem with a bounded minimum value, the minimizing argument is always $x = (b/a)$.*

a	b	Feasible region	Minimum value of cx when $c > 0$	Minimum value of cx when $c < 0$
> 0	> 0	$x \geq (b/a)$	(cb/a)	$-\infty$
> 0	< 0	$x \geq (b/a)$	(cb/a)	$-\infty$
< 0	> 0	$x \leq (b/a)$	$-\infty$	(cb/a)
< 0	< 0	$x \leq (b/a)$	$-\infty$	(cb/a)

$$\begin{aligned} &\min_x cx \\ &\text{s.t. } ax - b \geq 0, \end{aligned} \tag{4.80}$$

where $a, b, c \in \mathbb{R}$. Table 4.2 shows the solutions of the problem for the different possible combinations of signs of a, b, c, which are all assumed to be nonzero, in order to avoid trivial special cases.

To fix ideas, consider the problem in the first row of the table with a, b, and c all positive. The problem will now be solved using the penalty function method. Calling the positive penalty parameter $k > 0$, the penalized objective function or associated energy function is

$$E_k(x) := cx - k \min(0, ax - b). \tag{4.81}$$

Defining

$$r = ax - b, \tag{4.82}$$

this penalized objective function will now be minimized using gradient (i.e., derivative) descent, i.e., following the trajectories of the dynamical system

$$\dot{x} = -\frac{\mathrm{d}E_k(x)}{\mathrm{d}x} = -(c + ka \,\mathrm{hsgn}(r)). \tag{4.83}$$

Now observe that the time derivative of the energy function $E_k(x)$ is

$$\dot{E}_k = \frac{\mathrm{d}E_k}{\mathrm{d}x}\dot{x} = -\left(\frac{\mathrm{d}E_k}{\mathrm{d}x}\right)^2 = -(c + ka \,\mathrm{hsgn}(r))^2 \leq 0. \tag{4.84}$$

This equation shows that $\mathrm{d}E_k/\mathrm{d}t < 0$ almost everywhere, except when $k = (c/a)$ and $r = 0$, where $\mathrm{hsgn}(r)$ could assume the value -1.

Thus, the value of E_k decreases almost everywhere along trajectories of (4.83) which therefore converge to stationary points of E_k, i.e., points where $\mathrm{d}E_k/\mathrm{d}x = 0$.

As mentioned in section 4.3, from a basic theorem on exact penalty functions, it is known that if there exists a point x^* that satisfies the second-order sufficiency conditions for a local minimum of (4.80), then there exists a penalty parameter k for which x^* is also a local minimum of (4.81).

Since $\dot{r} = a\dot{x}$, (4.83) can be written in r-coordinates as

$$\dot{r} = -a(c + ka\,\mathrm{hsgn}(r)). \tag{4.85}$$

We first show that the phase space (which, for this example, is the real line $\mathbb{R}$) analysis of this system leads to a global stability condition for the equilibrium $r = 0$ of (4.85). Given the definition of hsgn$(\cdot)$, (4.85) can be written equivalently as

$$\dot{r} = \begin{cases} -ac & \text{if } r > 0, \\ -a(c - ka) & \text{if } r < 0. \end{cases} \tag{4.86}$$

Clearly, if

$$ka - c > 0, \tag{4.87}$$

then $\dot{r}$ is positive for $r < 0$ and negative for $r > 0$; i.e., $r(t)$ converges globally to the origin ($r = 0$) for all initial conditions. This confirms the gradient dynamical system analysis above, providing a condition for global stability of $r = 0$, i.e., of the solution $x = b/a$. Note that the condition $r\dot{r} < 0$ is satisfied; this will be identified, for higher dimensional systems encountered in what follows, as a condition that ensures the existence of a sliding mode (see Chapter 1).

This stability result is now investigated carrying out a Liapunov function analysis of this system, using the following nonsmooth Persidskii-type Liapunov function:

$$V_{rp}(r) = \int_0^r \mathrm{hsgn}(\tau)d\tau. \tag{4.88}$$

Notice that $V_{rp}(r) > 0$ whenever $r < 0$, $V_{rp}(r) = 0$ when $r \geq 0$, and V_{rp} is radially unbounded in the region $r < 0$. From this, it follows that this Liapunov function serves (only) to show that the feasible region ($r \geq 0$) is globally attractive, in the sense that all trajectories starting from initial conditions outside this region enter it, in finite time. In other words, the Liapunov function V_{rp} is suitable for reaching phase analysis.

The time derivative of (4.88) along the trajectories of (4.85) can be written as

$$\dot{V}_{rp} = -ac\,\mathrm{hsgn}(r) - ka^2\,\mathrm{hsgn}(r)^2. \tag{4.89}$$

The first term on the right-hand side of (4.89) attains a maximum value of ac (which is positive under the assumptions made), while the second term attains a minimum of ka^2, which shows that $\dot{V}_{rp} < 0$ under the condition $ka > c$, coinciding with the conclusion obtained from the direct phase space analysis above.

Equation (4.89) also allows us to understand the concept of *finite-time convergence*. Notice that

$$\dot{V}_{rp} \leq -\beta, \tag{4.90}$$

where

$$\beta = \begin{cases} 0 & \text{if } r > 0, \\ |ac - ka^2| & \text{if } r < 0. \end{cases} \tag{4.91}$$

In the reaching phase, when $r < 0$ and consequently $\beta > 0$, we can integrate the differential inequality (4.90) to get

$$V_{rp}(t) \leq -\beta t + \gamma, \tag{4.92}$$

where γ is an integration constant. This means that the value of the Liapunov function is bounded above by $-\beta t + \gamma$ (a straight line of negative slope), which will attain the value zero at the finite value $t_z = \gamma/\beta$. In other words, the Liapunov function, and hence the residue, must converge to zero in finite time, less than (γ/β) units of time. Whenever the use of variable structure control leads to a differential inequality of the type (4.90), finite-time convergence of the trajectories of the associated dynamical system can be concluded.

Finally, it is instructive to analyze the consequence of condition (4.87) on the shape of the energy function $E_k(\cdot)$. Observe that the energy function can be written as

$$E_k(x) = \begin{cases} cx & \text{if } x \geq (b/a), \\ (c - ka)x + kb & \text{if } x < (b/a), \end{cases} \tag{4.93}$$

which is clearly a convex function (V-shaped) with a minimum (i.e., (cb/a) at $x = (b/a)$), under the condition (4.87).

Most of the salient features of this example generalize to other situations in the rest of the chapter.

Example 4.10. Consider the following standard form linear programming problem in a single variable, with an inequality constraint as well as a nonnegativity constraint:

$$\begin{aligned} &\min_x cx \\ &\text{s.t. } x \geq 0, \ ax - b \geq 0, \end{aligned} \tag{4.94}$$

where $a, b, c \in \mathbb{R}$.

Assuming $a > 0$, the inequality constraint gives $x \geq (b/a)$. From the nonnegativity constraint on x, it follows that a and b must have the same sign, so b must be positive as well. If c is also assumed positive, then the solution of (4.94) is (cb/a), attained for the argument $x = (b/a)$.

For the sake of illustration, this problem is also solved using the penalty function method, assuming a, b, and c are all positive. Denoting the positive penalty parameters as k_1, k_2, the penalized objective function is

$$E_k(x) := cx - k_1 \min(0, x) - k_2 \min(0, ax - b). \tag{4.95}$$

Note that, for $k_1, k_2 > 0$, the second and third terms on the right-hand side of (4.95) are chosen such that, whenever constraints are violated, a positive quantity is added to the objective function cx.

Defining $r = ax - b$ as in (4.82), this penalized scalar objective function will now be minimized using "gradient" descent, i.e., using the scalar dynamical system:

$$\dot{x} = -\frac{\mathrm{d}E_k(x)}{\mathrm{d}x} = -[c - k_1(-\mathrm{hsgn}(x)) - k_2 a\,(-\mathrm{hsgn}(r))]. \tag{4.96}$$

Since the right-hand side of (4.96) contains both r and x, and $\dot{r} = a\dot{x}$, an augmented system can be written as follows:

$$\begin{aligned} \dot{x} &= -c - k_1 \mathrm{hsgn}(x) - ak_2 \mathrm{hsgn}(r), \\ \dot{r} &= -ac - ak_1 \mathrm{hsgn}(x) - a^2 k_2 \mathrm{hsgn}(r). \end{aligned} \tag{4.97}$$

For this augmented system of differential equations, it should be proved that both r and x eventually become nonnegative (meaning that the constraints are satisfied) and that the "equilibrium" $((b/a), 0)$ is eventually reached. It will be shown, in fact, that the point $((b/a), 0)$ is an equilibrium of the sliding mode type and that convergence to the feasible region $(x \geq 0, r \geq 0)$ occurs even if a consistent initial condition $r_0 = ax_0 - b$ is not satisfied.

Equation (4.97) can be written in vector form by defining the vector $\mathbf{z} = (x, r)$ as follows:

$$\dot{\mathbf{z}} = -\begin{bmatrix} c \\ ac \end{bmatrix} - \begin{bmatrix} 1 & a \\ a & a^2 \end{bmatrix} \begin{bmatrix} k_1 & 0 \\ 0 & k_2 \end{bmatrix} \begin{bmatrix} \text{hsgn}(x) \\ \text{hsgn}(r) \end{bmatrix}. \tag{4.98}$$

Introducing the notation

$$\bar{\mathbf{c}} = \begin{bmatrix} c \\ ac \end{bmatrix}, \quad \mathbf{S} = \begin{bmatrix} 1 & a \\ a & a^2 \end{bmatrix}, \quad \mathbf{K} = \text{diag}\,(k_1, k_2), \quad \mathbf{f}(\mathbf{z}) = \begin{bmatrix} \text{hsgn}(x) \\ \text{hsgn}(r) \end{bmatrix},$$

equation (4.98) can be written compactly as

$$\dot{\mathbf{z}} = -\bar{\mathbf{c}} - \mathbf{SKf}(\mathbf{z}). \tag{4.99}$$

Let $\mathbf{k} := (k_1, k_2)$ and

$$\mathbf{p}(\mathbf{z}) := \left(\int_0^x \text{hsgn}(\tau) \mathrm{d}\tau, \int_0^r \text{hsgn}(\tau) \mathrm{d}\tau \right). \tag{4.100}$$

A reaching phase candidate Persidskii-type Liapunov function is now defined as follows:

$$V_{rp}(\mathbf{z}) = V_{rp}((x, r)) := \mathbf{k}^T \mathbf{p}(\mathbf{z}). \tag{4.101}$$

Note that this Liapunov function is positive definite with respect to the feasible region in the sense that $V_{rp}(\mathbf{z}) > 0$ for all $\mathbf{z}$ outside the feasible region, i.e., when $x < 0$ or $r < 0$. On the other hand, if $\mathbf{z}$ is inside the feasible region, then both integrals in (4.100) evaluate to zero and $V_{rp}(\mathbf{z}) = 0$; this is the reason it is referred to as a reaching phase Liapunov function. Proceeding with the analysis, the time derivative of $V_{rp}(\cdot)$ along the trajectories is

$$\dot{V}_{rp} = (\nabla V_{rp})^T \dot{\mathbf{z}} = -\mathbf{f}(\mathbf{z})^T \mathbf{K}\bar{\mathbf{c}} - \mathbf{f}(\mathbf{z})^T \mathbf{KSKf}(\mathbf{z}). \tag{4.102}$$

Observe that $-\dot{V}_{rp}$ is the sum of a term $\mathbf{f}(\mathbf{z})^T \mathbf{K}\bar{\mathbf{c}}$, which is linear in $\mathbf{f}(\mathbf{z})$, and a term $\mathbf{f}(\mathbf{z})^T \mathbf{KSKf}(\mathbf{z})$, which is a quadratic form in $\mathbf{f}(\mathbf{z})$ in the positive semidefinite matrix $\mathbf{KSK}$. Our task is to find conditions that ensure that, in the reaching phase, the quadratic term dominates the linear term, so that a standard Liapunov argument will allow the conclusion that all trajectories of (4.96) enter the feasible region. The Liapunov analysis is carried out with the help of a phase line analysis.

Note that the vector $\mathbf{Kf}(\mathbf{z})$ assumes different values (see Figure 4.12) according to the region of the phase space: $(-k_1, -k_2)$ (region I); $(0, -k_2)$ (region II); $(0, k_2)$ (region III). We tabulate the values of the two terms on the right-hand side of (4.102) for these values of $\mathbf{Kf}(\mathbf{z})$, in order to obtain the conditions for $\dot{V} < 0$.

From Table 4.3, we conclude that the conditions on the penalty parameters to guarantee that the feasible region is reached are

$$ak_2 > c, \qquad k_1 + ak_2 > c. \tag{4.103}$$

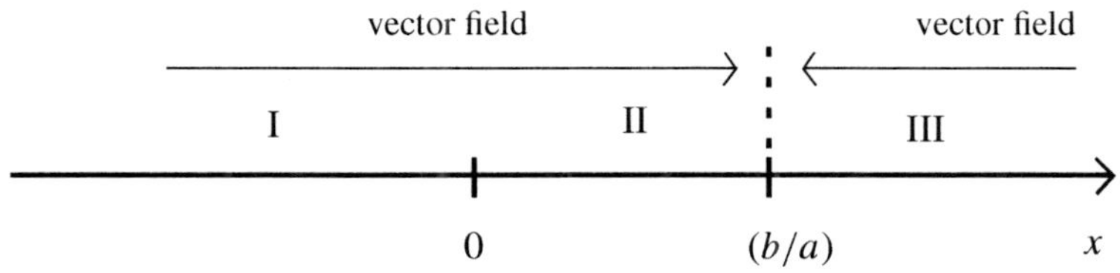

Figure 4.12. *Phase line for dynamical system* (4.96), *under the conditions* (4.103), *obtained from Table* 4.3.

Table 4.3. *Analysis of the Liapunov function. In the last row, it is assumed that* $(k_1 + ak_2) > 0$.

	Region I	Region II	Region III
$\mathbf{z}$	$(-k_1, -k_2)$	$(0, -k_2)$	$(0, k_2)$
$-\mathbf{f}(\mathbf{z})^T\mathbf{K}\bar{\mathbf{c}}$	$(k_1 + ak_2)c$	k_2ac	$-k_2ac$
$-\mathbf{f}(\mathbf{z})^T\mathbf{KSKf}(\mathbf{z})$	$-(k_1 + ak_2)^2$	$-k_2^2a^2$	$-k_2^2a^2$
Condition for $\dot{V} < 0$	$(k_1 + ak_2) > c$	$ak_2 > c$	None

Since $k_1 > 0$, the first condition is the only one required. Note that it also implies that the vector $\mathbf{Kf}(\mathbf{z})$ in not in the null space of $\mathbf{S}$.

An analysis of the dynamics of (4.96) on the phase line (Figure 4.12) shows that conditions (4.103) actually guarantee global convergence to the point $x^* = (b/a)$ which is the solution of (4.94). This shows that, in this simple one-dimensional example, the Liapunov function analysis actually gives conditions that are necessary and sufficient (i.e., not conservative) and ensure not only reaching phase convergence, but global convergence as well.

The strategy used in this example, namely, finding conditions that make the quadratic term dominate the linear term, will be followed in this section for general linear programming problems, although in some cases, it is convenient to make some majorizations instead of a complete combinatorial analysis of the type made in Table 4.3.

Note also that, with the choices of k_1, k_2 in (4.103), the penalized objective function (4.95) has a unique minimum at $x = (b/a)$, showing that, as expected, the penalty function is exact.

4.4.1 GDSs as linear programming solvers

It is well known that linear programming problems can be put into different, yet equivalent, formulations involving only inequality constraints or both equality and inequality constraints. We start with the first type of formulation, generalizing Example 4.9.

The linear programming problem in so-called canonical form I is defined as follows.

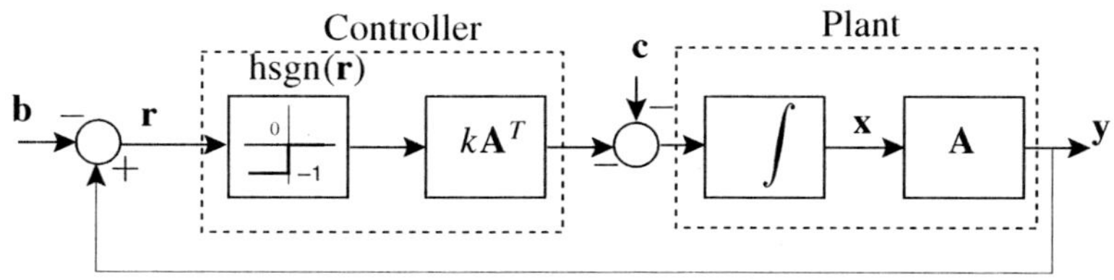

Figure 4.13. *Control system structure that solves the linear programming problem in canonical form I.*

Given $\mathbf{c}, \mathbf{x} \in \mathbb{R}^n$, $\mathbf{A} \in \mathbb{R}^{m\times n}$, with full row rank ($\text{rank}(\mathbf{A}) = m$) and $\mathbf{b} \in \mathbb{R}^m$:

$$\begin{aligned} &\min_{\mathbf{x}} \mathbf{c}^T\mathbf{x} \\ &\text{s.t. } \mathbf{Ax} - \mathbf{b} \geq \mathbf{0}. \end{aligned} \tag{4.104}$$

The following computational energy function (i.e., exact penalty function) is associated with this problem:

$$E(\mathbf{x}, k) = \mathbf{c}^T\mathbf{x} + k\ \min\{\mathbf{0}, \mathbf{Ax} - \mathbf{b}\} \quad (k > 0). \tag{4.105}$$

This energy function is minimized by the GDS $\dot{\mathbf{x}} = -\nabla E(\mathbf{x}, k)$, which can be written as

$$\dot{\mathbf{x}} = -\mathbf{c} - k\,\mathbf{A}^T\text{hsgn}(\mathbf{r}), \tag{4.106}$$

where $\mathbf{r} := \mathbf{Ax} - \mathbf{b}$.

Description of sliding mode equilibrium

The dynamical system (4.106) does not have an equilibrium solution in the classical sense because, when the trajectories enter the feasible region, the GDS (4.106) reduces to $\dot{\mathbf{x}} = -\mathbf{c}$. The way to understand this is to realize that solutions to the LP problem (4.104) occur at vertices of the feasible set. In fact, the trajectories of the dynamical system can be thought of as follows. The first term $\dot{\mathbf{x}} = -\mathbf{c}$ of (4.106) represents the movement of vector $\mathbf{x}$ in the direction of the vector $-\mathbf{c}$, which is the negative of the gradient of the objective function $\mathbf{f}(\mathbf{x}) = \mathbf{c}^T\mathbf{x}$. In other words, it represents gradient descent minimizing the objective function. The second term, to be understood as the control term (Figure 4.13) $-k\mathbf{A}^T\text{hsgn}(\mathbf{Ax} - \mathbf{b})$, "switches on" every time the vector leaves the feasible set and its function is to push the trajectory back into the feasible set. In a linear programming problem that has a unique solution, the trajectory defined in this manner ends up at a point (on the boundary of the feasible set) where the point $\mathbf{x}$ can retreat no further in the direction $-\mathbf{c}$ without leaving the feasible region. At this point the Karush–Kuhn–Tucker conditions for an optimum are satisfied. This point may be thought of as a dynamic equilibrium and is referred to in the control literature as a sliding mode equilibrium, in reference to the fact that such an equilibrium point may be approached by trajectories that "slide" along a (boundary) surface, being forced to display this behavior ("sliding mode") since the vector field on both sides of the boundary points towards it.

Convergence conditions for the linear programming problem in canonical form I

The convergence analysis is carried out by representing (4.106) in a Persidskii-type form obtained, in strict analogy with Example 4.9, by premultiplication of the system equation (4.106) by matrix $\mathbf{A}$, yielding

$$\dot{\mathbf{r}} = -\mathbf{Ac} - k\mathbf{AA}^T \text{hsgn}(\mathbf{r}). \tag{4.107}$$

The general result corresponding to the one obtained for one-dimensional Example 4.9 is as follows.

Theorem 4.11. *If the control gain (i.e., penalty parameter) k is chosen such that*

$$k > \frac{\|\mathbf{Ac}\|_1}{\lambda_{\min}(\mathbf{AA}^T)}, \tag{4.108}$$

then all trajectories of (4.106) *converge to the solution set of problem* (4.104).

Proof. As discussed in section 4.3.1, only the reaching phase is analyzed here. Consider the usual reaching phase diagonal Persidskii-type CLF that is associated with the system

$$V_{rp}(\mathbf{r}) = \sum_i \int_0^{r_i} \text{hsgn}(\tau)\, d\tau. \tag{4.109}$$

The time derivative of (4.109) along the trajectories of the system (4.107) is

$$\dot{V}_{rp} = \nabla V_{rp}^T\, \dot{\mathbf{r}} = -\text{hsgn}(\mathbf{r})^T \mathbf{Ac} - k\ \text{hsgn}(\mathbf{r})^T \mathbf{AA}^T \text{hsgn}(\mathbf{r}). \tag{4.110}$$

The largest value of the first term on the right-hand side is clearly $\|\mathbf{Ac}\|_1$, attained when all components of hsgn($\mathbf{r}$) take the value -1. On the other hand, in the reaching phase, at least one component of the vector hsgn($\mathbf{r}$) is equal to -1. Furthermore, the full rank assumption on $\mathbf{A}$ implies that the matrix $\mathbf{AA}^T$ is positive definite, so that the second term in (4.110) attains a minimum value of $k\,\lambda_{\min}(\mathbf{AA}^T)$, using Rayleigh's quotient estimate of the minimum value of the quadratic form $\text{hsgn}(\mathbf{r})^T \mathbf{AA}^T \text{hsgn}(\mathbf{r})$. Substituting these worst-case values in (4.110) proves that the condition (4.108) ensures $\dot{V}_{rp} < 0$, thus proving the theorem. □

Convergence results for the linear programming problem in canonical form II

The linear programming problem in canonical form II is defined as

$$\begin{aligned} &\min_{\mathbf{x}} \mathbf{c}^T\mathbf{x} \\ &\text{s.t. } \mathbf{x} \geq \mathbf{0}, \quad \mathbf{Ax} - \mathbf{b} \geq \mathbf{0}, \end{aligned} \tag{4.111}$$

where $\mathbf{A} \in \mathbb{R}^{m\times n}$, $\mathbf{c} \in \mathbb{R}^n$, $\mathbf{x} \in \mathbb{R}^n$, and $\mathbf{b} \in \mathbb{R}^m$, with $m < n$ and rank $(\mathbf{A}) = m$.

The associated computational energy function is

$$E(\mathbf{x}, k_1, k_2) = \mathbf{c}^T\mathbf{x} - k_1 \sum_{j=1}^{n} \min(0, x_j) - k_2 \sum_{i=1}^{m} \min(0, r_i), \tag{4.112}$$

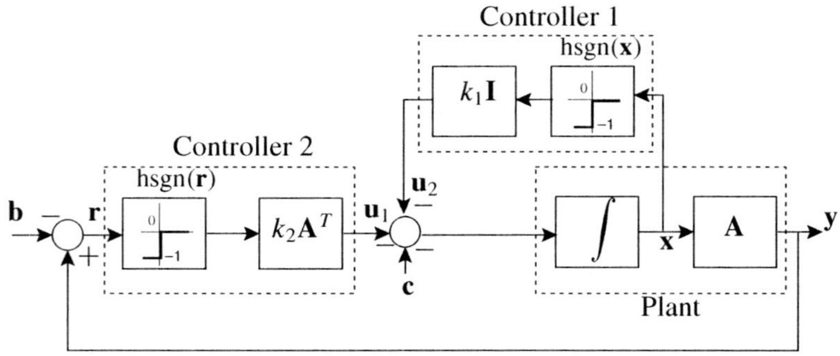

Figure 4.14. *Control system structure that solves the linear programming problem in canonical form* II.

where $r_i = \mathbf{a}_i^T\mathbf{x} - b_i$, the vectors $\mathbf{a}_i^T$ are the rows of matrix $\mathbf{A}$, and k_1, k_2 are control gains (i.e., penalty parameters) that will be determined by a CLF (4.120) that is actually the sum of the second and third terms on the right-hand side of (4.112).

The GDS that minimizes $E(\mathbf{x}, k_1, k_2)$ is

$$\dot{\mathbf{x}} = -\mathbf{c} - k_1\text{hsgn}(\mathbf{x}) - k_2\mathbf{A}^T\text{hsgn}(\mathbf{r}). \tag{4.113}$$

System (4.113) can be regarded from a control perspective (Figure 4.14) by writing it in the form

$$\dot{\mathbf{x}} = -\mathbf{c} + \mathbf{u}_1 + \mathbf{u}_2,$$

where $\mathbf{u}_1 = -k_1\text{hsgn}(\mathbf{x})$ represents a control that forces the trajectories into the set $\{\mathbf{x} : \mathbf{x} \geq 0\}$ and $\mathbf{u}_2 = -k_2\mathbf{A}^T\text{hsgn}(\mathbf{Ax} - \mathbf{b})$ represents a control that forces the trajectories into the set $\{\mathbf{x} : \mathbf{Ax} - \mathbf{b} \geq \mathbf{0}\}$ so that, overall, trajectories are maintained in the intersection set $\{\mathbf{x} : \mathbf{x} \geq 0\} \cap \{\mathbf{x} : \mathbf{Ax} - \mathbf{b} \geq \mathbf{0}\}$.

Premultiplying (4.113) by $\mathbf{A}$ yields

$$\dot{\mathbf{r}} = -\mathbf{Ac} - k_1\mathbf{A}\text{hsgn}(\mathbf{x}) - k_2\mathbf{AA}^T\text{hsgn}(\mathbf{r}). \tag{4.114}$$

In vector notation, (4.113) and (4.114) are written as

$$\dot{\mathbf{z}} = -\bar{\mathbf{c}} - \mathbf{SKf}(\mathbf{z}), \tag{4.115}$$

where $\dot{\mathbf{z}} := (\dot{\mathbf{x}}, \dot{\mathbf{r}})$, $\bar{\mathbf{c}} := (\mathbf{c}, \mathbf{Ac})$, $\mathbf{f}(\mathbf{z}) := (\text{hsgn}(\mathbf{x}), \text{hsgn}(\mathbf{r}))$, and

$$\mathbf{S} := \begin{bmatrix} \mathbf{I}_n & \mathbf{A}^T \\ \mathbf{A} & \mathbf{AA}^T \end{bmatrix}, \quad \mathbf{K} := \begin{bmatrix} k_1\mathbf{I}_n & \mathbf{0} \\ \mathbf{0} & k_2\mathbf{I}_m \end{bmatrix}. \tag{4.116}$$

Before stating a convergence result for this class of system, a lemma on the properties of the matrix $\mathbf{S}$ is needed.

Lemma 4.12. *The matrix* $\mathbf{S} \in \mathbb{R}^{(n+m)\times(n+m)}$ *has n positive eigenvalues, which are the eigenvalues of the symmetric positive definite matrix* $\mathbf{I} + \mathbf{A}^T\mathbf{A}$ *and m zero eigenvalues. The eigenvectors corresponding to the zero eigenvalues have the form* $(-\mathbf{A}^T\mathbf{y}, \mathbf{y})$, $\mathbf{y} \in \mathbb{R}^m$.

Proof. Writing the equation $\mathbf{Sv} = \lambda\mathbf{v}$, where $\mathbf{v} = (\mathbf{x}, \mathbf{y})$, yields the equations

$$\begin{aligned} \mathbf{x} + \mathbf{A}^T\mathbf{y} &= \lambda\mathbf{x}, \\ \mathbf{Ax} + \mathbf{AA}^T\mathbf{y} &= \lambda\mathbf{y}, \end{aligned} \tag{4.117}$$

whence, by elimination, it follows that $\lambda(\mathbf{y} - \mathbf{Ax}) = \mathbf{0}$ for all eigenvalues and eigenvectors. Specifically, for nonzero eigenvalues, this equation can be satisfied only if $\mathbf{y} = \mathbf{Ax}$. Substituting $\mathbf{y}$ by $\mathbf{Ax}$ in the first equation in (4.117) gives the equation $(\mathbf{I}_n + \mathbf{A}^T\mathbf{A})\mathbf{x} = \lambda\mathbf{x}$, proving the claim about the positive eigenvalues of $\mathbf{S}$. For the zero eigenvalues, it is enough to note that $\mathbf{v}$ is in the nullspace of $\mathbf{S}$ if and only if $\mathbf{v} = (-\mathbf{A}^T\mathbf{y}, \mathbf{y})$. □

The general convergence result corresponding to the one obtained for one-dimensional Example 4.10 is as follows.

Theorem 4.13. *If the control gains k_1 and k_2 satisfy the conditions*

$$(\min\{k_1, k_2\})^2 > \max\{k_1, k_2\}\|\bar{\mathbf{c}}\|_1 \tag{4.118}$$

and

$$\mathbf{Kf}(\mathbf{z}) = \begin{bmatrix} k_1\mathrm{hsgn}(\mathbf{x}) \\ k_2\mathrm{hsgn}(\mathbf{r}) \end{bmatrix} \neq \begin{bmatrix} -\mathbf{A}^T\mathbf{y} \\ \mathbf{y} \end{bmatrix} \textit{ for all } \mathbf{x} \in \mathbb{R}^n, \textit{ for all } \mathbf{r}, \mathbf{y} \in \mathbb{R}^m, \tag{4.119}$$

then all trajectories of (4.113) *converge to the solution set of problem* (4.111).

Proof. Let $\mathbf{k}$ denote the vector in $\mathbb{R}^{n+m}$ whose entries are the diagonal entries of the matrix $\mathbf{K}$ in (4.116) and let the Persidskii-type CLF be defined as

$$V_{rp} := \mathbf{k}^T\left(\int_0^{x_1}\mathrm{hsgn}(\tau)d\tau, \ldots, \int_0^{x_n}\mathrm{hsgn}(\tau)d\tau, \int_0^{r_1}\mathrm{hsgn}(\tau)d\tau, \ldots, \int_0^{r_m}\mathrm{hsgn}(\tau)d\tau\right). \tag{4.120}$$

As mentioned before, this CLF is exactly the sum of the second and third terms on the right-hand side of (4.112). Evaluating the time derivative of V_{rp} along the trajectories of (4.115) gives

$$\dot{V}_{rp} = -\mathbf{f}(\mathbf{z})^T\mathbf{K}\bar{\mathbf{c}} - \mathbf{f}(\mathbf{z})^T\mathbf{KSKf}(\mathbf{z}). \tag{4.121}$$

The first term on the right-hand side of (4.121) is clearly majorized by the expression $\max\{k_1, k_2\}\|\bar{\mathbf{c}}\|_1$. Observe that, by Lemma 4.12, condition (4.119) ensures that $\mathbf{Kf}(\mathbf{z}) \notin \mathcal{N}(\mathbf{S})$, where $\mathcal{N}(\mathbf{S})$ denotes the null-space of the matrix $\mathbf{S}$. Also, by Lemma 4.12, the smallest positive eigenvalue of $\mathbf{S}$ is 1, attained by a vector $\mathbf{Kf}(\mathbf{z})$. Since, in the reaching phase, at least one element of the vector $\mathbf{Kf}(\mathbf{z})$ is nonzero or, more specifically, equal to k_1 or k_2, the second term is lower bounded (by the Rayleigh quotient estimate) as $(\min\{k_1, k_2\})^2\lambda_{\mathrm{min,nz}}(\mathbf{S}) = (\min\{k_1, k_2\})^2$, since, by Lemma 4.12, the smallest nonzero eigenvalue of $\mathbf{S}$, denoted $\lambda_{\mathrm{min,nz}}(\mathbf{S})$, is 1. Substituting these bounds in (4.121) shows that, under the condition (4.118), $\dot{V}_{rp} < 0$, proving the theorem. □

Some comments on the conditions of Theorem 4.13 are appropriate. Given the universal quantifiers that occur in it, condition (4.119) looks difficult to satisfy, but it can be written as $k_1\mathrm{hsgn}(\mathbf{x}) \neq -k_2\mathbf{A}^T\mathrm{hsgn}(\mathbf{r})$. Now recall that, for all $\mathbf{x}$, $\mathbf{r}$, all components of the

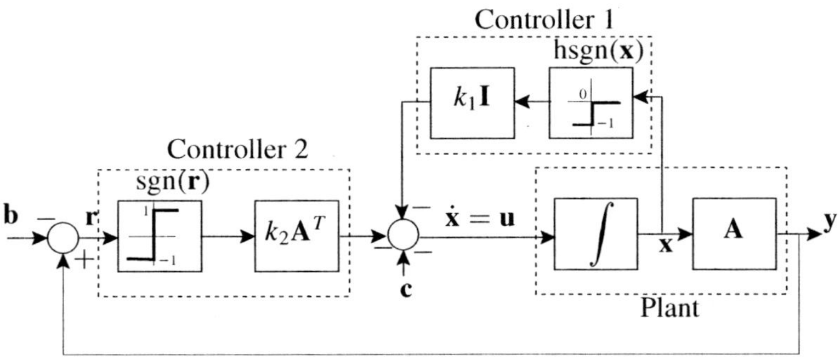

Figure 4.15. *Control system structure that solves the linear programming problem in standard form.*

vectors hsgn($\mathbf{x}$), hsgn($\mathbf{r}$) assume the values 0 or -1, and thus there are only a finite number of choices of k_1, k_2 that must be avoided. In other words, the condition is satisfied for almost all choices of k_1 and k_2. The condition (4.118) can also be easily satisfied in practice by specifying a relationship between k_1 and k_2. For example, from (4.118), choosing $k_1 = \alpha k_2$ with $\alpha > 1$ yields the bound $k_2^2 > \alpha k_2 \|\bar{\mathbf{c}}\|_1$, i.e.,

$$k_2 > \alpha \|\bar{\mathbf{c}}\|_1. \tag{4.122}$$

Other bounds can also be derived using different majorizations, and the reader is referred to [FKB02, FKB03] for examples of these.

Convergence results for the linear programming problem in standard form

Consider the linear programming problem in standard form,

$$\begin{aligned} &\min_{\mathbf{x}} \mathbf{c}^T\mathbf{x} \\ &\text{s.t. } \mathbf{A}\mathbf{x} = \mathbf{b}, \quad \mathbf{x} \geq \mathbf{0}, \end{aligned} \tag{4.123}$$

where $\mathbf{A} \in \mathbb{R}^{m\times n}$, $m < n$, rank($\mathbf{A}$) $= m$, $\mathbf{b} \in \mathbb{R}^m$, and $\mathbf{c} \in \mathbb{R}^n$.

The associated computational energy function, where $\mathbf{r}$ denotes $\mathbf{Ax} - \mathbf{b}$, is

$$E(\mathbf{x}, k_1, k_2) = \mathbf{c}^{\mathbf{T}}\mathbf{x} - k_1 \sum_{j=1}^{m} \min(0, x_j) + k_2\|\mathbf{r}\|_1. \tag{4.124}$$

The following GDS is associated with the energy function (4.124):

$$\dot{\mathbf{x}} = -\mathbf{c} - k_1 \text{hsgn}(\mathbf{x}) - k_2\mathbf{A}^T \text{sgn}(\mathbf{r}). \tag{4.125}$$

The system block diagram is given in Figure 4.15 and is similar to the one depicted in Figure 4.14.

As in the previous cases, writing the augmented system yields

$$\begin{bmatrix} \dot{\mathbf{x}} \\ \dot{\mathbf{r}} \end{bmatrix} = -\begin{bmatrix} \mathbf{c} \\ \mathbf{Ac} \end{bmatrix} - \begin{bmatrix} \mathbf{I}_n & \mathbf{A}^T \\ \mathbf{A} & \mathbf{AA}^T \end{bmatrix} \begin{bmatrix} k_1\mathbf{I}_n & \mathbf{0} \\ \mathbf{0} & k_2\mathbf{I}_m \end{bmatrix} \begin{bmatrix} \text{hsgn}(\mathbf{x}) \\ \text{sgn}(\mathbf{r}) \end{bmatrix}. \tag{4.126}$$

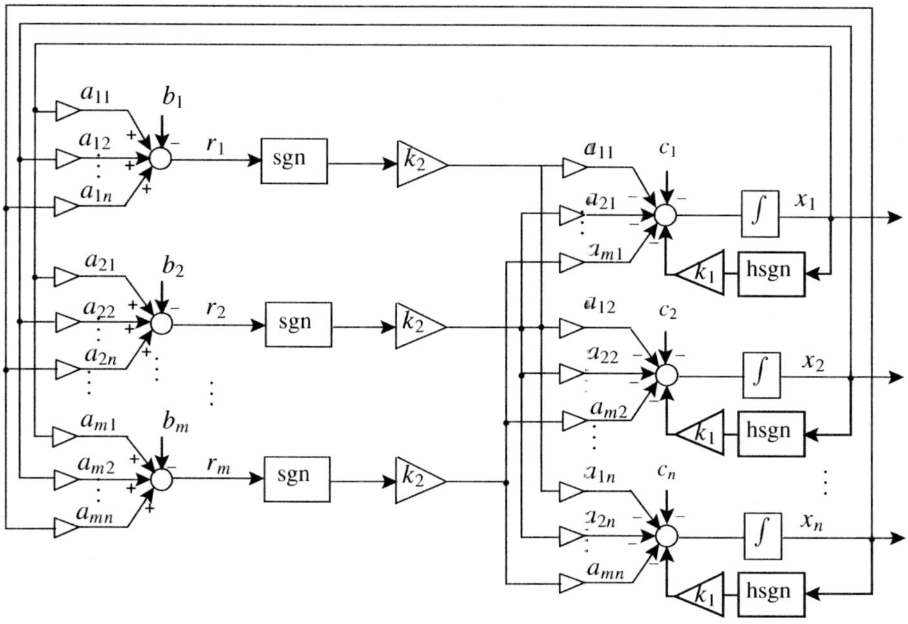

Figure 4.16. *The GDS* (4.125) *that solves standard form linear programming problems represented as a neural network.*

Consider the CLF that is associated to system (4.126):

$$V = k_1 \sum_{j=1}^{n} \int_0^{x_j} \mathrm{hsgn}(\tau)\, d\tau + k_2 \sum_{i=1}^{m} \int_0^{r_i} \mathrm{sgn}(\tau)\, d\tau. \tag{4.127}$$

The time derivative of (4.127) takes the form

$$\dot{V} = -\mathbf{f}(\mathbf{z})^T \mathbf{K}\bar{\mathbf{c}} - \mathbf{f}(\mathbf{z})^T \mathbf{K}\mathbf{S}\mathbf{K}\mathbf{f}(\mathbf{z}), \tag{4.128}$$

which has the same form as (4.121), with $\bar{\mathbf{c}}$, $\mathbf{K}$, $\mathbf{S}$ being defined as before and the only change being that $\mathbf{f}(\mathbf{z})$ is now defined as $\mathbf{f}(\mathbf{z}) := (\mathrm{hsgn}(\mathbf{x}), \mathrm{sgn}(\mathbf{r}))$. With this redefinition of $\mathbf{f}(\mathbf{z})$, it is easy to show that, for the standard form linear programming problem as well, the conditions of Theorem 4.13 ensure convergence of the trajectories to the solution set. As an illustration, the GDS (4.125) is represented as a neural network in Figure 4.16; the reader should be able to modify with no difficulty this figure for all the other GDSs encountered in this chapter.

Illustrative example of GDS linear programming solver trajectories

Example 4.14. A standard form linear programming problem, modeling a battery charger circuit with current constraints, is taken from [CHZ99], which in turn adapted it from

[Oza86, Chap. 1, pp. 39–40]. The matrices in (4.123) are given as

$$\mathbf{A} = \begin{bmatrix} 1 & -1 & -1 & 0 & 0 & 0 & 0 & 0 & 0 & 0 \\ 0 & 0 & 1 & -1 & -1 & 0 & 0 & 0 & 0 & 0 \\ 1 & 0 & 0 & 0 & 0 & 1 & 0 & 0 & 0 & 0 \\ 0 & 1 & 0 & 0 & 0 & 0 & 1 & 0 & 0 & 0 \\ 0 & 0 & 1 & 0 & 0 & 0 & 0 & 1 & 0 & 0 \\ 0 & 0 & 0 & 1 & 0 & 0 & 0 & 0 & 1 & 0 \\ 0 & 0 & 0 & 0 & 1 & 0 & 0 & 0 & 0 & 1 \end{bmatrix},$$

$$\mathbf{b} = (0, 0, 4, 3, 3, 2, 2),$$

$$\mathbf{c} = (0, -10, 0, -6, -20, 0, 0, 0, 0, 0),$$

and the solution can be calculated to be $\mathbf{x}^* = (4, 2, 2, 0, 2, 0, 1, 1, 2, 0)$.

To apply the condition (4.122) to this example, $\|\bar{\mathbf{c}}\|_1$ is calculated to be equal to 108. Choosing $\alpha = 1.2$ gives $k_2 > 129.6$. Thus, setting $k_2 = 130$ yields $k_1 = 1.2 \times 130 = 156$ in order to ensure convergence to the solution set of the linear programming problem (4.123).

The following alternative bounds on the control gains k_1, k_2 that ensure convergence to the solution set were obtained in [CHZ99, p. 2000]:

$$\begin{aligned} k_2 &> \|(\mathbf{A}\mathbf{A}^T)^{-1}\mathbf{A}\mathbf{c}\|_\infty + k_1 \|(\mathbf{A}\mathbf{A}^T)^{-1}\mathbf{A}\|_\infty, \\ k_1 &\geq \frac{\|\mathbf{c}\|_2}{\beta}, \end{aligned} \tag{4.129}$$

where

$$\beta = \inf_{\mathbf{y}} \{\|\mathbf{P}\mathrm{hsgn}(\mathbf{y})\|_2 : \mathbf{y} \in \partial g(\mathbf{x}), \mathbf{A}\mathbf{x} - \mathbf{b} = \mathbf{0}, x_j < 0 \text{ for some } j\}, \tag{4.130}$$

$\partial g(\mathbf{x})$ denotes the subdifferential of $g(\mathbf{x}) := \sum_{i=1}^n x_i^-$, where $x_i^- = -x_i$, if $x_i < 0$ and 0 otherwise, and the matrix $\mathbf{P} := \mathbf{I} - \mathbf{A}^T(\mathbf{A}\mathbf{A}^T)^{-1}\mathbf{A}$.

Comparison of the bound (4.122) with those in (4.129) is difficult since β in (4.130) is not easy to compute [CHZ99, p. 2002]; indeed, this is a significant advantage of the bounds obtained here. The trajectories of the GDS (4.125) are shown in Figure 4.17.

As pointed out in [CHZ99, Vid95], neural networks of the type considered in this chapter are related to continuous analogs of interior point methods for linear programming, and have the important feature that they do not need to be initiated within the feasible region.

As far as applications are concerned, it has been suggested in [CU93] that networks of the type treated in this section are actually implementable using switched capacitor technology and suitable for real-time problems. More recently, neural-gradient dynamical systems have also been proposed for adaptive blind signal and image processing applications [CA02], and the literature in this area is growing rapidly. More research that will lead to robust, reliable algorithms that can lead with large data sets is required in this area.

Finally, it should be pointed out that the CLF analysis above is quite general and can be applied in a variety of other situations, leading to other systems that can be interpreted as neural networks.

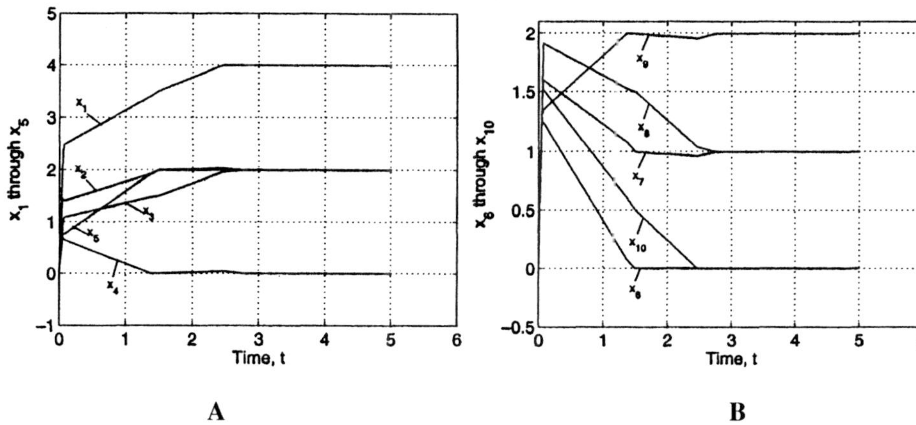

Figure 4.17. *Trajectories of the GDS* (4.125) *converging in finite time to the solution of the linear programming standard form, Example* 4.14. A: *Trajectories of the variables* x_1 *through* x_5. B: *Trajectories of the variables* x_6 *through* x_{10}.

4.5 Quadratic Programming and Support Vector Machines

This section describes a class of quadratic programming problems that can be treated in the framework of the generalized gradient Persidskii systems that are the main tool in this chapter. This is followed by a very brief introduction to support vector machines (SVMs), more specifically, to classes of SVMs that can be treated by the quadratic programming solving GDSs developed in this section.

Quadratic programming problems

Given $\mathbf{x} \in \mathbb{R}^n$, $\mathbf{Q} \in \mathbb{R}^{n\times n}$, $\mathbf{A} \in \mathbb{R}^{m\times n}$, $\mathbf{b} \in \mathbb{R}^m$, $\mathbf{H} \in \mathbb{R}^{p\times n}$, $\mathbf{c} \in \mathbb{R}^p$, consider the quadratic programming problem

$$\begin{aligned} &\min_{\mathbf{x}} \frac{1}{2}\mathbf{x}^T\mathbf{Q}\mathbf{x} \\ &\text{s.t. } \mathbf{A}\mathbf{x} = \mathbf{b} \text{ and } \mathbf{H}\mathbf{x} \geq \mathbf{c}. \end{aligned} \tag{4.131}$$

As discussed in the previous sections, the penalty function method applied to this problem leads to an energy function

$$E(\mathbf{x}, k_1, k_2) = \frac{1}{2}\mathbf{x}^T\mathbf{Q}\mathbf{x} - k_1\|\mathbf{A}\mathbf{x} - \mathbf{b}\|_1 - k_2\sum_{i=1}^{p}\min\{0, \mathbf{h}_i^T\mathbf{x} - c_i\}, \tag{4.132}$$

where $\mathbf{h}_i$ denotes the ith row of the matrix $\mathbf{H}$ and $\mathbf{c}_i$ is the ith component of the vector $\mathbf{c}$.

Introducing the notation

$$\mathbf{r}_1 := \mathbf{A}\mathbf{x} - \mathbf{b}, \quad \mathbf{r}_2 := \mathbf{H}\mathbf{x} - \mathbf{c}, \tag{4.133}$$

the associated GDS that minimizes $E(\cdot)$ can be written as

$$\dot{\mathbf{x}} = -\mathbf{Q}\mathbf{x} - k_1\mathbf{A}^T\text{sgn}(\mathbf{r}_1) - k_2\mathbf{H}^T\text{hsgn}(\mathbf{r}_2). \tag{4.134}$$

Noting that $\dot{\mathbf{r}}_1 = \mathbf{A}\dot{\mathbf{x}}$, $\dot{\mathbf{r}}_2 = \mathbf{H}\dot{\mathbf{x}}$, the dynamics of $\mathbf{r}_1$ and $\mathbf{r}_2$ can be written as

$$\begin{aligned}\dot{\mathbf{r}}_1 &= -\mathbf{A}\mathbf{Q}\mathbf{x} - k_1\mathbf{A}\mathbf{A}^T\text{sgn}(\mathbf{r}_1) - k_2\mathbf{A}\mathbf{H}^T\text{hsgn}(\mathbf{r}_2),\\ \dot{\mathbf{r}}_2 &= -\mathbf{H}\mathbf{Q}\mathbf{x} - k_1\mathbf{H}\mathbf{A}^T\text{sgn}(\mathbf{r}_1) - k_2\mathbf{H}\mathbf{H}^T\text{hsgn}(\mathbf{r}_2).\end{aligned} \tag{4.135}$$

Defining the vector $\mathbf{z} := (\mathbf{x}, \mathbf{r}_1, \mathbf{r}_2)$, and the vector $\mathbf{f}(\mathbf{z}) := (\mathbf{x}, \text{sgn}(\mathbf{r}_1), \text{hsgn}(\mathbf{r}_2))$, the gradient system dynamics can be written in generalized Persidskii form as follows:

$$\dot{\mathbf{z}} = -\mathbf{B}\mathbf{f}(\mathbf{z}) = -\begin{bmatrix} \mathbf{Q} & k_1\mathbf{A}^T & k_2\mathbf{H}^T \\ \mathbf{A}\mathbf{Q} & k_1\mathbf{A}\mathbf{A}^T & k_2\mathbf{A}\mathbf{H}^T \\ \mathbf{H}\mathbf{Q} & k_1\mathbf{H}\mathbf{A}^T & k_2\mathbf{H}\mathbf{H}^T \end{bmatrix}\mathbf{f}(\mathbf{z}). \tag{4.136}$$

The matrix $\mathbf{B}$ can clearly be factored as $\mathbf{SK}$, where

$$\mathbf{S} := \begin{bmatrix} \mathbf{I} & \mathbf{A}^T & \mathbf{H}^T \\ \mathbf{A} & \mathbf{A}\mathbf{A}^T & \mathbf{A}\mathbf{H}^T \\ \mathbf{H} & \mathbf{H}\mathbf{A}^T & \mathbf{H}\mathbf{H}^T \end{bmatrix}, \quad \mathbf{K} := \left[\begin{array}{c|cc} \mathbf{Q} & \mathbf{0} & \mathbf{0} \\ \hline \mathbf{0} & k_1\mathbf{I} & \mathbf{0} \\ \mathbf{0} & \mathbf{0} & k_2\mathbf{I} \end{array}\right], \tag{4.137}$$

where $\mathbf{K}$ is block diagonal, with the first block $\mathbf{K}_{11} := \mathbf{Q}$ a positive definite matrix and the second block $\mathbf{K}_{22} := \text{diag}\,(k_1\mathbf{I}, k_2\mathbf{I})$ a positive diagonal matrix.

Thus, Theorem 1.38 is applicable and allows for the conclusion that the reaching phase leads to convergence of the trajectories of (4.134) to the feasible set of the quadratic programming problem (4.131). Once inside the feasible set, convergence to the solution of the quadratic programming problem is ensured by the "pure" gradient dynamics $\dot{\mathbf{x}} = -\mathbf{Q}\mathbf{x}$, as explained in section 4.3.

Numerical example of sliding mode convergence for the GDS (4.131)

For illustrative purposes, consider the simple special case of (4.131) obtained by choosing

$$\mathbf{Q} = \begin{bmatrix} 1 & 0.5 \\ 0.5 & 2 \end{bmatrix}, \quad \mathbf{A} = [5 \;\; 9], \quad b = 3, \quad \mathbf{H} = -\mathbf{I}, \quad c = 0. \tag{4.138}$$

Figure 4.18 shows a trajectory of the GDS that solves the quadratic programming problem specified in (4.138).

Description of the two-class separation problem and SVMs

Given two classes A and B, the classical pattern recognition problem of finding the best surface that separates the elements of two given classes can be described as follows. Consider the training pairs

$$(y_1, \mathbf{z}_1), \ldots, (y_m, \mathbf{z}_m), \quad y_i \in \{-1, 1\}, \tag{4.139}$$

where the vectors $\mathbf{z}_i$ belong to the input space and the scalars y_i define the position of the vectors $\mathbf{z}_i$ in relation to the surface that separates the classes; i.e., if $y_i = +1$ the vector

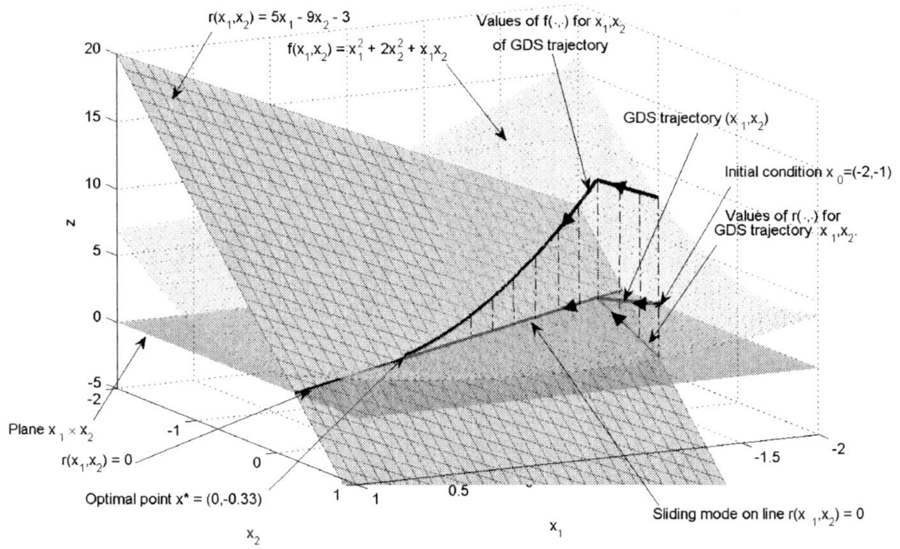

Figure 4.18. *Trajectories of the GDS* (4.131) *for the choices in* (4.138), *showing a trajectory that starts from an infeasible initial condition and converges, through a sliding mode, to the solution* (0, −0.33).

$\mathbf{z}_i$ is located above the separating surface and if $y_i = -1$, this vector is located below the separating surface. If given a set of pairs as in (4.139), a single hyperplane can be chosen such that for all i, $y_i = \pm 1$, then the set of points $\{\mathbf{z}_i\}_{i=1}^m$ is said to be *linearly separable*.

Consider two classes A and B, not necessarily linearly separable, identified as $y_A = +1$ and $y_B = -1$, respectively. The problem of finding the best hyperplane $\Pi := \{\mathbf{u} : \mathbf{u}^T\mathbf{z} + c = 0\}$ that separates the elements of classes A and B is modeled by the quadratic optimization problem [Vap98, CST00]

$$\min \xi(\mathbf{u}, \mathbf{e}, c) = \frac{1}{2}\mathbf{u}^T\mathbf{u} + b\sum_{k=1}^{N} e_k^p \tag{4.140}$$
$$\text{s.t. } y_i(\mathbf{u}^T\mathbf{z}_i + c) \geq 1 - e_i;\ e_i \geq 0, \quad i = 1, \ldots, m,$$

where p is a positive integer, $\mathbf{u}$, $\mathbf{z}_i \in \mathbb{R}^n$, and $e_i, \in \mathbb{R}$. The quantity $y_i(\mathbf{u}^T\mathbf{z} + c)$ is defined as the margin of the input $\mathbf{z}$ with respect to the hyperplane Π. The hyperplane Π that solves problem (4.140) gives the soft margin hyperplane, in the sense that the number of training errors is minimal [SS02, CST00]. The slack variables e_i are introduced in order to provide tolerance to misclassifications.

For nonlinear classification, a feature function ϕ that maps the input space into a higher dimensional space is introduced. In this case, the constraints of problem (4.140) become $y_i(\mathbf{u}^T\phi(\mathbf{z}_i) + c) \geq 1 - e_i, \quad i = 1, \ldots, m$. The traditional approach is to solve the dual of (4.140), since in this case, instead of the function ϕ, another class of functions, known as kernel functions and defined as $K(\mathbf{z}, \mathbf{z}_i) = \phi^T(\mathbf{z})\phi(\mathbf{z}_i)$, is used, with the advantage that it is not necessary to know the feature function ϕ. The feature function ϕ is defined implicitly by the kernel which is assumed to satisfy Mercer's theorem [SS02, CST00].

SVMs have been recently introduced for the solution of pattern recognition and function approximation problems. The so-called machine is actually a technique that consists of mapping the data into a higher dimensional input space and then constructs an optimal separating hyperplane in this space. The use of a technical result known as Mercer's theorem makes it possible to proceed without explicit knowledge of the nonlinear mapping and in this context is often called the kernel trick. The solution is written as a weighted sum of the data points. In the original formulation [Vap95], a quadratic programming problem is solved, leading to many zero weights. The data points corresponding to the nonzero weights are called *support vectors*, giving the machine technique its name. For more details on SVMs and classifiers, see [SS02, SS04].

4.5.1 ν-support vector classifiers for nonlinear separation via GDSs

There are several formulations for the nonlinear separation problem [CST00]. The ν-SVC formulation [SSWB00] fits into the general formulation of Theorem 1.38 and is modeled by the constrained optimization problem

$$\begin{aligned}&\min \tau(\mathbf{w}, \xi, \eta) = \frac{1}{2}\|\mathbf{w}\|^2 - \nu\eta + \frac{1}{m}\sum_{i=1}^{m} e_i \\ &\text{s.t. } y_i(\mathbf{w}^T\phi(\mathbf{z}_i) + c) \geq \eta - \xi_i, \text{ and } \eta \geq 0,\ e_i \geq 0,\end{aligned} \tag{4.141}$$

where ϕ is a feature map function, which provides the classifier with the ability to perform nonlinear discrimination of patterns. The additional parameter ν controls the number of margin errors and support vectors [SS02]. The dual of the constrained optimization problem (4.141) is as follows:

$$\min \xi_\nu(\boldsymbol{\alpha}) = \frac{1}{2}\boldsymbol{\alpha}^T\mathbf{Q}\boldsymbol{\alpha} \tag{4.142}$$

$$\text{s.t. } \mathbf{y}^T\boldsymbol{\alpha} = 0, \tag{4.143}$$

$$\boldsymbol{\alpha} - m^{-1}\mathbf{1}_m \leq \mathbf{0}_m, \text{ and} \tag{4.144}$$

$$\mathbf{B}\boldsymbol{\alpha} - \nu\mathbf{h} \geq \mathbf{0}_{m+1}, \tag{4.145}$$

where the column vectors $\mathbf{0}_m, \mathbf{1}_m \in \mathbb{R}^m$, $\mathbf{0}_{m+1} \in \mathbb{R}^{m+1}$, $\boldsymbol{\alpha} \in \mathbb{R}^m$, $\mathbf{Q} = q_{ij}$, where $q_{ij} = y_i y_j \phi(\mathbf{z}_i)^T\phi(\mathbf{z}_j)$, the matrix $\mathbf{B}^T := [\mathbf{1}_m\ \ \mathbf{I}_m]$, and the column vector $\mathbf{h}^T := [1\ \ \mathbf{0}_m]$. Notice that the objective function ξ_ν is homogeneously quadratic in $\boldsymbol{\alpha}$ and the constraints of the quadratic programming problem are linear. Let $\mathbf{r} := \mathbf{B}\boldsymbol{\alpha} - \nu\mathbf{h}$, $x = \mathbf{y}^T\boldsymbol{\alpha}$ and $\mathbf{v} := \boldsymbol{\alpha} - m^{-1}\mathbf{1}$. As in the previous cases, applying the exact penalty function method to (4.142)–(4.145) yields the energy function

$$E(\boldsymbol{\alpha}, k_1, k_2, k_3) := \frac{1}{2}\boldsymbol{\alpha}^T\mathbf{Q}\boldsymbol{\alpha} + k_1|x| - k_2\sum_{i=1}^{n}\min(0, r_i) + k_3\sum_{i=1}^{n}\max(0, v_i). \tag{4.146}$$

The gradient system $\dot{\boldsymbol{\alpha}} = -\nabla E(\boldsymbol{\alpha})$ associated to the unconstrained optimization problem (4.146) is given by

$$\dot{\boldsymbol{\alpha}} = -\mathbf{Q}\boldsymbol{\alpha} - k_1\,\mathbf{y}\,\mathrm{sgn}(x) - k_2\mathbf{B}^T\mathrm{hsgn}(\mathbf{r}) - k_3\mathrm{uhsgn}(\mathbf{v}). \tag{4.147}$$

Convergence analysis and determination of feedback gains

Define the function $\mathbf{f}$ and the vector $\boldsymbol{\theta}$ as

$$\mathbf{f}(\boldsymbol{\theta}) := (\boldsymbol{\alpha}^T, \mathrm{sgn}(x), \mathrm{hsgn}^T(\mathbf{r}), \mathrm{uhsgn}^T(\mathbf{v}))^T, \tag{4.148}$$

$$\boldsymbol{\theta} := (\boldsymbol{\alpha}, x, \mathbf{r}, \mathbf{v}). \tag{4.149}$$

An augmented dynamical system in the vector $\boldsymbol{\theta}$ can be written as

$$\dot{\boldsymbol{\theta}} = -\mathbf{A}\mathbf{f}(\boldsymbol{\theta}), \tag{4.150}$$

where the matrix $\mathbf{A}$ is:

$$\mathbf{A} = \begin{bmatrix} \mathbf{Q} & k_1\mathbf{y} & k_2\mathbf{B}^T & k_3\mathbf{I} \\ \mathbf{y}^T\mathbf{Q} & k_1\mathbf{y}^T\mathbf{y} & k_2\mathbf{y}^T\mathbf{B}^T & k_3\mathbf{y}^T \\ \mathbf{B}\mathbf{Q} & k_1\mathbf{B}\mathbf{y} & k_2\mathbf{B}\mathbf{B}^T & k_3\mathbf{B} \\ \mathbf{Q} & k_1\mathbf{y} & k_2\mathbf{B}^T & k_3\mathbf{I} \end{bmatrix}. \tag{4.151}$$

Matrix $\mathbf{A}$ can be factored into the form $\mathbf{A} = \mathbf{S}\mathbf{K}$, where

$$\mathbf{S} := \left[\begin{array}{c|ccc} \mathbf{I} & \mathbf{y} & \mathbf{B}^T & \mathbf{I} \\ \hline \mathbf{y}^T & \mathbf{y}^T\mathbf{y} & \mathbf{y}^T\mathbf{B}^T & \mathbf{y}^T \\ \mathbf{B} & \mathbf{B}\mathbf{y} & \mathbf{B}\mathbf{B}^T & \mathbf{B} \\ \mathbf{I} & \mathbf{y} & \mathbf{B}^T & \mathbf{I} \end{array}\right], \qquad \mathbf{K} := \left[\begin{array}{c|ccc} \mathbf{Q} & \mathbf{0} & \mathbf{0} & \mathbf{0} \\ \hline \mathbf{0} & k_1 & \mathbf{0} & \mathbf{0} \\ \mathbf{0} & \mathbf{0} & k_2\mathbf{I} & \mathbf{0} \\ \mathbf{0} & \mathbf{0} & \mathbf{0} & k_3\mathbf{I} \end{array}\right]. \tag{4.152}$$

After these preliminaries, the following theorem is straightforward.

Theorem 4.15. *If* $\mathbf{Q}$ *is positive definite then, for any positive gains* k_1, k_2, *and* k_3, *the trajectories of the system* (4.147) *converge to the solution of the dual quadratic programming problem* (4.142).

The proof of this theorem follows directly from Theorem 1.38, observing that the positive definiteness of matrix $\mathbf{Q}$, assumed in Theorem 4.15, is achieved by choosing positive-definite kernels in the implementations. Observe also that the convergence of the gradient system does not depend on the parameter ν. Numerical examples and further discussion of the choice of parameters in practical implementations of the GDS (4.147) can be found in [FKB04, FKB06b].

Support vector classifier for linear discrimination of two classes

Another type of SVM, also called a support vector classifier (SVC), for the linear discrimination of two classes is the so-called soft margin classifier, which is now described mathematically [CV95]. Define a quadratic objective function:

$$\xi(\mathbf{u}, \mathbf{e}, c) := \frac{1}{2}\left(\mathbf{u}^T\mathbf{u} + b\sum_{k=1}^{m} e_k^2\right). \tag{4.153}$$

In terms of this objective function, the SVC problem can be expressed as the following QP problem:

$$\begin{aligned} &\min \xi(\mathbf{u}, \mathbf{e}, c) \\ &\text{s.t. } \mathbf{u}^T\mathbf{z}_i + c \geq 1 - e_i,\ y_i = +1, \\ &\text{and } \mathbf{u}^T\mathbf{z}_i + c \leq -1 + e_i,\ y_i = -1, \end{aligned} \tag{4.154}$$

where m is the number of elements in the input space; $\mathbf{u}$, $\mathbf{z}_i \in \mathbb{R}^n$, and $e_i, c \in \mathbb{R}$. In order to proceed with the analysis, define

$$\mathbf{w} := (\mathbf{u} \vdots c)^T, \ \bar{\mathbf{z}}_i := (\mathbf{z}_i \vdots 1)^T,$$

where $\mathbf{w} \in \mathbb{R}^{n+1}$ and $\bar{\mathbf{z}}_i \in \mathbb{R}^{n+1}$, as well as the matrices

$$\mathbf{Z}_A := \begin{bmatrix} \mathbf{z}_{A_1}^T & 1 \\ \dots & \dots \\ \mathbf{z}_{A_p}^T & 1 \end{bmatrix} \text{ and } \mathbf{Z}_B := \begin{bmatrix} \mathbf{z}_{B_1}^T & 1 \\ \dots & \dots \\ \mathbf{z}_{B_q}^T & 1 \end{bmatrix}.$$

Observe that $\mathbf{Z}_A$ contains elements from class A, while $\mathbf{Z}_B$ contains elements from class B. Using exact penalization yields the energy function

$$E(\mathbf{w}, \mathbf{e}, k_1, k_2) = \frac{1}{2}\left(\mathbf{w}^T\mathbf{w} + b\sum_{k=1}^{m} e_k^2\right) - k_1\sum_{i=1}^{p}\min(0, r_{A_i}) + k_2\sum_{i=1}^{q}\max(0, r_{B_i}), \tag{4.155}$$

where $m = p + q$, $r_{A_i} = \mathbf{z}_{A_i}^T\mathbf{w} + e_{A_i} - 1$, $r_{B_i} = \mathbf{z}_{B_i}^T\mathbf{w} - e_{B_i} + 1$.

The GDS that minimizes (4.155) can now be written as

$$\begin{aligned} \dot{\mathbf{w}} &= -\mathbf{w} - k_1\mathbf{Z}_A^T\text{hsgn}(\mathbf{r}_A) - k_2\mathbf{Z}_B^T\text{uhsgn}(\mathbf{r}_B), \\ \dot{\mathbf{e}}_A &= -b\,\mathbf{e}_A - k_1\text{hsgn}(\mathbf{r}_A), \\ \dot{\mathbf{e}}_B &= -b\,\mathbf{e}_B + k_2\text{uhsgn}(\mathbf{r}_B), \end{aligned} \tag{4.156}$$

and put in standard form as

$$\dot{\mathbf{x}} = -\mathbf{A}\mathbf{f}(\mathbf{x}),$$

where

$$\begin{aligned} \mathbf{x} &= (\mathbf{w}, \mathbf{e}_A, \mathbf{e}_B, \mathbf{r}_A, \mathbf{r}_B), \\ \mathbf{f}(\mathbf{x}) &= (\mathbf{w}, \mathbf{e}_A, \mathbf{e}_B, \text{hsgn}(\mathbf{r}_A), \text{uhsgn}(\mathbf{r}_B)), \\ \mathbf{A} &= \mathbf{SK}. \end{aligned}$$

Application of Theorem 1.38 results in the following theorem.

Theorem 4.16. *Given any positive regularization parameter b, any positive penalty parameters k_1 and k_2, and any initial condition, trajectories of the dynamical system converge to the solution of the primal SVC quadratic programming problem.*

Note that convergence is independent of the penalty parameters and of the regularization parameter of the SVC. Numerical examples and further discussion of the choice of penalty parameters in practical implementations of the GDS (4.156) can be found in [FKB04, FKB06b].

4.6 Further Applications: Least Squares Support Vector Machines, K-Winners-Take-All Problem

This section gives two other applications of the generalized Persidskii theorem (Theorem 1.38) to problems that are of contemporary interest in the neural networks community, are related to the applications discussed earlier, and are amenable to treatment by the same techniques that have been used throughout this chapter.

4.6.1 A least squares support vector machine implemented by a GDS

The least squares support vector machine (LS-SVM) model is a modification of the original SVM model (4.140), in which the inequality constraints are replaced by equality constraints. The LS-SVM is modeled by the following constrained optimization problem [SGB$^+$02]:

$$\begin{aligned} \min\ \xi_{ls}(\mathbf{u}, \mathbf{e}, c) &= \frac{1}{2}\mathbf{u}^T\mathbf{u} + \frac{b}{2}\sum_{i=1}^{n} e_i \\ \text{s.t. } y_i(\mathbf{u}^T\phi(\mathbf{z}_i) + c) &= 1 - e_i, \qquad i = 1, \ldots, m. \end{aligned} \tag{4.157}$$

The dual problem of (4.157) is given by the following system of linear equations, also known as a KKT linear system [SGB$^+$02]:

$$\left[\begin{array}{c|c} 0 & \mathbf{y}^T \\ \hline \mathbf{y} & \mathbf{Q} + b^{-1}\mathbf{I} \end{array}\right] \left[\begin{array}{c} c \\ \hline \boldsymbol{\alpha} \end{array}\right] = \left[\begin{array}{c} 0 \\ \hline \mathbf{1} \end{array}\right], \tag{4.158}$$

where $\mathbf{Q}$ is a symmetric matrix given by $q_{ij} = y_i y_j K(\mathbf{z}_i, \mathbf{z}_j)$, K is defined by the kernel $K(\mathbf{z}, \mathbf{z}_j) = \phi^T(\mathbf{z})\phi(\mathbf{z}_j)$, and $\boldsymbol{\alpha}$ is the vector of dual variables. In the LS-SVM model, the problem of determining the best separating surface for classes A and B is reduced to solving the system of linear equations (4.158), which has a full rank coefficient matrix if $b^{-1} \neq -\lambda_i(\mathbf{Q})$ for all i. Thus by Theorem 4.2, the trajectories of the gradient system (4.46), with $\mathbf{A}$, $\mathbf{x}$, and $\mathbf{b}$ defined as in (4.158), converge in finite time to the solution of (4.157).

Implementation of this GDS to solve the LS-SVM problem and application examples are shown in [FKB05].

4.6.2 A GDS that solves the k-winners-take-all problem

The k-winners-takes-all problem is that of determining the k largest components of a given vector $\mathbf{c} \in \mathbb{R}^n$. This problem appears in decision making, pattern recognition, associative memories and competitive learning networks [Hay94, KK94, LV92]. Many networks have been proposed to solve this problem [UN95, MEAM89, WMA+91, CM00, YGC98], and they are referred to as KWTA networks.

This subsection proposes a new KWTA network by taking a previous proposal of Urahama and Nagao [UN95] as a starting point. The objective is to obtain advantages in implementation as well as performance. In fact, the integer programming formulation of the KWTA problem in [UN95] can be relaxed to a linear programming formulation with box constraints. As a result, it is possible to use the type of GDS analyzed in section 4.4 to solve the resulting linear programming problem, with advantages that will be discussed below. Urahama and Nagao [UN95] formulated the KWTA problem as the integer programming problem

$$\begin{aligned} &\max_{\mathbf{x}\in\{0,1\}^n} \ \mathbf{c}^T\mathbf{x} \\ &\text{s.t. } \mathbf{1}^T\mathbf{x} = k, \end{aligned} \tag{4.159}$$

converted it into a nonlinear programming problem, and solved it by minimizing an associated Lagrangian function. It can be shown that the integer programming problem above can be relaxed to the LP problem with bounded variables

$$\begin{aligned} &\max_{\mathbf{x}\in[0,1]^n} \ \mathbf{c}^T\mathbf{x} \\ &\text{s.t. } \mathbf{1}^T\mathbf{x} = k, \end{aligned} \tag{4.160}$$

where $\mathbf{c} = (c_1, \dots, c_n)$, $\mathbf{1} = (1, \dots, 1) \in \mathbb{R}^n$, $k \leq n \in \mathbb{N}$ is a nonnegative integer, and $\mathbf{x} \in \mathbb{R}^n$. The following proposition states that the integer programming problem (4.159) and its relaxed version (4.160) have the same solution $\mathbf{x}^*$.

Proposition 4.17 [FKB06a]. *Consider the linear programming problem* (4.160) *and let the components of vector* $\mathbf{c}$ *be distinct. Then, the solution of the linear programming problem* (4.160) *is unique and presents k components equal to* 1, *which, correspondingly, multiply the k largest components of vector* $\mathbf{c}$ *in the objective function, while the $n-k$ remaining components are equal to zero.*

In light of this proposition and the well-known fact that all linear programming problems, including one with bounded variables, as in (4.160), can be rewritten as one of the standard linear programming forms discussed above, it is now clear that the KWTA problem can be treated by the methods of section 4.4. However, since the rewriting into standard linear programming form involves the introduction of slack variables and, in general, an increase in the dimension of the linear programming problem, we show that the analysis developed in section 4.4 can be applied directly by a suitable definition of nonlinearity.

Consider the energy function associated with (4.160),

$$E(\mathbf{x}, k_1, k_2) := -\mathbf{c}^T\mathbf{x} - k_1\left(\sum_{j=1}^{n} \min\{0, x_j\} - \sum_{i=1}^{n} x_j^+\right) + k_2\left|\mathbf{1}^T\mathbf{x} - k\right|, \tag{4.161}$$

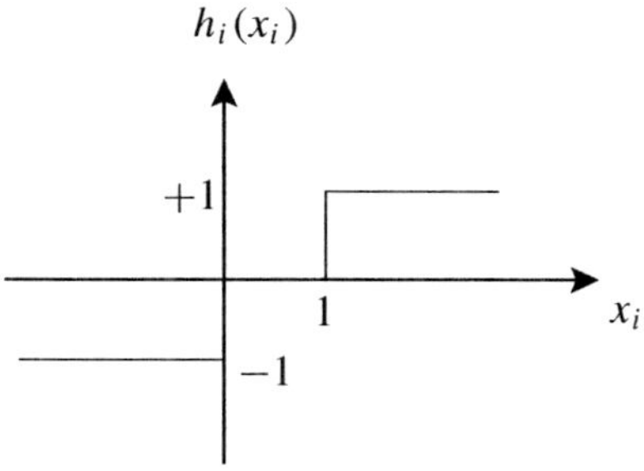

Figure 4.19. *The function $h_i(\cdot)$ defined in* (4.162) *is a first-quadrant–third-quadrant sector nonlinearity.*

where, for each j,

$$x_j^+ = \begin{cases} x_j - 1 & \text{if } x_j > 1, \\ 0 & \text{if } x_j \leq 1. \end{cases}$$

Observe that if

$$h_i := \min\{0, x_i\} - x_i^+, \tag{4.162}$$

then the graph of h_i is a first-quadrant–third-quadrant nonlinearity that satisfies a sector condition (Figure 4.19). Defining $\mathbf{h} := (h_1, \ldots, h_n)$, the gradient system $\dot{\mathbf{x}} = -\nabla E(\mathbf{x})$ that minimizes E is given by

$$\dot{\mathbf{x}} = \mathbf{c} - k_1\mathbf{h}(\mathbf{x}) - k_2\mathbf{1}\text{sgn}(\mathbf{1}^T\mathbf{x} - k). \tag{4.163}$$

Convergence results and determination of penalty parameters

As before, for the different versions of the linear programming problem, we give reaching phase conditions, i.e., conditions that ensure convergence to the feasible set of problem (4.160), which is given by the intersection

$$\Omega := \Pi \cap \Gamma, \tag{4.164}$$

where $\Pi := \{\mathbf{x} : \mathbf{1}^T\mathbf{x} - k = 0\}$ and $\Gamma := \{\mathbf{x} : x_j \in [0, 1],\ \text{for each } j\}$. Similarly to methods that use discontinuous switching functions (hsgn, uhsgn, sgn), the dynamical system (4.163) has the pleasant property of a finite-time reaching phase.

Defining $\dot{r} := \mathbf{1}^T\dot{\mathbf{x}}$ and writing the augmented equations in a manner similar to that of section 4.4 yields

$$\begin{bmatrix} \dot{\mathbf{x}} \\ \dot{r} \end{bmatrix} = \begin{bmatrix} \mathbf{c} \\ \mathbf{1}^T\mathbf{c} \end{bmatrix} - \begin{bmatrix} \mathbf{I}_n & \mathbf{1} \\ \mathbf{1}^T & \mathbf{1}^T\mathbf{1} \end{bmatrix} \begin{bmatrix} k_1\,\mathbf{I}_n & \mathbf{0}_{n\times 1} \\ \mathbf{0}_{1\times n} & k_2 \end{bmatrix} \begin{bmatrix} \mathbf{h}(\mathbf{x}) \\ \text{sgn}(r) \end{bmatrix}. \tag{4.165}$$

Defining

$$\mathbf{K} := \begin{bmatrix} k_1\,\mathbf{I}_n & \mathbf{0}_{n\times 1} \\ \mathbf{0}_{1\times n} & k_2 \end{bmatrix}, \quad \mathbf{S} := \begin{bmatrix} \mathbf{I}_n & \mathbf{1} \\ \mathbf{1}^T & \mathbf{1}^T\mathbf{1} \end{bmatrix}, \tag{4.166}$$

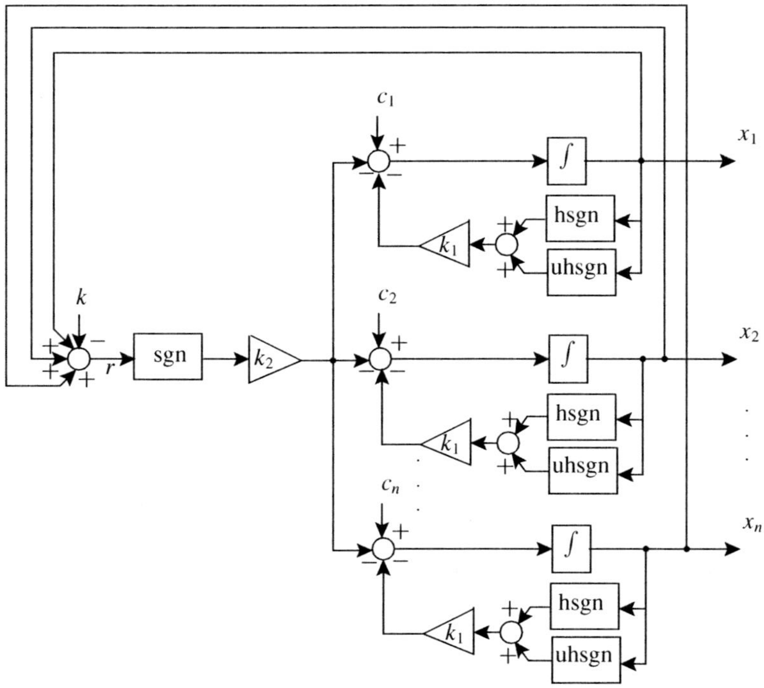

Figure 4.20. *The KWTA GDS represented as a neural network.*

using the CLF

$$V(\mathbf{x}, r) = k_1 \sum_{j=1}^{n} \int_0^{x_j} h(\tau)d\tau + k_2 \int_0^r \mathrm{sgn}(\tau)d\tau \tag{4.167}$$

as well as the simple structure of **S** and **K** matrices for the KWTA system, the following lemma is not difficult to prove [FKB06a].

Lemma 4.18. *Consider the system of ODEs* (4.163). *Provided that k_1 and k_2 satisfy one of the inequalities*

$$\min\left\{k_1^2, nk_2^2, n(k_1 - k_2)^2\right\} > \|\mathbf{c}\|_1(k_1 + k_2), \tag{4.168}$$

then, for any initial condition, the trajectories reach the feasible set Ω in finite time and remain in this set thereafter.

Assuming that $k_1 = 2k_2$, (4.168) yields a simple condition for KWTA behavior:

$$k_2 > \frac{3}{n}\|\mathbf{c}\|_1, \quad k_1 = 2k_2.$$

We close by showing the representation of the proposed GDS for the KWTA problem as a neural network (Figure 4.20).

4.7 Notes and References

GDSs for optimization

There is a vast literature on GDSs for optimization, so the references given here are the ones that most influenced our treatment, rather than an exhaustive list that seeks to establish priority. The solution of linear programming problems using GDSs was apparently first considered in [Pyn56], where a method for solving linear programming problems on an electronic analog computer was presented. Gradient methods for optimization were investigated for several mathematical programming problems in [AHU58]. Rybashov and coworkers, in a series of papers [Ryb65b, Ryb65a, Ryb69a, Ryb74, VR77], obtained several basic results about GDSs for optimization: These papers, especially [Ryb74], are precursors of the approach developed in this chapter, the additional ingredients in this book being the utilization of GDSs with discontinuous right-hand sides and the use of the associated Persidskii-type Liapunov function, leading in many cases to finite-time convergence results. A notable feature of the paper [Ryb74] is that, for the GDS proposed therein, it develops estimates for the size of basins of attraction of equilibria, as well as estimates for the rate of convergence.

Many neural-gradient dynamical systems were proposed in [CU93], which was mainly concerned with the properties of these systems as implementable circuits, rather than the theoretical analysis, and indeed, for this reason, served as the impetus for the development of part of the present chapter.

GDSs with discontinuous right-hand sides have also been studied in the literature on variable structure systems, specifically in the books [Utk92, Żak03] and the papers [GHŻ98, CHZ99], which provided theoretical justification of a discontinuous GDS. The textbook [Żak03] summarizes existing work on discontinuous GDSs and contains the theoretical background; it served as another of the sources of inspiration for the material presented in this chapter. A good survey of many approaches to optimization problems using continuous-time dynamical systems is [LQQ04]; this reference also introduces the term *neurodynamical optimization* and includes many references not cited here.

Feedback control for the design of GDSs in the solution of optimization problems arising in the study of equilibrium problems, such as Nash equilibria, has been considered by Antipin; a survey of this work can be found in [Ant00]. Another approach to convex programming using feedback control is described in [Kry98].

Authoritative references on optimization and, in particular, on gradient methods, are Polyak [Pol87] and Bertsekas [Ber99].

Global stability of neural-gradient dynamical systems

Stability analysis for neural-GDSs using Persidskii diagonal-type Liapunov functions was first carried out in [KB94]. The introductory discussion in section 4.1 is based on [KBC0].

SVMs

A comprehensive reference on SVMs and their relationship to optimization problems is [SS02]. A short introduction to the basics is [CST00]. A recent tutorial is [SS04].

Chapter 5

Control Tools in the Numerical Solution of Ordinary Differential Equations and in Matrix Problems

The control-based adaptivity (for automatic control and adaptive time-stepping for ODE integration methods) works because process models and controllers are mathematically analyzed and experimentally verified—*much less can be expected from a heuristic approach.*

—G. Söderlind [Söd02]

This chapter looks at some topics in the numerical solution of ODEs and in matrix theory from a control viewpoint.

We start with an application of control theory to the automatic stepsize control as well as optimal stepsize control of ODE integration methods; as pointed out in the preface, this is one of the success stories of the application of control ideas in numerical problems. Shooting methods for ODEs are given a feedback control formulation, and this leads to connections with the iterative methods discussed in the previous chapters as well as with a control technique called iterative learning control.

The matrix theory problems of diagonal preconditioning and D-stability are treated by using, respectively, the ideas of decentralized control and positive real systems. Finally, this chapter, as well as the book, closes with an application of the ideas of controllability and observability to the problem of finding common zeros of two polynomials in two variables.

5.1 Stepsize Control for ODEs

As pointed out by Söderlind [Söd02], a significant part of most modern software for initial value problems (IVPs) for ODEs is devoted to control logic and ancillary algorithms. Furthermore, in contrast to the heavily analyzed discretization methods, very little attention has been given to the analysis and design of control structure and logic, which have remained heuristic to a great extent. This situation is now being redressed by Söderlind and coworkers and this section makes an exposition of this approach, closely following [Söd02].

Consider the initial value problem

$$\dot{\mathbf{x}} = \mathbf{f}(\mathbf{x}(t)), \quad \mathbf{x}(0) = \mathbf{x}_0, \tag{5.1}$$

where $\mathbf{x}(\cdot) \in \mathbb{R}^n$ and $\mathbf{f} : \mathbb{R}^n \to \mathbb{R}^n$ is a Lipschitz-continuous function. The qualitative behavior of the solution $\mathbf{x}(\cdot)$ depends on the properties of the right-hand side of $\mathbf{f}$. The latter can be linear or nonlinear, and solutions can have very different properties with respect to sensitivity to perturbations, smoothness of solution, and so on. In some IVPs, high precision is required; in others, not as much. In other words, a general-purpose ODE solver must be able to deal with a wide variety of situations.

Properties such as efficiency of a stepsize method depend on the size of the IVP problem as well as the characteristics of the problem. The objective of the integration method is to attempt to compute a numerical solution $\mathbf{x}_k$ to (5.1) with minimum effort, subject to maintaining a *prescribed error tolerance*, `tol`. There is a tradeoff between computational effort and error. It is desired to have the global error $\|\mathbf{x}_k - \mathbf{x}(t_k)\|$ decrease as the error tolerance `tol` tends to 0; on the other hand, the computational effort increases.

The problem of minimizing the total computational effort subject to a bound on the global error can be viewed as a control problem or even an optimal control problem, and this approach will be detailed in section 5.1.2. An alternative approach is to argue that, since time-stepping methods are essentially local, an integration method for an IVP is a (sequential) procedure to compute the next state $\mathbf{x}(t+h)$ at the time step h units ahead. From this viewpoint, the size of the step h can be used to trade off accuracy and efficiency. In other words, the stepsize h is a control variable and can be used to keep the local error (per unit time) below the prespecified level `tol`. The rationale is that it can be shown that, under these conditions, the global error at time t is bounded by a term of the form $\alpha(t)\cdot$ `tol`, meaning that local error control indirectly influences global error. In addition, it is usually cheaper and simpler to control the local error.

A one-step method, given a step size h, can be thought of as a parameterized map $\Phi_h : \mathbb{R}^n \to \mathbb{R}^n$ such that

$$\mathbf{x}_{k+1} = \Phi_h(\mathbf{x}_k), \quad \mathbf{x}(0) = \mathbf{x}_0, \tag{5.2}$$

is a discrete-time dynamical system that approximates the flow of the continuous-time dynamical system (5.1) on $\mathbb{R}^n$; i.e., $\mathbf{x}_k$ approximates $\mathbf{x}(t_k)$.

A conceptual description of adaptive time-stepping for the efficient solution of IVPs for ODEs may be given as follows. In order to consider an adaptive or time-varying stepsize, it is necessary to introduce an additional map ("stepsize generator") $\sigma_{\mathbf{x}} : \mathbb{R} \to \mathbb{R}$ such that

$$h_{k+1} = \sigma_{\mathbf{x}}(h_k). \tag{5.3}$$

Note that the map $\sigma_{\mathbf{x}}(\cdot)$ uses information (feedback) about the state vector $\mathbf{x}$ and the current stepsize h_k in order to generate the next stepsize h_{k+1} (Figure 5.1).

From Figure 5.1, it is clear that the stability of an adaptive time-stepping method is equivalent to closed-loop stability of the feedback system depicted in the figure.

An important ingredient needed for control is a local error model. Assume that the integration method adopted also possesses a *reference method*, defined by a different quadrature formula, and let the reference values be denoted $\hat{\mathbf{x}}_k$. A local error estimate can then be defined as

$$\hat{\mathbf{e}}^{\ell}_{k+1} = \mathbf{x}_{k+1} - \hat{\mathbf{x}}_{k+1}. \tag{5.4}$$

Let $\hat{\mathbf{x}}(t; \tau, \boldsymbol{\eta})$ denote a solution of (5.1) with initial condition $\hat{\mathbf{x}}(\tau) = \boldsymbol{\eta}$. Then the *local error*, denoted $\mathbf{e}^{\ell}_k$ in a step from $\mathbf{x}_k$ to $\mathbf{x}_{k+1}$, can be written as

$$\mathbf{e}^{\ell}_{k+1} = \mathbf{x}_{k+1} - \hat{\mathbf{x}}(t_{k+1}; t_k, \mathbf{x}_k). \tag{5.5}$$

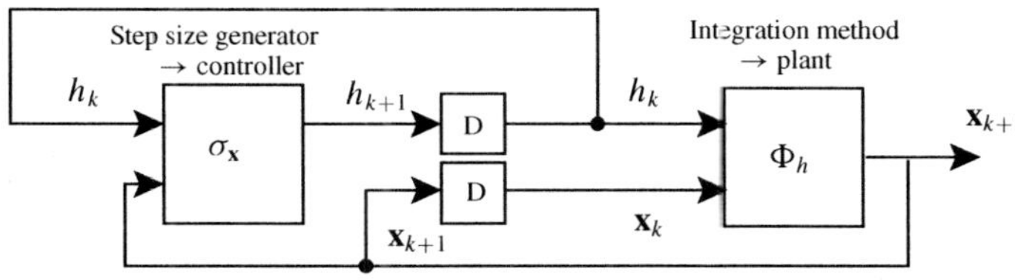

Figure 5.1. *Adaptive time-stepping represented as a feedback control system. The map $\Phi_h(\cdot)$ represents the discrete integration method (plant) and uses the current state ($\mathbf{x}_k$) and stepsize h_k to calculate the next state $\mathbf{x}_{k+1}$ which, in turn, is used by the stepsize generator or controller $\sigma_{\mathbf{x}}(\cdot)$ to determine the next stepsize h_{k+1}. The blocks marked D represent delays.*

By expanding the local error in an asymptotic series, it is possible to show that

$$\mathbf{e}_{k+1}^{\ell} = \Phi(t_k, \mathbf{x}_k)h^{p+1} + O(h^{p+2}), \tag{5.6}$$

where the term $\Phi(\cdot)$ is referred to as the *principal error function* and p is the *order* of the method. In an analogous fashion, the local error estimate is expressed as follows:

$$\hat{\mathbf{e}}_{k+1}^{\ell} = \hat{\Phi}(t_k, \mathbf{x}_k)h^{\hat{p}+1} + O(h^{\hat{p}+2}), \tag{5.7}$$

where $\hat{p}$ may be different from p in general but, for the purposes of this discussion, will be assumed to be equal to p.

Two error measures are usually of interest: the *local error per step (EPS)*, denoted $\hat{r}$ and defined as

$$\hat{r}_{k+1} := \|\hat{\mathbf{e}}_{k+1}^{\ell}\|, \tag{5.8}$$

and the *local error per unit step (EPUS)*, defined as

$$\hat{r}_{k+1} := \|\hat{\mathbf{e}}_{k+1}^{\ell}/h_k\|. \tag{5.9}$$

In practical computation, an asymptotic error estimate is usually available, and from the preceding discussion, in the asymptotic limit as $h \to 0$, the *stepsize–error relation* can be written as

$$\hat{r}_{k+1} = \hat{\varphi}_k h_k^{\overline{p}}, \tag{5.10}$$

where $\hat{r}_n$ is the norm of the local error estimate, $\hat{\varphi}$ is the norm of the principal error function, and $\overline{p} = p$ (EPUS) or $\overline{p} = p + 1$ (EPS), where p is the order of the integration method [Söd98, DB02, HNW93]. In fact, the model of this stepsize–error relation determines the design of the adaptive time step generator. A popular candidate for a local error control law is [Gea71]:

$$h_{k+1} = \left(\frac{\varepsilon}{\hat{r}_{k+1}}\right)^{1/\overline{p}} h_k, \tag{5.11}$$

where ε is a fraction of the local error tolerance. The rationale behind this choice of control law, as will be seen below, is that it eliminates error between the error norm $\hat{r}_k$ and the error tolerance ε in one step, provided that the assumptions underlying (5.10) hold, namely, that the method is operating in the asymptotic regime and the principal error function $\hat{\varphi}$ is slowly varying (i.e., $\hat{\varphi}_{k+1} \approx \hat{\varphi}_k$). In these circumstances, if there is a deviation between ε and $\hat{r}_{k+1}$, then the choice of h_{k+1} as in (5.11) clearly results in the error norm estimate becoming equal to the tolerance ε at the next step, as the following calculation shows:

$$\hat{r}_{k+2} = \hat{\varphi}_{k+1} h_{k+1}^{\overline{p}} \approx \hat{\varphi}_k \left(\left(\frac{\varepsilon}{\hat{r}_{k+1}} \right)^{1/\overline{p}} h_k \right)^{\overline{p}} = \hat{\varphi}_k \left(\left(\frac{\varepsilon}{\hat{\varphi}_k h_k^{\overline{p}}} \right)^{1/\overline{p}} h_k \right)^{\overline{p}} = \varepsilon.$$

It is, of course, well known [HW96, Gus91, Söd02] that, in practice, the assumptions underlying (5.10) may not hold. The first assumption, namely, that the method is operating in steady state, can become false because a stiff ODE is being integrated with a value of stepsize which is outside the regime for a mode that decays fast [HW96] or because an explicit method with a bounded stability region is being used [Gus91]. The second assumption, regarding slow variation of the estimated error norm, is tantamount to asserting that the function $\mathbf{f}$ and/or the argument $\mathbf{x}$ presents small variation during a time interval of length h and is rarely true [Gus94, Söd02]. Nevertheless, this so-called elementary error control has shown itself to be very useful in practical computation, basically because it is quite efficient as a feedback mechanism that prevents numerical instability. In colloquial terms, this feedback mechanism can be described as follows.

For an explicit method, the control law (5.11) allows the stepsize to grow, as long as the error is small (and the computed $\mathbf{x}_k$ is "smooth"). If the value of h_k increases beyond the numerical stability limit, the resulting nonsmooth behavior of $\mathbf{x}_k$ causes an increase in the estimated error norm $\hat{r}_k$, causing h_k to decrease. This verbal description seems to suggest oscillation around the largest stepsize that preserves numerical stability. In fact, oscillations of this nature are observed in practice and have motivated two different approaches to their removal.

In the first approach, the oscillation of stepsizes is seen as an additional problem, and new discretization methods that maintain the stepsize control law (5.11), but reduce stepsize oscillations, are constructed [HH90]. The second approach, which is the focus of this section, is to regard the choice of the law as the cause of the stepsize oscillations and to design new stepsize control laws that work together with the integration method of choice as well as suppress or reduce oscillations. In a succinct phrase that underlines the difference between the two approaches, Söderlind says, "instead of constructing methods that match the control law (5.11), the controller is designed to match the method." The second approach is, of course, the control approach that is being promoted in this book. In this particular situation, the control approach can also be regarded as a more natural approach, since one stays with the familiar integration methods, adding theoretically justifiable control logic to the method.

5.1.1 Stepsize control as a linear feedback system

At first sight, the control law (5.11) looks anything but linear. However, taking logarithms gives

$$\log h_{k+1} = \log h_k + \frac{1}{\overline{p}}(\log \varepsilon - \log \hat{r}_{k+1}). \tag{5.12}$$

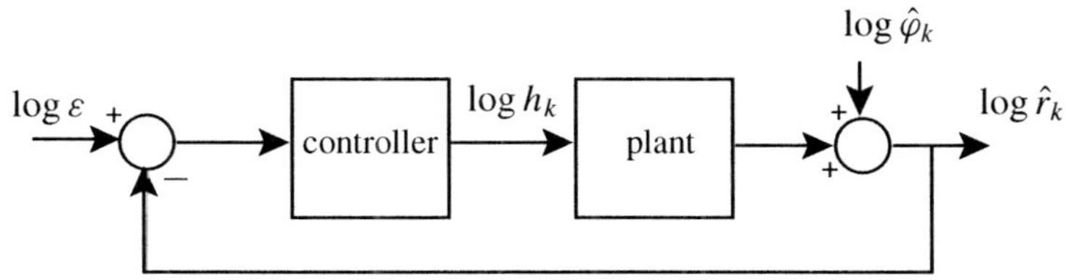

Figure 5.2. *The stepsize control problem represented as a plant P and controller C in standard unity feedback configuration.*

Comparing (5.12) with (1.10), the logarithmic form of the stepsize control law can clearly be identified as an integral controller of the type discussed in section 1.1.2. The block diagram corresponding to (5.12) is shown in Figure 5.2. Similarly, taking logarithms of the relation between stepsize and error estimate (5.10) yields the process or plant dynamics

$$\log \hat{r}_{k+1} = \overline{p} \log h_k + \log \hat{\varphi}_k. \tag{5.13}$$

The closed-loop dynamics of the system in Figure 5.2 is then easily found by substituting (5.13) in (5.12) and gives

$$\log h_{k+1} = \frac{1}{\overline{p}}(\log \varepsilon - \log \hat{\varphi}_k). \tag{5.14}$$

This is recognizable as *deadbeat* dynamics from the discussion in section 1.1.2 and resulted from the particular choice of gain $(1/\overline{p})$ in (5.12). Replacing this particular choice by a gain k_I yields instead the closed-loop dynamics

$$\log h_{k+1} = (1 - \overline{p}k_I) \log h_k + k_I(\log \varepsilon - \log \hat{\varphi}_k). \tag{5.15}$$

As discussed in section 1.1.2, the characteristic equation is now $q = 1 - \overline{p}k_I$. The important point to notice is that the solution of (5.15) can be written explicitly as the discrete convolution:

$$\log h_k = (1 - \overline{p}k_I)^k \log h_0 + k_I \sum_{m=1}^{k} (1 - \overline{p}k_I)^{k-m} (\log \varepsilon - \log \hat{\varphi}_{m-1}). \tag{5.16}$$

This has an interesting interpretation when $\overline{p}k_I$ lies in the interval (0, 1), since the term $(1 - \overline{p}k_I)^{n-m}$ is a "forgetting factor" that reduces the effect of variations in $\log \hat{\varphi}_n$, thus leading to smoother stepsize sequences.

With the change to the gain k_I, the general integral controller can be written as

$$h_{k+1} = \left(\frac{\varepsilon}{\hat{r}_{k+1}}\right)^{k_I} h_k, \tag{5.17}$$

so that the change with regard to (5.11) is that $(1/\overline{p})$ has been replaced by k_I. This is a very significant change in the sense that it has a theoretical basis that implies smoother stepsize sequences for appropriate choices of k_I. In addition, it assumes only that the asymptotic error model that defines the plant is (approximately) correct—there is no need for the assumption of slow variation. Finally, this simple change from $(1/\overline{p})$ to k_I, and the subsequent realization that a general integral controller results, opens the door to the utilization of other controllers. In particular, the proportional-integral (PI) and proportional-integral-derivative (PID) controllers are very popular in control theory and practice and are used to improve closed-loop performance.

In terms of the logarithm of the stepsize, a PI controller consists of two terms: one proportional to the control error and the other proportional to the summation or discrete integral of the control error:

$$\log h_k = \log h_0 + k_I \sum_{m=1}^{k} (\log \varepsilon - \log \hat{r}_m) + k_P (\log \varepsilon - \log \hat{r}_k). \tag{5.18}$$

This controller can also be written in recursive form, and thence in multiplicative form, as

$$h_{k+1} = \left(\frac{\varepsilon}{\hat{r}_{k+1}} \right)^{k_I} \left(\frac{\hat{r}_k}{\hat{r}_{k+1}} \right)^{k_P} h_k, \tag{5.19}$$

which is a modification of (5.11) that is easy to implement. The new factor, which because of the form of the third term on the right-hand side of (5.18) is referred to as the *proportional factor*, can be interpreted by observing that it is greater than 1 if the error is decreasing ($\hat{r}_{k+1} < \hat{r}_k$) and less than 1 if the error is increasing. This means that increasing error will result in faster stepsize reduction, and decreasing error in faster stepsize increase, relative to the use of a purely integral controller. As shown in section 1.1.2, the closed-loop dynamics are now determined by the roots of the characteristic equation

$$q^2 - (1 - \overline{p}k_I - \overline{p}k_P)q - \overline{p}k_P = 0. \tag{5.20}$$

In this simple case of a static plant (constant gain) and constant output disturbance ($\log \hat{\varphi}_k$), it is a well-known result in control theory (cf. section 1.1.2) that a controller with integral action is both necessary and sufficient for zero steady state regulation error. The assumptions, of the validity of the asymptotic model and almost constant disturbance, are reasonable for most nonstiff computation, but there are also many cases where a better (i.e., more robust) controller is required, and here again, one has to resort to existing tools of controller synthesis.

The details of controller parameterization and good choices of these parameters are very lucidly explained in [Söd02], to which the reader is referred. A mixture of control theory and extensive experimentation was used by Gustafsson [Gus91] to arrive at good starting choices of $(\overline{p}k_I, \overline{p}k_P) = (0.3, 0.4)$ in order to subsequently fine tune these and other parameters required in practice for each individual integration method.

In order to emphasize the contribution of control, the steps taken in the analysis and design of the PI controller detailed above are now outlined in the following form of a design procedure:

(i) An integration method is chosen. This implies that an underlying asymptotic error model is chosen and this defines the process or plant. The regulator problem is to be solved for this plant; i.e., it is required to find a controller that makes the output (estimated error) equal to the reference input (specified tolerance).

(ii) A control structure is chosen. This involves choosing a type of controller (e.g., P, PI, or PID) as well as free or design parameters that are chosen to give desired performance (e.g., the ability to regulate "robustly," in the presence of errors, either as disturbances or in the hypothesized model).

(iii) The closed-loop dynamics are found and the control parameters chosen in step (ii) adjusted so that the desired performance is obtained.

(iv) If the desired performance is not obtained, it is necessary to return to a previous step, (i) or (ii), and redesign.

Both modeling (step (i)) and selection of control structure are nontrivial tasks and require detailed knowledge of the problem to be solved (plant) in order to arrive at an accurate and reasonably simple model as well as controller. In the application considered here, design of a PI controller for a constant gain plant, one can use difference equation techniques, which are familiar to numerical analysts.

In general, for the IVP for ODEs, the control-based algorithms are efficient and, from a qualitative point of view, lead to smoother stepsize sequences and fewer rejected steps, as shown in numerical studies by Söderlind and coworkers [Gus91, Gus94, GS97, Söd98, Söd02, Söd03].

However, as pointed out by Söderlind (himself a numerical analyst, in addition to being one of the discoverers and leading researchers of the material of this section), "mathematically, ... it is more elegant (and less cumbersome) to follow the control theoretic practice [and, furthermore] the efficiency gain is in terms of qualitative improvement and increased computational stability."

5.1.2 Optimal stepsize control for ODEs

This section, based on the work of [UTK96], takes up the question of global error control, formulating the problem as an optimal control problem. The essential argument in favor of looking at optimal stepsize control is that, unlike the local method considered in the previous section, global methods can seek to minimize accumulated error or error incurred at the end of integration. Another interesting feature of this approach is that it becomes possible to prove that certain strategies are optimal (e.g., constant stepsize for constant coefficient linear problems). There are disadvantages as well, principally increased difficulty in computation of the optimal controller, and this will limit the treatment to simple illustrative examples in what follows.

Recall that, in section 3.3, in the context of optimization, the question of local versus global methods was discussed.

Error dynamics for scalar ODEs

Consider the scalar IVP

$$\dot{x} = f(x,t), \quad x(0) = x_0, \text{ given} \tag{5.21}$$

with $x(t) \in \mathbb{R}$. In any numerical method used to integrate (5.21), an error in x generated after a step $\Delta t = h$ is given by (5.10) and depends on both the right-hand side f and the integration method utilized. The integer p is the order of the method; thus $p = 1$ for the Euler method and $p = 4$ for the standard fourth-order Runge–Kutta method. Writing the variation of x to the first order as δx, from (5.1), it follows that

$$\frac{\mathrm{d}\delta x}{\mathrm{d}t} = \frac{\partial f}{\partial x}\delta x + \hat{\varphi}h^p, \tag{5.22}$$

where the *error generating coefficient* $\hat{\varphi}$ can be calculated in principle. As an example, consider the forward Euler method:

$$\hat{\varphi}_{\text{Euler}} = \frac{1}{2}\left(\frac{\partial f}{\partial t} + \frac{\partial f}{\partial x}f\right). \tag{5.23}$$

For the fourth-order Runge–Kutta method, the expression for $\hat{\varphi}_{\text{RK4}}$ is more complicated, although its numerical value can be calculated in a simple manner [Gea71, HNW87, DB02].

State variable description of error dynamics

In order to provide a state variable description of the error dynamics, as well as to introduce error measures of interest in this problem, the following state variables are introduced:

$$\begin{aligned} z_1(t) &:= \textstyle\int_0^t \delta x(t)\mathrm{d}t, \\ z_2(t) &:= \delta x(t), \\ z_3(t) &:= \textstyle\int_0^t (1/h(\tau))\mathrm{d}\tau. \end{aligned} \tag{5.24}$$

To simplify notation, denote the partial derivative $\frac{\partial f}{\partial x}$ in (5.22) by a, i.e.,

$$a := \frac{\partial f}{\partial x}, \tag{5.25}$$

and, dropping the subscripts on $\hat{\varphi}$, the error dynamics can be written in state space form as

$$\begin{aligned} \dot{z}_1 &= z_2, \\ \dot{z}_2 &= az_2 + \hat{\varphi}h^p, \\ \dot{z}_3 &= (1/h), \end{aligned} \tag{5.26}$$

which can be written in vector notation as

$$\dot{\mathbf{z}} = \mathbf{f}(\mathbf{z}, h), \quad \mathbf{z}(0) \text{ given}, \tag{5.27}$$

where $\mathbf{z} = (z_1, z_2, z_3)$ and

$$\mathbf{f}(\mathbf{z}, h) := \begin{bmatrix} 0 & 1 & 0 \\ 0 & a & 0 \\ 0 & 0 & 0 \end{bmatrix}\mathbf{z} + \begin{bmatrix} 0 \\ \hat{\varphi}h^p \\ (1/h) \end{bmatrix} =: \mathbf{F}\mathbf{z} + \mathbf{g}(h). \tag{5.28}$$

Choice of error measures for ODE integration methods

Two simple error measures that are "global" in the sense discussed above are as follows. The first measures the error $\delta x(\cdot) =: z_2(\cdot)$ at the end of the interval of integration:

$$z_2(T) = \delta x(T). \tag{5.29}$$

The second measures the accumulated error, i.e., the integral of the error on the interval $[0, T]$:

$$z_1(T) = \int_0^T \delta x(\tau)d\tau. \tag{5.30}$$

Note that the error and therefore the state variables z_1 and z_2 decrease as the stepsize h approaches zero. Clearly, other measures of error, such as the integral of $|\delta x|$ or of $(\delta x)^2$, are better choices but are more difficult to calculate: for simplicity, therefore, the error measures $z_1(T)$ and $z_2(T)$ will be considered.

The computational effort is proportional to the number of steps N. The number of steps N increases as a function of $1/h$ and, in fact, from the definition of the state variable z_3 it follows that

$$z_3(T) = N. \tag{5.31}$$

Choice of a cost function for an ODE integration method

Since we want to reduce error and keep the computational effort as low as possible, it is reasonable to define a general cost function that is a linear combination of the final values of the three state vector components

$$J(\mathbf{z}(T)) = \alpha_1 z_1(T) + \alpha_2 z_2(T) + \alpha_3 z_3(T) = \boldsymbol{\alpha}^T \mathbf{z}(T). \tag{5.32}$$

Two special cases of the cost function J that are used in what follows are

$$J_1 = z_2(T) + \alpha_3 z_3(T) \tag{5.33}$$

$$J_2 = z_1(T) + \alpha_3 z_3(T). \tag{5.34}$$

In the first cost function J_1, a compromise is sought between final error and number of steps, whereas, in the second cost function J_2, a compromise between accumulated error and number of steps is sought.

Stepsize control formulated and solved as an optimal control problem

With these preliminaries, the stepsize control problem can be formulated as the following optimal control problem:

$$\begin{aligned} &\min J(\mathbf{z}(T)) =: S(\mathbf{z}(T)) = \boldsymbol{\alpha}^T \mathbf{z}(T) \\ &\text{s.t. } \dot{\mathbf{z}} = \mathbf{f}(\mathbf{z}, h), \quad \mathbf{z}(0) \text{ given,} \end{aligned} \tag{5.35}$$

where J is defined in (5.32) and the dynamics for $\mathbf{z}$ is defined in (5.27). In optimal control terminology, this is a fixed final time, free final state problem, and indeed, the objective is to choose the control (i.e., stepsize $h(t)$) over the fixed horizon $[0, T]$ in such a way that the free final state $\mathbf{z}(T)$ minimizes the cost J.

Denoting the costate vector as $\boldsymbol{\lambda} = (\lambda_1, \lambda_2, \lambda_3)$, the Hamiltonian H is defined as

$$H := \boldsymbol{\lambda}^T \mathbf{f} = \lambda_1 z_2 + \lambda_2(a z_2 + \hat{\varphi} h^p) + \lambda_3 h^{-1}. \tag{5.36}$$

The costate or adjoint equations are calculated from $\dot{\boldsymbol{\lambda}} = -(\partial H / \partial \mathbf{z})$ as

$$\begin{aligned} \dot{\lambda}_1 &= 0, \\ \dot{\lambda}_2 &= -(\lambda_1 + a\lambda_2), \\ \dot{\lambda}_3 &= 0. \end{aligned} \tag{5.37}$$

From optimal control theory, for the optimal costate trajectory $\boldsymbol{\lambda}^*(\cdot)$:

$$\boldsymbol{\lambda}^*(T) = \left.\frac{\partial S}{\partial \mathbf{z}}\right|_T = \boldsymbol{\alpha}. \tag{5.38}$$

In particular, the terminal conditions for the optimal cost functions J_1^* and J_2^* can be tabulated as follows:

Table 5.1. *Choices of costate terminal conditions for the optimal cost functions J_1^* and J_2^*.*

	$\lambda_1(T)$	$\lambda_2(T)$	$\lambda_3(T)$
J_1^*	0	1	α_3
J_2^*	1	0	α_3

From (5.37) it follows that λ_1 and λ_3 are constant and hence determined by the terminal conditions (5.38). This allows for the simplification of the equation for λ_2, given the terminal values specified in Table 5.1. For the cost function J_1, the equation for λ_2 simplifies to

$$\dot{\lambda}_2 = -a\lambda_2, \qquad \lambda_2(T) = 1, \tag{5.39}$$

while, for the cost function J_2, the equation for λ_2 becomes

$$\dot{\lambda}_2 = -1 - a\lambda_2, \qquad \lambda_2(T) = 0. \tag{5.40}$$

The solution of (5.39) is

$$\lambda_2(t) = \exp \int_t^T a(\tau) \mathrm{d}\tau = b(T)/b(t), \tag{5.41}$$

where

$$b(t) := \int_0^t a(\tau) \mathrm{d}\tau. \tag{5.42}$$

The solution of (5.40) is

$$\lambda_2(t) = \frac{\int_t^T b(\tau) \mathrm{d}\tau}{b(t)}. \tag{5.43}$$

Substituting these results into the expression for the Hamiltonian yields the following expression, for cost function J_1:

$$H_1 = \lambda_2(az_2 + \hat{\varphi}h^p) + \alpha_3 h^{-1}, \tag{5.44}$$

and for cost function J_2:

$$H_2 = z_2 + \lambda_2(az_2 + \hat{\varphi}h^p) + \alpha_3 h^{-1}. \tag{5.45}$$

The optimal control $h^*(t)$ minimizes the Hamiltonian and, since H_1 and H_2 differ only in a term that does not involve h, $\partial H_1/\partial h = \partial H_2/\partial h$. Setting either of the latter partial derivatives to zero yields the optimal control (i.e., stepsize), for both cost functions J_1 and J_2, as

$$h^*(t) = \left(\frac{\alpha_3}{p\lambda_2\hat{\varphi}}\right)^{\frac{1}{p+1}}. \tag{5.46}$$

This expression can be written as $\beta(\lambda_2\hat{\varphi})^{-\frac{1}{p+1}}$, where the constant $\beta^{p+1} = \alpha_3/p$ can be determined from the specified total number of steps N as follows:

$$N = \int_0^T \left(\frac{1}{h(\tau)}\right) \mathrm{d}\tau = \left(\frac{1}{\beta}\right) \int_0^T (\lambda_2\hat{\varphi})^{\frac{1}{p+1}} \mathrm{d}\tau, \tag{5.47}$$

whence

$$\beta = \frac{1}{N} \int_0^T (\lambda_2\hat{\varphi})^{\frac{1}{p+1}} \mathrm{d}\tau. \tag{5.48}$$

Simple theoretical results on optimal stepsize control

The optimal control approach outlined above allows for the deduction of optimal stepsize choices, in a precise sense, for some simple classes of problems.

For instance, given a set of linear differential equations:

$$\dot{\mathbf{x}} = \mathbf{F}\mathbf{x}, \quad \mathbf{x}(0) = \mathbf{x}_0 \in \mathbb{R}^n, \ \mathbf{F} \in \mathbb{R}^{n\times n} \ \text{ given.} \tag{5.49}$$

Denoting the error by $\delta\mathbf{x} =: \mathbf{z}$, in analogy with the scalar case (5.22), the error dynamics can be written as

$$\dot{\mathbf{z}} = \mathbf{F}\mathbf{z} + \frac{\mathbf{F}^{p+1}\mathbf{x}}{(p+1)!} h^p, \quad \mathbf{z}(0) = \mathbf{z}_0. \tag{5.50}$$

For this problem, the corresponding optimal control problem for stepsize choice can be formulated as follows.

Let a cost function (i.e., figure of merit for error) be defined as

$$J_3 = \mathbf{w}^T \delta\mathbf{x}(T) + \alpha N = \mathbf{w}^T \mathbf{z}(T) + \alpha N, \tag{5.51}$$

where $\mathbf{w} = (w_1, \ldots, w_n)$ is a vector containing the weights w_i for each error $\delta\mathbf{x}_i$. Since the number of steps N can be written as

$$N = \int_0^T (1/h(\tau)) \mathrm{d}\tau, \tag{5.52}$$

the cost function can be rewritten as

$$J_3 = \mathbf{w}^T \mathbf{z}(T) + \alpha \int_0^T (1/h(\tau))\mathrm{d}\tau. \tag{5.53}$$

An optimal control problem is then defined as

$$\begin{aligned} &\min J_3 \\ &\text{s.t. the dynamics (5.50).} \end{aligned} \tag{5.54}$$

For (5.54), the following theorem holds.

Theorem 5.1. *The strategy of constant stepsize is optimal, in the sense of problem* (5.54), *for integration of a constant coefficient linear ODE* (5.49).

Proof. The Hamiltonian H_3 corresponding to the optimal control problem (5.54) can be written as

$$H_3 = \boldsymbol{\lambda}^T \mathbf{F}\mathbf{z} + \frac{\boldsymbol{\lambda}^T \mathbf{F}^{p+1}\mathbf{x}}{(p+1)!} h^p + \alpha h^{-1}. \tag{5.55}$$

The costate dynamics is given by

$$\dot{\boldsymbol{\lambda}} = -\mathbf{F}^T \boldsymbol{\lambda}, \quad \boldsymbol{\lambda}(T) = \mathbf{w}. \tag{5.56}$$

The optimal h minimizes the Hamiltonian H_3 and is therefore a solution of $\partial H_3/\partial h = 0$, i.e.,

$$\frac{p\boldsymbol{\lambda}^T \mathbf{F}^{p+1}\mathbf{x} h^{p-1}}{(p+1)!} - \frac{\alpha}{h^2} = 0,$$

which can be rearranged as

$$h^{p+1} = \frac{(p+1)!\alpha}{p(\boldsymbol{\lambda}^T \mathbf{F}^{p+1}\mathbf{x})}. \tag{5.57}$$

Calculation shows that the time derivative of the denominator term $\boldsymbol{\lambda}^T \mathbf{F}^{p+1}\mathbf{x}$ is zero:

$$\begin{aligned} \mathrm{d}(\boldsymbol{\lambda}^T \mathbf{F}^{p+1}\mathbf{x})/\mathrm{d}t &= \dot{\boldsymbol{\lambda}}^T \mathbf{F}^{p+1}\mathbf{x} + \boldsymbol{\lambda}^T \mathbf{F}^{p+1}\dot{\mathbf{x}} \\ &= -\boldsymbol{\lambda}^T \mathbf{F}^{p+2}\mathbf{x} + \boldsymbol{\lambda}^T \mathbf{F}^{p+2}\mathbf{x} = 0, \end{aligned}$$

implying that the optimal choice of h is a constant. □

Another simple theorem that is deduced directly from (5.46) is as follows.

Theorem 5.2. *For the problem $\dot{x} = f(t)$, a strategy of constant error generation per time step is optimal.*

Proof. For this problem, the solution of which is simple integration of $f(\cdot)$, the associated optimal control problem has cost function J_1, with $a = 0$ and $\lambda_2 = 1$. Thus, from (5.46),

$$h = c\hat{\varphi}^{-\frac{1}{p+1}}. \tag{5.58}$$

Since the error generation per step is $\hat{\varphi}h^{p+1}$ (5.9), the theorem follows. □

The optimal stepsize control procedure just outlined is clearly not applicable to problems in which $\lambda\hat{\varphi}$ changes sign in the time interval of interest. Basically, this is due to the use of cost functions like J in (5.32) in which positive and negative errors cancel out, leading to an ill-posed optimization problem. One remedy is to use an integral squared error criterion instead, e.g.,

$$J_4 = \int_0^T \sum_i w_i \delta x_i^2(\tau) \mathrm{d}\tau + \alpha N.$$

The drawback is that the resulting system of equations to be solved in order to find the optimal stepsize is very complicated, perhaps as complicated as the original ODE which it is supposed to solve. This means that the optimal control method, although it provides theoretical insight, will not always be practical in applications to general nonlinear ODEs.

A general conclusion about optimal stepsize is that

$$h_{\text{opt}} = c(\lambda\hat{\varphi})^{-\frac{1}{p+1}}, \tag{5.59}$$

where c is a proportionality constant. This choice of stepsize is applicable to both linear and nonlinear problems and to many kinds of cost functions. Although (5.59) has been derived and written in different forms only for some specific classes of problems ((5.46) and (5.58)), it is a simple, general, and theoretically useful expression, derived from a natural optimal control problem formulation.

5.2 A Feedback Control Perspective on the Shooting Method for ODEs

> *Trifft's ist's gut; trifft's nicht, ist die moralische Wirkung eine ungeheure.* (If it hits, that's good; if it misses, the moral impact will be immense.)
>
> —Motto of the imperial Austrian field artillery, quoted in [DB02].

In this section a discrete state space representation of an ODE is used to provide a feedback control formulation for the iterative shooting method used to solve two-point boundary value problems (BVPs) in ODEs; with this formulation in hand, convergence conditions for these iterative methods can be obtained.

The solution of an nth-order ODE depends on the specification of n boundary conditions. Unlike the IVP, where these conditions are specified at one point, BVPs are subject to conditions that are given at different points. The most common case is the two-point boundary value problem (TPBVP), where the n boundary conditions are split in two; i.e., m conditions are given at the initial point and $m - n$ at the final point. An IVP can be solved by numerical integration, and many methods like Euler, Runge–Kutta, etc. have been proposed (cf. section 5.1). To take advantage of these numerical integration methods in the case of a TPBVP, one approach is to use the so-called (iterative) shooting method, which is the focus of this section.

The objective of this section is to express the shooting method for the solution of TPBVPs for *linear* ODEs in the mathematical framework of state space linear system realizations and feedback systems. The motivation for choosing linear ODEs, instead of nonlinear ODEs, which is the main application of shooting methods, is that this allows for the establishment of relations between the concepts of shooting method, feedback control, and the iterative solution of algebraic linear systems in the simplest case. Of course, all that is said can be suitably extended to the nonlinear case as well.

It is demonstrated how finite-difference approximations associated with state space representations enable the shooting iterative method to be interpreted from a feedback perspective. For linear ODEs, convergence conditions and error analysis are also discussed.

A linear-time invariant nth-order ODE can always be discretized using finite-difference approximation methods and thus can be represented as a classical discrete-time state-space model (cf. (1.1)),

$$\begin{aligned} \mathbf{z}_{k+1} &= \mathbf{F}\mathbf{z}_k + \mathbf{G}\mathbf{u}_k, \\ \mathbf{y}_k &= \mathbf{H}\mathbf{z}_k + \mathbf{J}\mathbf{u}_k, \end{aligned} \tag{5.60}$$

where $k = 0, 1, 2, \ldots$. In (5.60), $\mathbf{z}_k$ is the state vector of dimension n. The matrices $\mathbf{F}, \mathbf{G}, \mathbf{H}$, and $\mathbf{J}$ have dimensions $n \times n$, $n \times p$, $r \times n$, and $r \times p$, respectively; $\mathbf{u}_k$ is the input function; and $\mathbf{y}_k$ is the output of the system. In the case of dynamical systems, the independent variable is the time t. Therefore, when we proceed to the discretization of the corresponding ODE, this independent variable is consequently also discretized. In order to discretize the independent variable t, a constant stepsize h is chosen, and the relationship between the continuous- and discrete-time variables is given by $t = kh$, where k is the iteration counter.

In what follows, a TPBVP and the associated shooting procedure are formulated, taking as a reference the discretized system (5.60). This is done with the corresponding boundary conditions $\mathbf{y}_0$ (vector of initial conditions) and $\mathbf{y}_N = \boldsymbol{\alpha}$ (vector of final conditions).

5.2.1 A state space representation of the shooting method

Although the shooting method is mainly used for the solution of nonlinear ODEs, this section takes the case of linear ODEs as the starting point, beginning with a simple example.

Example 5.3. Consider a typical TPBVP for the linear ODE:

$$\frac{d^2 y(t)}{dt^2} = y(t) + f(t) \tag{5.61}$$

with the *boundary conditions* $y(a) = y_0$ and $y(b) = \alpha$, the domain of interest for the independent variable being the continuous closed interval $[a, b]$. Note that, for the solution of the IVP associated with (5.61), it is necessary to specify the values of the *initial conditions* $y(a)$ and $\dot{y}(a)$.

Using a central difference approximation [KH83] with a grid spacing (h) for the domain $[a, b]$, such that $h = (b - a)/N$, where $N + 1$ is the number of points in the grid (discrete domain of interest) $[0, 1, 2, \ldots, N]$, we get the scalar difference equation

$$y_{k+2} - \left(2 + h^2\right) y_{k+1} + y_k = h^2 f_{k+1} \tag{5.62}$$

with the boundary conditions y_0 and y_N.

A discrete state space representation, or realization, in the form (5.60), corresponding to this difference equation is straightforward. Defining the vector $\mathbf{z}_k = (z_k^1, z_k^2) = (y_k, y_{k+1})$

and $u_k = f_{k+1}$ yields

$$\begin{bmatrix} z_{k+1}^1 \\ z_{k+1}^2 \end{bmatrix} = \begin{bmatrix} 0 & 1 \\ -1 & (2+h^2) \end{bmatrix} \begin{bmatrix} z_k^1 \\ z_k^2 \end{bmatrix} + \begin{bmatrix} 0 \\ h^2 \end{bmatrix} u_k, \tag{5.63}$$

$$y_k = [1 \quad 0] \begin{bmatrix} z_k^1 \\ z_k^2 \end{bmatrix}, \tag{5.64}$$

where (5.63) and (5.64) represent the discretization of (5.61). The boundary conditions given can be represented as follows: $\mathbf{z}(0) = (z_0^1, z_0^2) = (y_0, \boxed{?})$, *and* $\mathbf{z}(N) = (z_N, z_{N+1}) = (y_N, \boxed{?})$, where $\boxed{?}$ represents unknown entries. Thus if z_0^2 is specified, the initial condition $\mathbf{z}(0)$ is completely known and (5.63) and (5.64) define a discrete-time IVP. When the boundary conditions y_0 and y_N are specified, both the initial and final conditions $\mathbf{z}(0)$ and $\mathbf{z}(N)$ are only partially known, and (5.63) and (5.64) define a TPBVP.

For the IVP associated with (5.60), the state $\mathbf{z}_k$ and output $\mathbf{y}_k$ trajectories are given by the closed-form expressions

$$\begin{aligned} \mathbf{z}_k &= \mathbf{F}^k \mathbf{z}_0 + \sum_{j=0}^{k-1} \mathbf{F}^{k-j-1} \mathbf{G} \mathbf{u}_j, \\ \mathbf{y}_k &= \mathbf{H} \mathbf{z}_k + \mathbf{J} \mathbf{u}_k. \end{aligned} \tag{5.65}$$

For Example 5.3, $\mathbf{y}_k = y_k$, and from (5.63), (5.64), these matrices are

$$\mathbf{F} = \begin{bmatrix} 0 & 1 \\ -1 & (2+h^2) \end{bmatrix}, \quad \mathbf{G} = \begin{bmatrix} 0 \\ h^2 \end{bmatrix}, \quad \mathbf{H} = [1 \quad 0], \quad \mathbf{J} = 0. \tag{5.66}$$

One way to solve a TPBVP is to transform it into an IVP by arbitrarily guessing the values for the unknown components of $\mathbf{z}_0$. In order to organize the calculations, the initial state vector $\mathbf{z}_0$ is partitioned into two subvectors as follows:

$$\mathbf{z}_0 = \begin{bmatrix} \mathbf{z}_0^g \\ \mathbf{z}_0^a \end{bmatrix}, \tag{5.67}$$

where the subvectors $\mathbf{z}_0^g$ and $\mathbf{z}_0^a$ correspond, respectively, to the given initial values and the unknown values. In the case of Example 5.3, these values are: $\mathbf{z}_0^g = z_0^1 = y_0$ and $\mathbf{z}_0^a = z_0^2 = y_1$.

An iterative shooting method is described as follows. After N steps, using (5.65) either explicitly or implicitly (by iterating the corresponding state space equation (5.60)), the final vector $\mathbf{y}_N$ is obtained. The calculated final components are compared with the corresponding given final conditions. The error (or discrepancy) vector between these two values is fed back to apply a correction to the guessed initial conditions (in the case of Example 5.3, the value of $z_0^2 = y_1$ is adjusted).

Using (5.65), after N steps we have

$$\mathbf{z}_N = \mathbf{F}^N \mathbf{z}_0 + \sum_{j=0}^{N-1} \mathbf{F}^{N-j-1} \mathbf{G} \mathbf{u}_j, \tag{5.68}$$

$$\mathbf{y}_N = \mathbf{H} \mathbf{z}_N + \mathbf{J} \mathbf{u}_N. \tag{5.69}$$

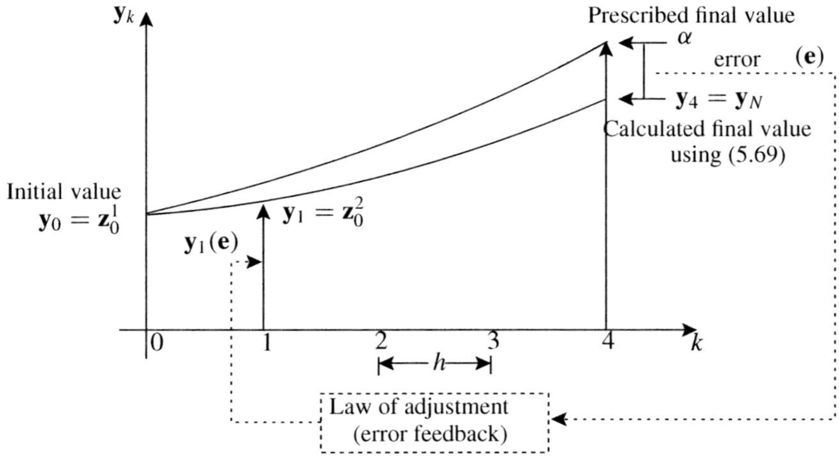

Figure 5.3. *Pictorial representation of a shooting method based on error feedback.*

Since the set of initial conditions is incomplete, some elements have been arbitrarily assigned, and consequently, the calculated value of $\mathbf{y}_N$ is not the correct one. With the discrepancy between the calculated value of $\mathbf{y}_N$ and the value given by the boundary condition α, the elements of $\mathbf{z}_0^a$ have to be modified in order to reduce this discrepancy in the next iteration (entry z_0^2 in Example 5.3). The basic idea, for Example 5.3, is represented graphically in Figure 5.3.

Each time a modification of the initial condition $\mathbf{z}_0$ is carried out, and $\mathbf{y}_N = \mathbf{H}z_N + \mathbf{J}\mathbf{u}_N$ is evaluated, we say that one *shooting iteration* has been completed. Given an initial condition $\mathbf{z}_0$ it is necessary to perform N steps, or iterations of (5.60), or, equivalently, to set $k = N$ in (5.65) in order to get $\mathbf{y}_N$. Thus one shooting iteration corresponds to N iterations of (5.60), which is, in turn, equivalent to setting the counter $k = N$ in (5.65). A new counter m is associated with each shooting iteration.

In order to complete the state space description of the iterative shooting method, the following variables are defined as

$$\mathbf{q}_m := \mathbf{z}_{N,m}, \qquad \mathbf{w}_m := \mathbf{z}_{0,m}, \tag{5.70}$$

where $\mathbf{q}_m$ represents the vector of $\mathbf{z}_N$ at the mth shooting iteration (i.e., corresponds to the mth application of the expression (5.68)), and similarly $\mathbf{w}_m$ represents the vector $\mathbf{z}_0$ at the mth shooting iteration. Noting that the second term in (5.68) is constant for each iteration of the shooting, it can be represented by a constant vector $\overline{\mathbf{g}}$, where

$$\overline{\mathbf{g}} := \sum_{j=0}^{N-1} \mathbf{F}^{N-j-1}\mathbf{G}\mathbf{u}_j. \tag{5.71}$$

From definitions (5.70) and (5.71), we can write the relationships

$$\begin{aligned} \mathbf{q}_m &= \mathbf{F}^N \mathbf{w}_m + \overline{\mathbf{g}}, \\ \mathbf{y}_{N,m} &= \mathbf{H}\mathbf{q}_m, \end{aligned} \tag{5.72}$$

where $\mathbf{F}^N$ represents the Nth power of matrix $\mathbf{F}$ and, without loss of generality, matrix $\mathbf{J}$ from (5.69) is set to zero.

As can be observed, the value of $\mathbf{y}_{N,m} = \mathbf{H}\mathbf{q}_m$ is associated with a given value of $\mathbf{z}_0$ or equivalently to $\mathbf{w}_m$ at the mth iteration of the shooting. The error in each iteration, associated with the iteration counter m, is defined by the discrepancy between the prescribed or correct vector $\boldsymbol{\alpha}$ and the calculated vector $\mathbf{y}_{N,m}$, i.e.,

$$\mathbf{e}_m = \boldsymbol{\alpha} - \mathbf{y}_{N,m} = \boldsymbol{\alpha} - \mathbf{H}\mathbf{q}_m, \tag{5.73}$$

where $\mathbf{e}_m \in \mathbb{R}^r$. In order to complete the state space description, it is necessary to define an update law for the subvector of arbitrated initial conditions as a function of the error $\mathbf{e}_m$. This can be achieved by the use of a dynamic feedback controller of the form

$$\begin{aligned} \boldsymbol{\delta}_{m+1} &= \boldsymbol{\delta}_m + \mathbf{G}_c\mathbf{K}\mathbf{e}_m, \\ \mathbf{w}_m &= \mathbf{H}_c\boldsymbol{\delta}_m + \mathbf{v}, \end{aligned} \tag{5.74}$$

where $\boldsymbol{\delta} := \mathbf{z}_0^a$ is the state variable vector which corresponds to the subvector of $\mathbf{z}_0$ that needs to be arbitrated, and $\mathbf{v} := (\mathbf{z}_0^g, 0)$ is a constant vector which corresponds to the part of the initial value vector $\mathbf{z}_0$ that is given. From this definition of $\mathbf{v}$, it is clear that the $n \times r$ matrix $\mathbf{H}_c$ should be chosen as

$$\mathbf{H}_c = \begin{bmatrix} \mathbf{0} \\ \mathbf{I} \end{bmatrix}. \tag{5.75}$$

The matrix $\mathbf{K}$ expresses a feedback gain matrix with adequate dimensions and the matrices $\mathbf{G}_c$ and $\mathbf{H}_c$ play, respectively, the roles of input and output coupling matrices of the dynamic controller.

Combining (5.72), (5.73), and (5.74), the following closed-loop equation for the tandem connection of the controller (5.74) and plant (5.72) dynamics is obtained:

$$\begin{aligned} \boldsymbol{\delta}_{m+1} &= \left(\mathbf{I} - \mathbf{G}_c\mathbf{K}\mathbf{H}\mathbf{F}^N\mathbf{H}_c\right)\boldsymbol{\delta}_m + \mathbf{G}_c\mathbf{K}\left(\boldsymbol{\alpha} - \mathbf{H}\mathbf{F}^N\mathbf{v} - \mathbf{H}\overline{\mathbf{g}}\right), \\ \mathbf{w}_m &= \mathbf{H}_c\boldsymbol{\delta}_m + \mathbf{v}. \end{aligned} \tag{5.76}$$

In this case, the dynamics of this iterative system is governed by the feedback law described by (5.74) and the error, in each iteration, is defined by (5.73).

Therefore, (5.76) is a state space representation of the iterative shooting method. In terms of a control perspective, this equation corresponds to a closed-loop dynamical system with an output feedback law given by (5.74), where $\mathbf{K}$ is the corresponding feedback gain matrix. This perspective is represented in Figure 5.4. The controller (5.74) is of the class of integral controllers that also appeared in section 5.1. Once again, a control interpretation opens up the possibility of using controllers with different dynamics such as the class of PI controllers, or even more general classes of controllers.

5.2.2 Error dynamics of the iterative shooting scheme

Considering the error variable defined in (5.73) at the $(m+1)$th iteration of the shooting iterative scheme (5.76), one has

$$\mathbf{e}_{m+1} = \boldsymbol{\alpha} - \mathbf{H}\mathbf{F}^N\mathbf{w}_{m+1} - \mathbf{H}\overline{\mathbf{g}}, \tag{5.77}$$

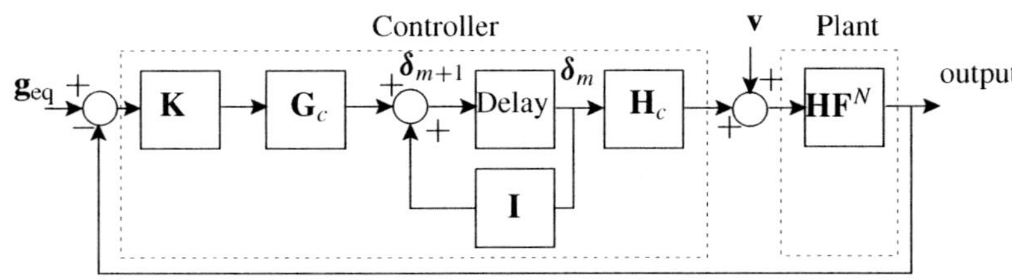

Figure 5.4. *The shooting method represented as a feedback control system in the standard configuration. The vector* $\mathbf{g}_{\text{eq}}$ *is given by* $\boldsymbol{\alpha} - \mathbf{H}\overline{\mathbf{g}}$.

or equivalently,

$$\mathbf{e}_{m+1} = \boldsymbol{\alpha} - \mathbf{HF}^N(\mathbf{H}_c\boldsymbol{\delta}_m + \mathbf{v}) - \mathbf{HF}^N\mathbf{H}_c\mathbf{G}_c\mathbf{K}\mathbf{e}_m - \mathbf{H}\overline{\mathbf{g}}. \tag{5.78}$$

Using (5.72) we get

$$\mathbf{e}_{m+1} = \left(\mathbf{I} - \mathbf{HF}^N\mathbf{H}_c\mathbf{G}_c\mathbf{K}\right)\mathbf{e}_m. \tag{5.79}$$

Equation (5.79) is called the *error equation* for the closed-loop dynamical (or iterative) system (5.76). In order to ensure the convergence of this iterative shooting method, it is necessary to ensure the Schur stability (see Chapter 1) of the $r \times r$ matrix $\mathbf{S}$ defined as

$$\mathbf{S} := \left(\mathbf{I} - \mathbf{HF}^N\mathbf{H}_c\mathbf{G}_c\mathbf{K}\right). \tag{5.80}$$

The Schur stability criterion requires matrix $\mathbf{S}$ to have all the eigenvalues less than 1 in modulus, or equivalently that all eigenvalues lie inside the unit circle (see Chapter 1). Since $\mathbf{K}$ is an arbitrary matrix, one can resort to the eigenvalue placement for the system (5.79) using well-known methods in control theory (see, e.g., [ÅW84, Kai80]). Consequently, the correct placement of the eigenvalues of matrix $\mathbf{S}$ will provide the desired rate of convergence for the shooting method.

The eigenvalues of matrix $\mathbf{S}$ in (5.80), in terms of the matrix $\mathbf{K}$, are the roots of the characteristic equation

$$\det\left(\mathbf{I}\lambda - \mathbf{I} + \mathbf{HF}^N\mathbf{H}_c\mathbf{G}_c\mathbf{K}\right) = 0. \tag{5.81}$$

It is easily seen that if $\mathbf{K}$ equals $\left(\mathbf{HF}^N\mathbf{H}_c\mathbf{G}_c\right)^{-1}$, then all the eigenvalues of $\mathbf{S}$ are equal to zero; such a choice is referred to as *deadbeat control* [ÅW84].

Clearly, when the error defined in (5.73) equals zero, this means that the iterative scheme (5.76) associated with the shooting method has reached the desired solution, or equivalently $\boldsymbol{\delta}_{m+1} = \boldsymbol{\delta}_m = \boldsymbol{\delta}$, and by (5.78),

$$\boldsymbol{\alpha} = \mathbf{HF}^N\mathbf{H}_c\boldsymbol{\delta} + \mathbf{HF}^N\mathbf{v} + \mathbf{H}\overline{\mathbf{g}}, \tag{5.82}$$

where $\mathbf{HF}^N\mathbf{H}_c$ is an $r \times r$ square matrix and $\boldsymbol{\delta}$ represents the vector of the values that need to be arbitrated in order to transform the TPBVP into an IVP (in Example 5.3, the variable y_1). Solving (5.82) for $\boldsymbol{\delta}$ is equivalent to solving a linear system of algebraic equations,

$$\mathbf{A}_s\mathbf{x} = \mathbf{b}, \tag{5.83}$$

where $\mathbf{A}_s = \mathbf{HF}^N\mathbf{H}_c$, $\mathbf{b} = \boldsymbol{\alpha} - \mathbf{HF}^N\mathbf{v} - \mathbf{H}\overline{\mathbf{g}}$, and $\mathbf{x} = \boldsymbol{\delta}$.

Equation (5.83) can be solved, for example, by an iterative method, such as Jacobi or Gauss–Seidel [Var00], and once the solution of this equation ($\boldsymbol{\delta}$) is found, then we have determined $\mathbf{w} = \mathbf{z}_0$, which is the complete initial value vector corresponding to the solution of the TPBVP. Clearly (5.83) can also be (and usually is) solved by a direct method such as LU decomposition [AMR95].

We have shown that the so-called iterative shooting method for solving TPBVPs in ODEs can be represented by an iterative feedback description, thus making explicit the error feedback scheme inherent to this method.

In fact, in the shooting method as well as in the classical iterative methods used to solve an algebraic linear system of equations, obtaining a suitable feedback gain matrix $\mathbf{K}$ is equivalent to obtaining a suitable preconditioner matrix [SK01]. Finding an optimal preconditioner, i.e., one that minimizes the number of iterations of the iterative method, is equivalent to finding an optimal feedback gain matrix that induces suitable eigenvalue placement of an associated state space discrete-time dynamical system.

The above relationships, derived in [SK01], show the usefulness of feedback control techniques in the design of efficient iterative numerical algorithms to solve sets of algebraic and differential equations, both linear and nonlinear, and also make explicit the limitations of these methods. Applications of this control approach to the design of iterative methods for the solution of fluid dynamics problems, modeled by partial differential equations, are reported in [SKM04], where they were used, in conjunction with multilevel Schwarz schemes, to solve incompressible fluid flow problems.

Connection between shooting and ILC

Another conceptual connection, mentioned earlier (section 2.3.2) and emerging from the control interpretation of the shooting method, is its equivalence to the ILC scheme. ILC can be viewed as a rediscovery of shooting in a "control" context. To substantiate this claim, the ILC scheme for a linear-time invariant system or plant is outlined, following [SO91]. Consider the dynamical system

$$\begin{aligned} \dot{\mathbf{x}}(t) &= \mathbf{F}\mathbf{x}(t) + \mathbf{G}\mathbf{u}(t), \\ \mathbf{y}(t) &= \mathbf{H}\mathbf{x}(t) + \mathbf{J}\mathbf{u}(t), \end{aligned} \tag{5.84}$$

$$\mathbf{x}(T) = e^{t\mathbf{F}}\mathbf{x}_0 + \int_0^T e^{(t-\tau)\mathbf{F}}\mathbf{G}\mathbf{u}(\tau)\mathrm{d}\tau, \tag{5.85}$$

where $\mathbf{F} \in \mathbb{R}^{n\times n}$, $\mathbf{G} \in \mathbb{R}^{n\times p}$, $\mathbf{H} \in \mathbb{R}^{m\times n}$, $\mathbf{J} \in \mathbb{R}^{m\times p}$ are constant matrices. Suppose further that the dynamical system (5.84) can be operated repeatedly over a finite-time interval $[0, T]$ with the same initial condition $\mathbf{x}(0) = \mathbf{x}_0$. A desired output $\mathbf{y}_d(t)$, over the whole interval $[0, T]$, is specified in advance, and the objective is to obtain a control input sequence $\mathbf{u}_k(t)$ that generates the desired output exactly over this time interval. The ILC scheme arrives at such a control input iteratively as follows:

$$\mathbf{e}_k(t) = \mathbf{y}_d(t) - \mathbf{y}_k(t), \tag{5.86}$$

$$\dot{\mathbf{v}}_k(t) = \mathbf{F}_c(t)\mathbf{v}_k(t) + \mathbf{G}_c(t)\mathbf{e}_k(t), \tag{5.87}$$

$$\mathbf{w}_k(t) = \mathbf{H}_c(t)\mathbf{v}_k(t) + \mathbf{J}_c(t)\mathbf{e}_k(t), \tag{5.88}$$

$$\mathbf{u}_{k+1}(t) = \mathbf{u}_k(t) + \mathbf{w}_k(t). \tag{5.89}$$

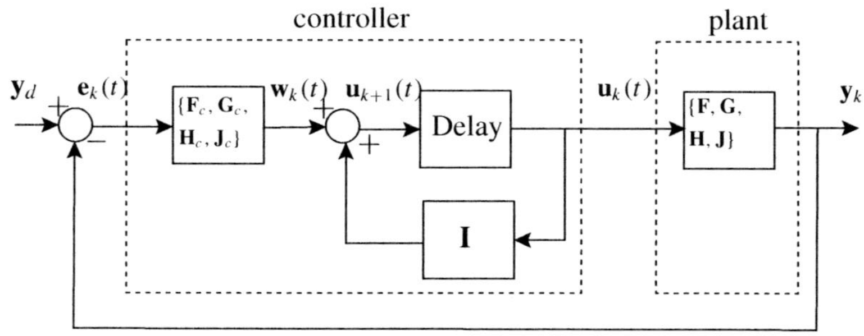

Figure 5.5. *Iterative learning control (ILC) represented as a feedback control system. The plant, the object of the learning scheme, is represented by the dynamical system* (5.84)*, while the (dynamic) controller is represented by* (5.87)–(5.88).

Here $\mathbf{u}_k(t)$ is the input at the kth iteration, $\mathbf{y}_k(t)$ and $\mathbf{e}_k(t)$ are the corresponding output and error vectors, and $\mathbf{v}_k(t)$ and $\mathbf{w}_k(t)$ are the controller state and controller output vectors, respectively, that are used to generate the new control input $\mathbf{u}_{k+1}$ from the previous control input and from the error vector sequences $\{\mathbf{u}_k(t), \mathbf{y}_k(t) : t \in [0, T]\}$. The initial state of the controller $\mathbf{v}_k(0)$ is assumed to be zero. The resulting learning control system is shown in block diagram form in Figure 5.5, which is similar to the linear iterative method depicted in Figure 5.4. One of the basic results of ILC theory is stated as follows.

Theorem 5.4 [SO91]. *Suppose that the learning control scheme* (5.86)–(5.89) *is applied to the plant* (5.84)*. Then, if*

$$\|\mathbf{I}_m - \mathbf{J}\mathbf{J}_c(t)\| < 1, \quad \textit{for all } t \in [0, T] \tag{5.90}$$

holds, then the learning process is convergent in the sense that there exist numbers $0 \le b_0 < 1$ *and* $q \ge 0$ *such that*

$$\|\mathbf{e}_{k+1}(t)\|_q \le b_0 \|\mathbf{e}_k(t)\|_q \tag{5.91}$$

holds for any k, which implies that $\mathbf{e}_k(t) \to \mathbf{0}$ *as* $k \to \infty$.

From this theorem, it is not difficult to show that it is possible to obtain a control input that yields an arbitrary desired output by the ILC scheme *if and only if* the plant is (right) invertible [SO91].

From Figure 5.5 it becomes clear that ILC, as well as the shooting method, can be easily recast in an iterative feedback control framework (see Figure 2.12b). Furthermore, in order to connect the above results with linear iterative methods, consider a static plant $\mathbf{y} = \mathbf{D}\mathbf{u}$ and the iterative method

$$\mathbf{u}_{k+1} = \mathbf{u}_k + \mathbf{D}_c(\mathbf{y} - \mathbf{D}\mathbf{u}_k), \tag{5.92}$$

where $\mathbf{D}_c$ is chosen appropriately.

Notice that (5.92) is exactly the standard linear iterative method for solving the linear equation $\mathbf{D}\mathbf{u} = \mathbf{y}$ (compare with (2.109) in section 2.3, replacing $\mathbf{u}$ with $\mathbf{x}$, $\mathbf{y}$ with $\mathbf{b}$, and $\mathbf{D}$ with $\mathbf{A}$).

Compare the block diagram of Figure 5.4 (depicting shooting) with that of Figure 5.5 (depicting ILC). It becomes clear that ILC has the same structure as the well-known shooting method.

This connection should certainly enable the use of techniques from the vast literature on shooting to improve ILC schemes and vice versa. Specifically, "ILC-inspired shooting" can be used to extend the shooting method presented in this section to the case of nonlinear ODEs.

Another connection deserves comment. Assume that the ILC method is applied to a discrete-time plant (i.e., suppose that the dynamics in (5.84) are discrete time) and recall that the shooting method will integrate the given ODE using some discretization method, resulting in discrete-time dynamics. Thus, both the shooting and ILC methods can be thought of as involving two discrete dynamical systems, one being the plant and the other being the controller. Let the "time" variable of the plant be denoted t and that of the controller denoted k. Since the controller is thought of as belonging to an "outer loop" involving the iterative scheme, we are naturally led to the idea of using the so-called two-dimensional discrete-time systems (which are discrete versions of PDEs), evolving two time counters t and k [KZ93], and once again there is scope for a mutually beneficial interaction between control and numerical techniques, since there is extensive literature on the control and system theory of multidimensional systems [Bos03].

5.3 A Decentralized Control Perspective on Diagonal Preconditioning

The condition number $\kappa(\mathbf{A})$ of a matrix $\mathbf{A}$, measured in the l_2-norm, is the ratio of the largest singular value of $\mathbf{A}$ to the smallest:

$$\kappa(\mathbf{A}) = \|\mathbf{A}\|\|\mathbf{A}^{-1}\| = \sigma_{max}(\mathbf{A})/\sigma_{min}(\mathbf{A}).$$

It is an important quantity in the sensitivity and convergence analysis of many problems in numerical linear algebra [Dem97, GL89, Dat95]. The sensitivity of a linear system and the convergence of the conjugate gradient method are two of the best known examples where the condition number plays a determining role. The latter method is an important motivation for the conditioning problem, for if the condition number of a given matrix can be made closer to unity by pre- and postmultiplying it by the same diagonal matrix, then faster convergence of the conjugate gradient method is assured [GL89].

The optimal condition number of a matrix $\mathbf{A}$ is the minimum, over all positive diagonal matrices $\mathbf{P}$, of $\kappa(\mathbf{PA})$.

In this section we interpret the problem of finding the optimal preconditioner $\mathbf{P}$ that minimizes $\kappa(\mathbf{PA})$ as the equivalent problem of maximally clustering the poles of a suitably defined dynamical system by the choice of a positive diagonal stabilizing feedback matrix $\mathbf{K}(= \mathbf{P}^2)$. This allows us to give a control-theoretic proof of a characterization of perfect preconditioners, thereby making connections between the Hadamard and Wielandt inequalities and the condition number, and to use results on constrained linear quadratic (LQ) optimal control to give a control interpretation for optimal preconditioners.

In this section we consider only one-sided diagonal conditioning (for the most part, preconditioning, which corresponds to scaling the rows of a given matrix by the respective diagonal elements of the preconditioning matrix). Much work has already been done on theoretical and practical aspects of the preconditioning problem (also known as the scaling problem). Here we confine ourselves to referencing a few of the theoretical papers most relevant to this subject [Bau63, FS55, GV74, MS73, Sha82, vdS69, BM94b].

Let us define perfect and optimal diagonal preconditioners for a real nonsingular matrix $\mathbf{A}$. Though much of the discussion in this section is valid for complex matrices as well, we will restrict ourselves to real matrices.

Definition 5.5. *A real diagonal matrix* $\mathbf{D}$ *is said to be a* perfect preconditioner *for* $\mathbf{A}$ *if* $\kappa(\mathbf{DA}) = 1$.

It is clear that a perfect preconditioner may not exist for an arbitrary $\mathbf{A}$, so we are led to the following definition.

Definition 5.6. *A diagonal matrix* $\mathbf{D}^{\text{opt}}$ *is said to be an* optimal preconditioner *for* $\mathbf{A}$ *if* $\kappa(\mathbf{D}^{\text{opt}}\mathbf{A})$ *is the infimum, over all diagonal matrices* $\mathbf{D}$, *of* $\kappa(\mathbf{DA})$. *The* optimal condition number $c(\mathbf{A})$ *is defined as* $\kappa(\mathbf{D}^{\text{opt}}\mathbf{A})$.

McCarthy and Strang [MS73] survey the available results for the l_2-norm and find upper and lower bounds for the optimal condition number $c(\mathbf{A})$ of a matrix $\mathbf{A}$, which can also be defined as the infimum, over all diagonal matrices $\mathbf{D}$, of $\|\mathbf{DA}\|\|(\mathbf{DA})^{-1}\|$. These bounds are expressed as

$$b(\mathbf{A}) \leq c(\mathbf{A}) \leq 2b(\mathbf{A}),$$

where $b(\mathbf{A})$ is the supremum, over diagonal unitary matrices $\mathbf{U}$, of $\|\mathbf{U}^{-1}\mathbf{AU}\|$. The lower bound was obtained in [Bau63] and is simple to prove; however, establishing the upper bound is nontrivial. It is also worthwhile to point out that the results of McCarthy and Strang do not provide a computational technique to find a diagonal matrix $\mathbf{D}$ that achieves the optimum or, indeed, one that reduces the condition number. Computational procedures based on convex optimization are discussed in [BM94b].

We reinterpret a geometric version of Wielandt's inequality as a characterization of perfectly preconditionable matrices—those that can be diagonally preconditioned so as to have a condition number equal to unity. We also discuss, in the 2×2 case, the problem of optimal preconditioning; i.e., minimization of the condition number and a geometric argument using Hadamard's inequality allows us to characterize optimal 2×2 preconditioners. We also interpret the preconditioning problem as a pole allocation problem via decentralized feedback applied to a suitably defined dynamical system. This allows us to rederive the characterization result via control theory and, moreover, leads naturally to a formulation of the optimal preconditioner problem as the problem of determining an optimal diagonal feedback, which is an LQ problem with the feedback matrix constrained to be diagonal.

Perfect diagonal conditioning

The characterization of the class of matrices that can be perfectly conditioned by a diagonal matrix is essentially contained in Wielandt's inequality. A geometric version of this inequality is stated below to derive an explicit characterization of perfect conditionability.

Theorem 5.7 (Wielandt's inequality). *Let* $\mathbf{A} \in \mathbb{R}^{n\times n}$ *be a given nonsingular matrix with condition number* $\kappa(\mathbf{A})$*, and define the angle* $\theta(\mathbf{A})$ *in the first quadrant by*

$$\theta(\mathbf{A}) = 2\cot^{-1}[\kappa(\mathbf{A})]. \tag{5.93}$$

Then

$$|\langle \mathbf{Ax}, \mathbf{Ay}\rangle| \leq \|\mathbf{Ax}\|\|\mathbf{Ay}\|\cos[\theta(\mathbf{A})] \tag{5.94}$$

for every pair of orthogonal vectors $\mathbf{x}, \mathbf{y} \in \mathbb{R}^n$*, where* $\langle \mathbf{u}, \mathbf{v}\rangle := \mathbf{v}^T\mathbf{u}$ *denotes the Euclidean inner product and* $\|\mathbf{u}\| = (\mathbf{u}^T\mathbf{u})^{1/2}$ *denotes the Euclidean norm. Moreover, there exists an orthonormal pair of vectors* $\mathbf{x}, \mathbf{y} \in \mathbb{R}^n$ *for which equality holds in* (5.94).

Wielandt's inequality (5.94) leads to a geometrical interpretation of the angle $\theta(\mathbf{A})$: *The minimum angle between* $\mathbf{Ax}$ *and* $\mathbf{Ay}$*, as* $\mathbf{x}$ *and* $\mathbf{y}$ *range over all possible orthonormal pairs of vectors, is given by* $\theta(\mathbf{A}) = 2\cot^{-1}[\kappa(\mathbf{A})]$.

A proof of Wielandt's inequality, discussion of the geometrical interpretation, and other useful information can be found in [HJ88].

In light of the preceding interpretation, perfect diagonal conditionability is characterized as follows.

Proposition 5.8 (characterization of perfect diagonal conditionability). *Given* $\mathbf{A} \in \mathbb{R}^{n\times n}$

$$\exists \mathbf{P} \in \mathcal{D} \text{ such that } \kappa(\mathbf{PA}) = 1 \Leftrightarrow \mathbf{AA}^T \in \mathcal{D}^+, \tag{5.95}$$

$$\exists \mathbf{Q} \in \mathcal{D} \text{ such that } \kappa(\mathbf{AQ}) = 1 \Leftrightarrow \mathbf{A}^T\mathbf{A} \in \mathcal{D}^+, \tag{5.96}$$

where $\mathcal{D}$ $(\mathcal{D}^+)$ *is defined as the class of real diagonal (respectively, positive diagonal) matrices of appropriate dimension.*

Proof. The proofs of (5.95) and (5.96) are dual to each other; the proof of ($\Rightarrow$) in (5.96) is immediate from two observations: (a) for a matrix $\mathbf{A}$ to be perfectly conditioned, we must have $\theta(\mathbf{A}) = 2\cot^{-1}(1) = \pi/2$; i.e., the minimum angle between $\mathbf{Ax}$ and $\mathbf{Ay}$ must equal 90 degrees as $\mathbf{x}$ and $\mathbf{y}$ range over all pairs of orthonormal vectors, and since this is also the maximum possible first quadrant angle between any pair of vectors $\mathbf{Ax}$ and $\mathbf{Ay}$, we can conclude that the columns of $\mathbf{A}$ are mutually orthogonal (choose $\mathbf{x} = \mathbf{e}_i$, $\mathbf{y} = e_j$, $i \neq j, i, j = 1, \ldots, n$); i.e., $\mathbf{A}^T\mathbf{A}$ is positive diagonal. (b) Postmultiplication of a matrix $\mathbf{A}$ by a diagonal matrix $\mathbf{Q}$ does not change the angle between the columns of $\mathbf{A}$, and thus the columns of $\mathbf{AQ}$ are orthogonal if and only if the columns of $\mathbf{A}$ are. In other words, for all $\mathbf{Q} \in \mathcal{D}$, $(\mathbf{AQ})^T\mathbf{AQ} \in \mathcal{D}^+$ if and only if $\mathbf{A}^T\mathbf{A} \in \mathcal{D}^+$.

($\Leftarrow$) Let $\mathbf{A}^T\mathbf{A} = \mathbf{D} = \text{diag}\,(d_1, \ldots, d_n)$ with $d_i > 0$, for all i. Then, choosing $\mathbf{Q} = \mathbf{Q}^T = \mathbf{D}^{-1/2}$ gives

$$\mathbf{Q}^T\mathbf{A}^T\mathbf{AQ} = \mathbf{D}^{-1/2}\mathbf{D}\mathbf{D}^{-1/2} = \mathbf{I},$$

which implies that $\kappa(\mathbf{AQ}) = 1$. □

Proposition 5.8 immediately raises the question of optimal diagonal conditioning; namely, if the rows (respectively, columns) of a given matrix $\mathbf{A}$ are not orthogonal, then what is the diagonal pre- (respectively, post-) conditioner $\mathbf{P}$ (respectively, $\mathbf{Q}$) that minimizes $\kappa(\mathbf{PA})$ (respectively, $\kappa(\mathbf{AQ})$)? We turn to this question now.

Optimal diagonal preconditioning: The second-order case

Since pre- and postconditioning are dual problems (preconditioners for $\mathbf{A}$ are postconditioners for $\mathbf{A}^T$) in what follows we will deal only with preconditioning and give a geometric interpretation of the optimal preconditioner for a 2×2 matrix.

If

$$\mathbf{A} = (a_{ij}) = \begin{bmatrix} \mathbf{a}_{r_1} \\ \mathbf{a}_{r_2} \end{bmatrix} \in \mathbb{R}^{2\times 2},$$

where $\mathbf{a}_{r_i} = [\ a_{i1} \quad a_{i2}\ ], i = 1, 2$, and we set $\mathbf{K} = \text{diag}\,(k_1, k_2) = \text{diag}\,(p_1^2, p_2^2) = \mathbf{P}^2$, then it is easy to show by calculation that

$$\text{spec}((\mathbf{PA})^T\mathbf{PA}) = \text{spec}(\mathbf{A}^T\mathbf{KA}) = \left\{ \frac{(a+b) \pm \sqrt{(a+b)^2 - 4c}}{2} \right\}, \tag{5.97}$$

where

$$a := k_1\|\mathbf{a}_{r_1}\|^2, \ b := k_2\|\mathbf{a}_{r_2}\|^2, \ c := k_1k_2 \det \mathbf{A}^2. \tag{5.98}$$

Hadamard's inequality for $\mathbf{A}$ is

$$\|\mathbf{a}_{r_1}\|^2\|\mathbf{a}_{r_2}\|^2 \geq \det \mathbf{A}^2, \tag{5.99}$$

and moreover, equality is attained if and only if $\mathbf{a}_{r_1}$ is orthogonal to $\mathbf{a}_{r_2}$. Since k_1, k_2 are positive, from (5.99),

$$k_1k_2\|\mathbf{a}_{r_1}\|^2\|\mathbf{a}_{r_2}\|^2 \geq k_1k_2 \det \mathbf{A}^2;$$

i.e.,

$$ab \geq c. \tag{5.100}$$

Moreover, equality in (5.100) is attained, making $\kappa(\mathbf{PA}) = 1$, exactly when the row vector $\mathbf{a}_{r_1}$ is orthogonal to the row vector $\mathbf{a}_{r_2}$, as we saw in Proposition 5.8: The additional geometric interpretation of the equality attained at orthogonality in (5.99) is that the area enclosed by $\mathbf{a}_{r_1}$ and $\mathbf{a}_{r_2}$ (i.e., $\det \mathbf{A}$) is maximized when $\mathbf{a}_{r_1}$ is orthogonal to $\mathbf{a}_{r_2}$.

If the rows of $\mathbf{A}$ are not orthogonal, we also know from Hadamard's inequality that

$$c = ab - \epsilon^2 \quad \text{for some} \quad \epsilon^2 > 0. \tag{5.101}$$

It is clear that it is no longer possible to make the eigenvalues of $\mathbf{A}^T\mathbf{KA}$ equal, but inspection of (5.97) together with the fact that

$$(a+b)^2 - 4c = (a+b)^2 - 4(ab - \epsilon^2) = (a-b)^2 + 4\epsilon^2$$

shows that $\kappa(\mathbf{PA})$ is minimized when k_1 and k_2 (i.e., p_1^2 and p_2^2) are chosen such that $a = b$. In other words, the optimal preconditioner scales the rows of the matrix so that both rows have norm equal to 1, for this minimizes ϵ and makes $a = b = 1$, which in turn minimizes the ratio of the eigenvalues of $\mathbf{A}^T\mathbf{KA}$ which are now $1 \pm \epsilon$, so that, denoting the optimal condition number by κ_{opt}, we have

$$\kappa_{opt} = \sqrt{\frac{1+\epsilon}{1-\epsilon}}. \tag{5.102}$$

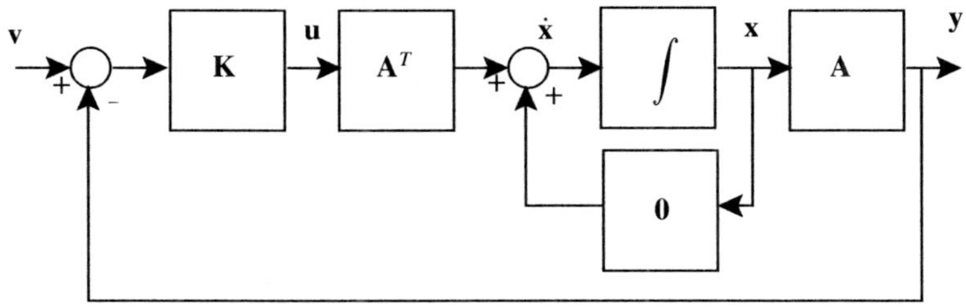

Figure 5.6. *Minimizing the condition number of the matrix* **A** *by choice of diagonal preconditioner* **P** *(i.e., minimizing* $\kappa(\mathbf{PA})$*) is equivalent to clustering closed-loop poles by decentralized (i.e., diagonal) positive feedback* $\mathbf{K} = \mathbf{P}^2$.

We will not elaborate further on the above because the obvious conjecture generalizing this (true) observation to matrices of dimension greater than or equal to 3 is false. The following counterexample suffices to show this:

$$\mathbf{A} = \begin{bmatrix} 1 & 2 & 3 \\ 2 & 1 & 3 \\ 0 & 3 & 2 \end{bmatrix} = \begin{bmatrix} \mathbf{a}_{r_1} \\ \mathbf{a}_{r_2} \\ \mathbf{a}_{r_3} \end{bmatrix}; \ \mathbf{P} = \text{diag}\left[\|\mathbf{a}_{r_1}\|^{-1}, \|\mathbf{a}_{r_2}\|^{-1}, \|\mathbf{a}_{r_3}\|^{-1}\right].$$

For this matrix **A**, $\kappa(\mathbf{A}) = 25.573$; however, $\kappa(\mathbf{PA}) = 25.892$. Note, however, that McCarthy and Strang [MS73] state that the above conjecture is true if the condition number is measured in the Frobenius norm.

5.3.1 Perfect diagonal preconditioning

In this section we rederive the characterization of Proposition 5.8 by the alternate route of pole-zero interlacing and *root-locus* arguments that are routine analysis tools used in control. The justification for this is to close the circle of ideas relating Wielandt's inequality, Hadamard's inequality, Cauchy's interlacing inequalities, and feedback control.

The starting point of the analysis is the fact that the dynamical system $\mathcal{S}^0$ represented by the triple $\{\mathbf{0}, \mathbf{A}^T, \mathbf{A}\}$, $\mathbf{A} \in \mathbb{R}^{n\times n}$, subject to output feedback through the matrix $\mathbf{K} \in \mathcal{D}^+$, is transformed to the dynamical system $\mathcal{S}^{\mathbf{K}} = \{-\mathbf{A}^T\mathbf{K}\mathbf{A}, \mathbf{A}^T, \mathbf{A}\}$. Since the eigenvalues of $\mathbf{A}^T\mathbf{K}\mathbf{A}$ are the squares of the singular values of **PA** (where $\mathbf{P} = \mathbf{K}^{1/2}$), it follows that the problem of minimizing $\kappa(\mathbf{PA})$ is equivalent to the problem of minimizing the distance between the smallest and the largest closed-loop poles of $\mathcal{S}^{\mathbf{K}}$ (i.e., eigenvalues of $-\mathbf{A}^T\mathbf{K}\mathbf{A}$, which are all real and negative) by choice of an appropriate $\mathbf{K} \in \mathcal{D}^+$. Since **K** is restricted to be diagonal, this last problem can be regarded as one of decentralized feedback in the following manner. The output matrix (**A**) in $\mathcal{S}$ is of dimension $n \times n$, where n is the dimension of the system $\mathcal{S}^0$; thus we are considering an output feedback problem where there are as many outputs as states and where the feedback gain matrix is constrained to be diagonal (see Figure 5.6).

To formalize the preceding discussion we make some preliminary observations. Under the change of variable $\mathbf{z} = \mathbf{Ax}$, the system $\mathcal{S}^0$ is transformed (by similarity) to

$$\tilde{\mathcal{S}}^0 = \{\mathbf{0}, \mathbf{A}\mathbf{A}^T, \mathbf{I}\}.$$

Since the output matrix of $\tilde{\mathcal{S}}^0$ is the identity, output feedback through a positive diagonal $\mathbf{K}$ for $\tilde{\mathcal{S}}^0$ is equivalent to a decentralized state feedback.

The poles of $\mathcal{S}^{\mathbf{K}} = \{-\mathbf{A}^T\mathbf{K}\mathbf{A}, \mathbf{A}^T, \mathbf{A}\}$ are defined as the eigenvalues of the matrix $-\mathbf{A}^T\mathbf{K}\mathbf{A}$ and hence are all real and negative. We denote them as

$$\lambda_n \le \lambda_{n-1} \le \cdots \le \lambda_1 < 0. \tag{5.103}$$

Since the poles of $\mathcal{S}^0 = \{\mathbf{0}, \mathbf{A}^T, \mathbf{A}\}$ are all zero, the matrix $\mathbf{K}$ is a stabilizing feedback for $\mathcal{S}^0$ and leads to a stable closed-loop system $\mathcal{S}^{\mathbf{K}}$. Thus we may reformulate the perfect and optimal conditioning problems, abbreviated as **P** and **O**, as follows:

(P) *Given* $\mathcal{S}^0 = \{\mathbf{0}, \mathbf{A}^T, \mathbf{A}\}$, *find a positive diagonal stabilizing feedback matrix* $\mathbf{K}$ *that makes all the closed-loop poles coincide; i.e.,* $\lambda_1 = \lambda_2 = \cdots = \lambda_n < 0$ *are the poles of* $\mathcal{S}^{\mathbf{K}} = \{-\mathbf{A}^T\mathbf{K}\mathbf{A}, \mathbf{A}^T, \mathbf{A}\}$.

(O) *Given* $\mathcal{S}^0 = \{\mathbf{0}, \mathbf{A}^T, \mathbf{A}\}$, *find a positive diagonal stabilizing feedback* $\mathbf{K}$ *such that the system* $\mathcal{S}^{\mathbf{K}} = \{-\mathbf{A}^T\mathbf{K}\mathbf{A}, \mathbf{A}^T, \mathbf{A}\}$ *has* $|\lambda_1| = 1$ *(normalization) and such that* $|\lambda_n|$ *is minimized (clearly this minimizes the ratio* $|\lambda_n|/|\lambda_1| = \kappa(\mathbf{PA})$, *where* $\mathbf{P} = \mathbf{K}^{1/2}$*).*

Transmission zeros and interlacing theorems

We start by motivating the idea of a transmission zero. Given a dynamical system $\mathcal{S} = \{\mathbf{F}, \mathbf{G}, \mathbf{H}\}$, we define the notion of a transfer function, for which the starting point is to take the Laplace transforms of the dynamical system equations. Let the Laplace-transformed variables be denoted with the same letter as the original variable, adorned with a hat, and let the standard letter s be used for the complex variable. Thus, e.g., the Laplace transform of the vector $\mathbf{x}(t)$ is denoted $\hat{\mathbf{x}}(s)$. Then, considering zero initial conditions (for simplicity), the Laplace transform of the state space dynamical system equations for the system $\mathcal{S}$ is

$$s\hat{\mathbf{x}}(s) = \mathbf{F}\hat{\mathbf{x}}(s) + \mathbf{G}\hat{\mathbf{u}}(s); \quad \hat{\mathbf{y}}(s) = \mathbf{H}\hat{\mathbf{x}}(s), \tag{5.104}$$

whence it follows that

$$\hat{\mathbf{y}}(s) = \mathbf{H}(s\mathbf{I} - \mathbf{F})^{-1}\mathbf{G}\hat{\mathbf{u}}(s) =: \hat{\mathbf{W}}(s)\hat{\mathbf{u}}(s). \tag{5.105}$$

The matrix $\hat{\mathbf{W}}(s)$ that relates the transform of the output to that of the input is known as a *transfer function* and, roughly speaking, values of s that cause the transfer function matrix to have a nontrivial null space are referred to as *transmission zeros*, since for these values of s, a nonzero sinusoidal input signal can lead to a zero output signal. As this brief digression shows, the utility of the Laplace transform approach is that it enables an algebraic approach to the calculation of the output of the system $\mathcal{S}$ given the input.

For the purposes of this section, we adopt a specific definition of transmission zero, in terms of the matrices $\mathbf{F}, \mathbf{G}, \mathbf{H}$ (without explicit reference to the underlying transfer function, which was introduced only to provide motivation).

Definition 5.9. *The complex number* $z_0 \in \mathbb{C}$ *is a* transmission zero *of* $\mathcal{S} = \{\mathbf{F}, \mathbf{G}, \mathbf{H}\}$ *if and only if*

$$\det \begin{bmatrix} z_0\mathbf{I} - \mathbf{F} & -\mathbf{G} \\ \mathbf{H} & \mathbf{0} \end{bmatrix} = 0. \tag{5.106}$$

It is straightforward to observe that, equivalently, the complex number z_0 must also satisfy

$$\det \begin{bmatrix} z_0\mathbf{I} - \mathbf{F} & \mathbf{G} \\ \mathbf{H} & \mathbf{0} \end{bmatrix} = 0. \tag{5.107}$$

The following result on the transmission zeros of a special class of dynamical systems $\mathcal{S} = \{\mathbf{F}, \mathbf{G}, \mathbf{H}\}$ for which $\mathbf{H} = \mathbf{G}^T$ and $\mathbf{F}$ is symmetric negative definite will be needed below.

Lemma 5.10 [Bha86]. *Let $\mathbf{F} = \mathbf{F}^T \in \mathbb{R}^{n\times n}$ be negative definite and $\mathbf{G} \in \mathbb{R}^{n\times m}$ and consider the dynamical system $\mathcal{S}^L = \{\mathbf{F}, \mathbf{G}, \mathbf{G}^T\}$. $\mathcal{S}^L$ has $(n-m)$ real, negative transmission zeros z_i, $1 \le i \le n-m$, given by*

$$z_i = \frac{\mathbf{v}_i^T \mathbf{F}\, \mathbf{v}_i}{\mathbf{v}_i^T \mathbf{v}_i}, \quad \textit{where } \mathbf{v}_i \in \mathcal{N}(\mathbf{G}^T),\ \mathbf{v}_i \neq \mathbf{0} \in \mathbb{R}^n.$$

Perfect diagonal preconditioning and decentralized feedback

We define a family of single-input, single-output systems derived from $\mathcal{S}^{\mathbf{K}} = \{-\mathbf{A}^T\mathbf{K}\mathbf{A}, \mathbf{A}^T, \mathbf{A}\}$ by considering that all loops, except the ith, have been closed with fixed values of the feedback gains and that one loop, the ith, is left open, to be analyzed as it is closed with the ith feedback gain k_i. The single-input, single-output system referred to is the dynamical system $\mathcal{S}^{(i)} = \{\mathbf{F}^{(i)}, \mathbf{g}^{(i)}, \mathbf{h}^{(i)}\}$ obtained from $\mathcal{S}^0 = \{\mathbf{0}, \mathbf{A}^T, \mathbf{A}\}$ by closing $(n-1)$ loops with the feedback gains k_j, $j \neq i$, and considering u_i as input and y_i as output.

To simplify the notation required to identify the matrices $\mathbf{F}^{(i)}, \mathbf{g}^{(i)}, \mathbf{h}^{(i)}$ and to state properties of $\mathcal{S}^{(i)}$ that we need, let us introduce the following.

Definition 5.11.

$$\mathbf{u}^{(i)} := [u_1, u_2, \ldots, u_{i-1}, u_{i+1}, \ldots, u_n] \in \mathbb{R}^{n-1}, \tag{5.108}$$

$$\mathbf{y}^{(i)} := [y_1, y_2, \ldots, y_{i-1}, y_{i+1}, \ldots, y_n] \in \mathbb{R}^{n-1}, \tag{5.109}$$

$$\mathbf{A}^{(i)} := [\mathbf{a}_{r_1}^T, \mathbf{a}_{r_2}^T, \ldots, \mathbf{a}_{r_{i-1}}^T, \mathbf{a}_{r_{i-1}}^T, \ldots, \mathbf{a}_{r_n}^T]^T \in \mathbb{R}^{(n-1)\times n}, \tag{5.110}$$

where $\mathbf{a}_{r_k}$ is the kth row of $\mathbf{A} \in \mathbb{R}^{n\times n}$,

$$\mathbf{K}^{(i)} := \operatorname{diag}(k_1, k_2, \ldots, k_{i-1}, k_{i+1}, \ldots, k_n) \in \mathbb{R}^{(n-1)\times(n-1)}. \tag{5.111}$$

Lemma 5.12. *For $i = 1, \ldots, n$, the system $\mathcal{S}^{(i)} = \{\mathbf{F}^{(i)}, \mathbf{g}^{(i)}, \mathbf{h}^{(i)}\}$ is defined as*

$$\mathbf{F}^{(i)} := -\mathbf{A}^{(i)T}\mathbf{K}^{(i)}\mathbf{A}^{(i)}, \tag{5.112}$$

$$\mathbf{g}^{(i)} := \mathbf{a}_{r_i}^T, \tag{5.113}$$

$$\mathbf{h}^{(i)} := \mathbf{a}_{r_i}, \tag{5.114}$$

where $\mathbf{a}_{r_i}$ is the ith row of $\mathbf{A} \in \mathbb{R}^{n\times n}$ and $\mathbf{A}^{(i)}$ is $\mathbf{A}$ with the ith row, $\mathbf{a}_{r_i}$, deleted (similarly for $\mathbf{K}^{(i)}$, see (5.110), (5.111)). Furthermore, the $(n-1)$ transmission zeros, $\{z_j^{(i)}\}_{j=1}^{n-1}$, and n poles $\{p_j^{(i)}\}_{j=1}^{n}$ of $\mathcal{S}^{(i)}$ are real, negative, and nonstrictly interlaced (i.e., $p_n^{(i)} \le z_{n-1}^{(i)} \le$

$p_{n-1}^{(i)} \le z_{n-2}^{(i)} \le \cdots \le p_2^{(i)} \le z_1^{(i)} \le p_1^{(i)} = 0$*), starting with a pole at the origin* ($p_1^{(i)} = 0$), *provided that the matrix* $\mathbf{K}$ *is positive diagonal.*

Proof. Since $\mathcal{S}^{(i)}$ is derived from $\mathcal{S}^0 = \{\mathbf{0}, \mathbf{A}^T, \mathbf{A}\}$ under the feedback $\mathbf{K}^{(i)}$, we write

$$\dot{\mathbf{x}} = \mathbf{A}^T\mathbf{u} \tag{5.115}$$

or, making the input u_i to $\mathcal{S}^{(i)}$ explicit,

$$\dot{\mathbf{x}} = \mathbf{A}^{(i)T}\mathbf{u}^{(i)} + \mathbf{a}_{r_i}^T u_i. \tag{5.116}$$

The feedback is given by

$$\mathbf{u}^{(i)} = -\mathbf{K}^{(i)}\mathbf{y}^{(i)}, \tag{5.117}$$

where $\mathbf{y}^{(i)}$ in turn is

$$\mathbf{y}^{(i)} = \mathbf{A}^{(i)}\mathbf{x}. \tag{5.118}$$

Finally, the output y_i of $\mathcal{S}^{(i)}$ is

$$y_i = \mathbf{a}_{r_i}\mathbf{x}. \tag{5.119}$$

Substituting (5.117), (5.118) in (5.116) gives

$$\dot{\mathbf{x}} = -\mathbf{A}^{(i)T}\mathbf{K}^{(i)}\mathbf{A}^{(i)}\mathbf{x} + \mathbf{a}_{r_i}^T u_i, \tag{5.120}$$

and consequently (5.119), (5.120) imply (5.112), (5.113), and (5.114). From the definitions of $\mathbf{F}^{(i)}$ and $\mathbf{A}^{(i)}$, it is clear that

(i) $\mathbf{F}^{(i)}$ is symmetric negative semidefinite, which implies that all poles of $\mathcal{S}^{(i)}$ are real and negative as claimed, and

(ii) the dimension of the null space of $\mathbf{F}^{(i)}$ is 1: the matrix $\mathbf{F}^{(i)}$ has one zero eigenvalue; i.e., the system $\mathcal{S}^{(i)}$ has one pole at the origin. The statement about the zeros follows immediately from Lemma 5.12 and Cauchy's interlacing theorem (with $m = n - 1$). □

We are now ready to state and prove the control-theoretic equivalent of Proposition 5.8, as follows.

Proposition 5.13 (control version of Proposition 5.8). *There exists a stabilizing positive diagonal feedback matrix* $\mathbf{K}$ *for the system* $\mathcal{S}^0 = \{\mathbf{0}, \mathbf{A}^T, \mathbf{A}\}$ *that makes all the poles* $\{\lambda_i\}_{i=1}^n$ *of the closed-loop system* $\mathcal{S}^{\mathbf{K}} = \{-\mathbf{A}^T\mathbf{K}\mathbf{A}, \mathbf{A}^T, \mathbf{A}\}$ *coincide (i.e.,* $\lambda_1 = \lambda_2 = \cdots = \lambda_n < 0$*) if and only if the rows of the matrix* $\mathbf{A}$ *are orthogonal to each other, i.e.,* $\mathbf{A}\mathbf{A}^T \in \mathcal{D}^+$.

Proof. We know that there exists a feedback matrix $\mathbf{K}$ corresponding to each distribution of closed-loop poles. Thus we may study a given distribution of closed-loop poles that is achieved by setting $\mathbf{K} = \mathbf{K}^0 = \text{diag}\,(k_1^0, k_2^0, \ldots, k_n^0)$ as follows:

(i) Close all loops j, except the ith (i.e., $j \neq i$) with the feedback gain set to k_j^0. This results in the single-input, single-output system $\mathcal{S}^{(i)}$ described completely above, in Lemma 5.12.

(ii) Feedback gain k_i is applied to $\mathcal{S}^{(i)}$ and we analyze the displacement of the poles as k_i varies through all positive values (thus through k_i^0 in particular). Clearly any distribution of closed-loop poles that cannot be attained for any value of k_i is also an unattainable distribution for $\mathcal{S}^{\mathbf{K}}$.

Since, for any i, $\mathcal{S}^{(i)}$ is a single-input, single-output system with poles and zeros nonstrictly interlaced on the negative real axis, classical root-locus properties [Tru55] tell us that, as k_i is increased from zero to infinity, the poles tend towards the zeros along the negative real axis but never actually cross them for any finite value of k_i. Thus coincidence of poles is ruled out unless all poles and zeros coincide (i.e., cancel) (recall that the interlacing is nonstrict). Given a minimal realization of a single-input, single-output dynamical system (i.e., one that is controllable and observable), we know that each cancellation of a pole-zero pair is equivalent to a drop in rank of unity for the controllability or observability matrix.

Denoting the observability matrix of $\mathcal{S}^{(i)}$ as $O_{\mathcal{S}^{(i)}}$, we have

$$O_{\mathcal{S}^{(i)}} := \begin{bmatrix} \mathbf{a}_{r_i} \\ \mathbf{a}_{r_i}\mathbf{A}^{(i)T}\mathbf{K}^{(i)}\mathbf{A}^{(i)} \\ \vdots \\ \mathbf{a}_{r_i}(\mathbf{A}^{(i)T}\mathbf{K}^{(i)}\mathbf{A}^{(i)})^{n-1} \end{bmatrix}. \tag{5.121}$$

Inspection of (5.121) and nonsingularity of $\mathbf{A}$ ($\Rightarrow \mathbf{a}_{r_i} \neq \mathbf{0} \in \mathbb{R}^{1\times n}$, for all i) make it clear that $O_{\mathcal{S}^{(i)}}$ can drop rank by $(n-1)$ (i.e., rank $(O_{\mathcal{S}^{(i)}}) = 1$) if and only if

$$\mathbf{a}_{r_i}(\mathbf{A}^{(i)T}\mathbf{K}^{(i)}\mathbf{A}^{(i)})^k = \mathbf{0}, \quad k = 1, \ldots, n-1. \tag{5.122}$$

Equation (5.122) is equivalent to

$$\mathbf{a}_{r_i}\mathbf{A}^{(i)T}\mathbf{K}^{(i)}\mathbf{A}^{(i)} = \mathbf{0}. \tag{5.123}$$

Clearly, if the rows of $\mathbf{A}$ are orthogonal to each other, then (5.123) is satisfied and the poles of $\mathcal{S}^{\mathbf{K}}$ coincide, proving one direction of the proposition.

To see that the orthogonality of the rows of $\mathbf{A}$ is a necessary condition for the coincidence of the poles of $\mathcal{S}^{\mathbf{K}}$, observe that (5.123) may be written as

$$\sum_{j\neq i} k_j \langle \mathbf{a}_{r_i}, \mathbf{a}_{r_j}\rangle \mathbf{a}_{r_j} = \mathbf{0}; \tag{5.124}$$

i.e., a linear combination of the linearly independent row vectors $\mathbf{a}_{r_j}$ sums to zero. Since the k_j's are positive, this implies that

$$\text{for all } j \neq i, \ \langle \mathbf{a}_{r_i}, \mathbf{a}_{r_j}\rangle = 0, \tag{5.125}$$

which completes the proof, since (5.125) must hold for all i. □

From Proposition 5.13 we conclude that the natural interpretation of the preconditioning problem as the problem (**P**) of decentralized feedback stabilization (more accurately, pole clustering) of a dynamical system leads to another proof of the characterization of perfect diagonal preconditioners. From the point of view of control, the proof also offers insight into the mechanism of diagonal preconditioning and an understanding of its complexity.

5.3.2 LQ perspective on optimal diagonal preconditioners

The formulation of the preconditioning problem in the previous section as a feedback stabilization problem, under the constraint that the feedback matrix be positive diagonal, leads us naturally to an optimal control formulation of the problem, known as the *fixed structure* or *structurally constrained linear quadratic (LQ) problem.* We describe this perspective briefly below.

Consider the system $\mathcal{S} = \{\mathbf{F}, \mathbf{G}, \mathbf{H}\}$. The classical linear quadratic regulator problem (see, e.g., [KS72]) is that of finding a control $\mathbf{u}(\cdot)$ that minimizes the *quadratic performance index* or *loss function* J:

$$J := \int_0^\infty \{\mathbf{x}(t)^T \mathbf{Q}\mathbf{x}(t) + \mathbf{u}(t)^T \mathbf{R}\mathbf{u}(t)\}\mathrm{d}t, \tag{5.126}$$

where $\mathbf{R} \in \mathbb{R}^{m\times m}$ is a positive definite matrix, and $\mathbf{Q} \in \mathbb{R}^{n\times n}$ is a positive semidefinite matrix. The solution of this problem turns out to be a linear state feedback law

$$\mathbf{u}^{\text{opt}}(t) = -\mathbf{K}^{\text{opt}}\mathbf{x}(t) \tag{5.127}$$

with the state feedback matrix $\mathbf{K}^{\text{opt}}$ given as

$$\mathbf{K}^{\text{opt}} = \mathbf{R}^{-1}\mathbf{G}^T\mathbf{P}^{\text{opt}}, \tag{5.128}$$

where $\mathbf{P}^{\text{opt}} \in \mathbb{R}^{n\times n}$ is the symmetric positive semidefinite solution of the following continuous-time *algebraic Riccati equation*:

$$\mathbf{F}^T\mathbf{P} + \mathbf{P}\mathbf{F} + \mathbf{Q} - \mathbf{P}\mathbf{G}\mathbf{R}^{-1}\mathbf{G}^T\mathbf{P} = \mathbf{0}, \tag{5.129}$$

and the minimum value of the index J is

$$J^{\text{opt}} = \mathbf{x}_0^T\mathbf{P}^{\text{opt}}\mathbf{x}_0. \tag{5.130}$$

The resulting closed-loop system is asymptotically stable; i.e., the eigenvalues of the matrix $(\mathbf{F} - \mathbf{G}\mathbf{K})$ are contained in the left half of the complex plane.

Structurally constrained optimal control

In the usual LQR problem, no restrictions are imposed on the structure of the feedback matrix $\mathbf{K}$. The reformulation of the optimal conditioning problem as a structurally constrained problem requiring that the feedback matrix $\mathbf{K}$ be positive diagonal is also known as the *decentralized feedback control problem.*

A general formulation is given in terms of the following definition.

Definition 5.14. *The matrix* $\mathbf{K}$ *is said to satisfy a* decentralization constraint *if*

$$\Phi(\mathbf{K}) := \mathbf{K} - \text{diag}\,(\mathbf{K}_1, \ldots, \mathbf{K}_r) = \mathbf{0}, \tag{5.131}$$

where the $\mathbf{K}_i$*'s are square matrices and the sum of their dimensions is that of* $\mathbf{K}$.

Remark. If $\mathbf{K} \in \mathbb{R}^{n\times n}$ is required to be positive diagonal, it suffices to choose $r = n$ so that each $\mathbf{K}_i$ is a positive number. The general formulation of (5.131) will be useful if one wishes to go beyond simple diagonal preconditioners to more complex block-diagonal preconditioners.

There are three main approaches to the minimization of the performance index (5.126) subject to the dynamics specified by the triple $\{\mathbf{F}, \mathbf{G}, \mathbf{H}\}$ and subject to the decentralization constraint (5.131). A popular approach is to parameterize the feedback matrix in some way and then use gradient descent to minimize the loss function or index (5.126). Other techniques involve using a homotopy method in conjunction with a projection operator, as well as linear matrix inequalities. It would take us too far afield to describe any one of these methods in greater detail. Furthermore, all three approaches currently involve a large amount of computation, rendering all unsuitable for cheap calculation of the optimal preconditioner. Thus we refer the reader to [GYA89, KBR95] and merely present an illustrative numerical example here, in order to show that the optimal preconditioner can indeed be calculated using decentralized feedback.

Numerical example of optimal diagonal preconditioner

Consider the following 9×9 symmetric matrix $\mathbf{A}$ that arises in the structural analysis of a loaded beam:

$$\mathbf{A} = \begin{bmatrix} 0.6 & 0 & 0 & -0.18 & 0 & 0 & 0 & 0 & 0 \\ 0 & 0.6 & 0 & 0 & -0.18 & 0.009 & 0 & 0 & 0 \\ 0 & 0 & 0.002 & 0 & -0.009 & 0.0003 & 0 & 0 & 0 \\ -0.18 & 0 & 0 & 0.6 & 0 & 0 & -0.18 & 0 & 0 \\ 0 & -0.18 & -0.009 & 0 & 0.6 & 0 & 0 & -0.18 & 0.009 \\ 0 & 0.009 & 0.0003 & 0 & 0 & 0.002 & 0 & -0.009 & 0.0003 \\ 0 & 0 & 0 & -0.18 & 0 & 0 & 0.3 & 0 & 0 \\ 0 & 0 & 0 & 0 & -0.18 & -0.009 & 0 & 0.3 & -0.015 \\ 0 & 0 & 0 & 0 & 0.009 & 0.0003 & 0 & -0.015 & 0.001 \end{bmatrix}.$$

The condition number of $\mathbf{A}$, $\kappa(\mathbf{A})$, is 3508.5. An algorithm proposed in [GYA89] was implemented in the context of the preconditioning problem in [KBR95] and yielded the optimal diagonal preconditioner

$$\mathbf{P}^{\text{opt}} = 10^{-4} \times \text{diag}\,[5\ \ 28\ \ 1306\ \ 6\ \ 27\ \ 1839\ \ 10\ \ 524\ \ 9728],$$

and the condition number of the optimally preconditioned matrix, $\mathbf{P}^{\text{opt}}\mathbf{A}$, is

$$\kappa(\mathbf{P}^{\text{opt}}\mathbf{A}) = 200.71.$$

For comparison, consider two other commonly used preconditioners: $\mathbf{P}_{\text{eq}}$, which scales the rows so as to equalize the row 2-norms, and $\mathbf{P}_{\text{diag}}$, which normalizes the diagonal elements to unity. We calculate $\kappa(\mathbf{P}_{\text{eq}}\mathbf{A})$ and $\kappa(\mathbf{P}_{\text{diag}}\mathbf{A})$ as follows:

$$\mathbf{P}_{\text{eq}} = [1.596\ \ 1.596\ \ 108.4\ \ 1.534\ \ 1.534\ \ 77.57\ \ 2.858\ \ 2.855\ \ 57.06];$$

$$\kappa(\mathbf{P}_{\text{eq}}\mathbf{A}) = 203.07;$$

$$\mathbf{P}_{\text{diag}} = [1.667\ \ 1.667\ \ 500\ \ 1.667\ \ 1.667\ \ 500\ \ 3.333\ \ 3.333\ \ 1000];$$

$$\kappa(\mathbf{P}_{\text{diag}}\mathbf{A}) = 1260.7.$$

The control-theoretic approach used in this section showed that the problem of finding the optimal diagonal preconditioner $\mathbf{P}^{\text{opt}}$ that minimizes the condition number $\kappa(\mathbf{P}^{\text{opt}}\mathbf{A})$ is equivalent to the problem of allocating the poles of a suitably defined dynamical system by feedback since this is, in turn, equivalent to allocating the singular values of **PA**.

The fact that a matrix **A** possesses a perfect preconditioner if and only if it has mutually orthogonal rows is given a control-theoretic interpretation and proof by use of the pole-zero interlacing property of certain single-input, single-output systems associated with $\mathcal{S}^{\mathbf{K}}$ (all poles coincide when **A** has mutually orthogonal rows).

The pole-allocation interpretation of the preconditioning problem provides a clear picture of the mechanisms and strong restrictions involved in the problem of finding diagonal preconditioners. The formulation of the optimal diagonal preconditioner problem as a decentralized or fixed-structure LQR problem also shows that the well-known numerical difficulties associated with the various techniques for finding fixed-structure or decentralized controllers have their counterparts in the equally well-known matrix-theoretic difficulty of finding optimal diagonal preconditioners.

5.4 Characterization of Matrix D-Stability Using Positive Realness of a Feedback System

A matrix **A** is called *D-stable* if **DA** is Hurwitz stable for all positive diagonal $\mathbf{D} \in \mathbb{R}^{n\times n}$. The concept of D-stability was introduced in the 1950s, and since then, innumerable sufficient conditions for it have been presented in the literature [Her92, Joh74b]. For dimensions 2, 3, and 4, algebraic characterizations are known [KB00], and it is probable that such characterizations do not exist for higher dimensions, although there has been some recent progress in computational tests for D-stability [OP05]. Note that, since the matrix **D** is positive diagonal, it is nonsingular, so that **DA** is similar to $\mathbf{D}^{-1}(\mathbf{DA})\mathbf{D} = \mathbf{AD}$, which means that the concept of D-stability could be equally well defined in terms of postmultiplication by a positive diagonal matrix **D**.

The objective of this section is to study the D-stability problem using the well-known concept of strictly positive real functions widely used in control and circuit theory. In particular, simple algebraic characterizations of D-stability for matrices of order 2 and 3 are given using this approach, and in addition, a research direction for the general problem is opened up.

There are two ways to view the D-stability problem as a problem of stability of a standard feedback system $S(\mathcal{P}, \mathcal{C})$. The first is to define a feedback system $S_1(\mathcal{P}(\mathbf{A}), \mathcal{C}(\mathbf{D}))$, with $\mathcal{P}(\mathbf{A}) := \{\mathbf{0}, \mathbf{I}, -\mathbf{A}, \mathbf{0}\}$ and $\mathcal{C}(\mathbf{D}) := \{\mathbf{0}, \mathbf{0}, \mathbf{0}, \mathbf{D}\}$, which has the closed-loop dynamics represented by $\dot{\mathbf{x}} = \mathbf{DAx}$. Thus the family of dynamical systems $S_1(\mathcal{P}(\mathbf{A}), \mathcal{C}(\mathbf{D}))$ is stable for all $\mathcal{C}(\mathbf{D})$, $\mathbf{D} \in \mathcal{D}^+$, if and only if the matrix **DA** is Hurwitz stable for all $\mathbf{D} \in \mathcal{D}^+$, i.e., if and only if the matrix **A** is D-stable. There is a strong similarity between this representation of the D-stability problem and that of diagonal preconditioning presented in the previous section, as can be seen by comparing Figures 5.6 and the figure corresponding to the feedback system $S(\mathcal{P}(\mathbf{A}), \mathcal{C}(\mathbf{D}))$, which the reader can easily sketch.

In this section, however, since we are concerned only with stability under decentralized feedback, and not with optimality, we use an alternative representation in order to make connections with, as well as use, the ideas of integral control and strict positive realness.

Consider, therefore, the feedback system $S_2(\mathcal{P}(\mathbf{A}), \mathcal{C}(\mathbf{D}))$, where $\mathcal{P}(\mathbf{A}) := \{\mathbf{0}, \mathbf{0}, \mathbf{0}, -\mathbf{A}\}$ and $\mathcal{C}(\mathbf{D}) := \{\mathbf{0}, \mathbf{D}, \mathbf{I}, \mathbf{0}\}$. The reader can easily verify that the closed-loop dynamics, once again, is given by $\dot{\mathbf{x}} = \mathbf{DAx}$, so, as before, the family of dynamical systems $S_2(\mathcal{P}(\mathbf{A}), \mathcal{C}(\mathbf{D}))$ is stable for all $\mathcal{C}(\mathbf{D})$, $\mathbf{D} \in \mathcal{D}^+$ if and only if the matrix $\mathbf{A}$ is D-stable. This feedback system representation of the D-stability problem has been studied in control theory as the so-called problem of decentralized integral control, since the controller $\mathcal{C}(\mathbf{D}) := \{\mathbf{0}, \mathbf{D}, \mathbf{I}, \mathbf{0}\}$ can be viewed as a set of integral controllers (i.e., integrators), each one integrating the ith input of the controller and then multiplying with gain d_i, where d_i is the ith diagonal element of the diagonal matrix $\mathbf{D}$. The objective of this decentralized integral control is to stabilize the (closed-loop) feedback system $S_2(\mathcal{P}(\mathbf{A}), \mathcal{C}(\mathbf{D}))$. The reader should be alerted to the fact that, strictly speaking, the concept of decentralized integral control allows for zero integrator gains; i.e., the corresponding input is disconnected. In this case, stability in the face of disconnection of some of the inputs means that any subset of integral controllers can be disconnected while maintaining the stability of the system. Conversely, if a system is decentralized integral controllable, then it can be stabilized by adjusting (i.e., tuning) each integral controller separately, in any order. For more details on the control and theoretical aspects of this problem, the reader is referred to [MZ89, YF90, HCB92].

In order to bring the well-developed theory of positive real functions to bear on this problem, the additional observation required is that it is possible to reduce the D-stability problem to one of stability of a family of linear plants parameterized by $n-1$ of the n elements of the diagonal of the matrix $\mathbf{D}$, with an integral controller of gain equal to the remaining diagonal element of $\mathbf{D}$, in the standard feedback configuration. The theory of positive real functions can then be used to obtain sufficient conditions for D-stability.

The main objectives of this section are to use the connections between matrix D-stability, suitably chosen standard feedback systems, and strictly positive real functions, in order to present a new sufficient condition for D-stability for matrices of order greater than 3 and to prove that this sufficient condition is also necessary for matrices of orders 2 and 3, leading to a conjecture that this condition is also necessary for matrices of any order.

A definition of a matrix class that is useful in the D-stability problem is as follows.

Definition 5.15. *A matrix* $\mathbf{A}$ *belongs to* class $\mathcal{P}_0^+$ *if*

(i) *for all* $k = 1, \ldots, n$, *all* $k \times k$ *principal minors of* $\mathbf{A}$ *are nonnegative;*

(ii) *at least one principal minor of each order* k *is positive.*

A well-known necessary condition for D-stability is given in terms of the class $\mathcal{P}_0^+$ in the theorem below.

Theorem 5.16 [Joh74b]. *If* $\mathbf{A}$ *is D-stable, then* $-\mathbf{A} \in \mathcal{P}_0^+$.

For 2×2 matrices it is known, in addition, that the condition of Theorem 5.16 is also sufficient for D-stability [Joh74b]. Thus, we have the following theorem.

Theorem 5.17 [Joh74b]. $\mathbf{A} \in \mathbb{R}^{2\times 2}$ *is D-stable if and only if* $-\mathbf{A} \in \mathcal{P}_0^+$.

Strictly positive real functions and stability

Some preliminaries on strictly positive real functions and related definitions are required in what follows. Recall from Chapter 1 that, given a quadruple $\mathcal{P} := \{\mathbf{F}, \mathbf{g}, \mathbf{h}, 0\}$ that describes a linear dynamical system, there exists an associated rational function, called a *transfer function*, denoted as $H(s) = n(s)/d(s)$, and defined as

$$H(s) = \mathbf{h}^T(s\mathbf{I} - \mathbf{F})^{-1}\mathbf{g} = \mathbf{h}^T \operatorname{adj}(s\mathbf{I} - \mathbf{F})\mathbf{g}/\det(s\mathbf{I} - \mathbf{F}) =: n(s)/d(s),$$

where $n(s)$ and $d(s)$ are the numerator and denominator polynomials of the rational function $H(s)$, adj denotes the classical adjoint matrix, and $\mathcal{P}$ is called a realization of the transfer function $H(s)$. The concepts of *positive real* and *strictly positive real* functions originated in circuit theory [Bru31, Val60]. A rational transfer function is the driving point impedance of a passive network if and only if it is positive real. Similarly, it is the driving point impedance of a dissipative network if and only if it is strictly positive real. The formal definitions are as follows.

Definition 5.18. *A rational function $H : \mathbb{C} \to \mathbb{C}$ of the complex variable $s = \sigma + j\omega$ is positive real if $H(\sigma) \in \mathbb{R}$ for all $\sigma \in \mathbb{R}$ and its real part $Re(H(\sigma + j\omega)) \geq 0$ for all $\sigma > 0$ and $\omega \geq 0$. It is strictly positive real, if, for some $\epsilon > 0$, $H(s - \epsilon)$ is positive real.*

Theorem 5.19. *Consider the rational function $H(s) = n(s)/d(s)$, where $n(s)$ and $d(s)$ are polynomials that have no zeros on the imaginary axis in common, and also do not have zeros in the right half complex plane. If $H(s)$ is strictly positive real, $\mathcal{P}_H$ is a minimal realization of $H(s)$, and $\mathcal{C}_k := \{0, k, 1, 0\}$, then the feedback system $S(\mathcal{P}_H, \mathcal{C}_k)$ is asymptotically stable for all $k \in (0, \infty)$.*

Proof. The proof is based on the polar plot of $H(s)$. If $H(s)$ is strictly positive real, the phase of $H(j\omega)$ lies between $-90°$ and $+90°$, so that the phase of the loop transfer function $(k/j\omega)H(j\omega)$ does not attain $\pm 180°$. Consequently, the polar plot does not cross or touch the negative semi-axis; by the Nyquist criterion [GGS01], it may now be concluded that the feedback system $S(\mathcal{P}_H, \mathcal{C}_k)$ is stable for all $k \in (0, \infty)$. □

Note that the controller $\mathcal{C}_k := \{0, k, 1, 0\}$ is an integral controller with gain k. Thus, in control terminology, Theorem 5.19 says that strict positive realness of the transfer function of the dynamical system $\mathcal{P}_H$ guarantees the stability of the feedback system $S(\mathcal{P}_H, \mathcal{C}_k)$ for all integrator gains k.

Proposition 5.20. *Let $H(s)$, $\mathcal{P}_H$, and $\mathcal{C}_k$ be as defined in Theorem 5.19. If there exists $\omega^* > 0$ such that $Re(H(j\omega^*)) \leq 0$ and $Im(H(j\omega^*)) < 0$, then the closed-loop system $S(\mathcal{P}_H, \mathcal{C}_k)$ is not asymptotically stable.*

Proof. From the hypotheses, the phase of $H(j\omega^*) \geq 90°$. Thus the phase of $(1/j\omega^*)H(j\omega^*)$ $\geq 180°$. Then, by the Nyquist criterion, there exists $k^* \in (0, \infty)$ such that $S(\mathcal{P}_H, \mathcal{C}_k)$ is not asymptotically stable. □

5.4.1 A feedback control approach to the D-stability problem via strictly positive real functions

The matrix $\mathbf{A}$ is said to be D-stable if all possible choices of $\mathbf{D}$ in $\mathcal{D}^+$ result in a stable matrix $\mathbf{AD}$, or equivalently, in a stable polynomial $\det(s\mathbf{I} - \mathbf{AD})$.

The following lemma shows the equivalence between a standard feedback system $S(\mathcal{P}, \mathcal{C})$, where the dynamical systems $\mathcal{P}$ and $\mathcal{C}$ are suitably chosen, and the D-stability of the matrix $\mathbf{A}$, thus allowing the use of system-theoretic tools in the analysis of D-stability.

Lemma 5.21. *Let $\mathbf{A} \in \mathbb{R}^{n\times n}$ and a positive diagonal $\mathbf{D} \in \mathbb{R}^{n\times n}$ be conformally partitioned as*

$$\mathbf{A} = \begin{bmatrix} \mathbf{A}_1 & \mathbf{B} \\ \mathbf{C} & \mathbf{A}_2 \end{bmatrix}, \quad \mathbf{D} = \begin{bmatrix} \mathbf{D}_1 & \mathbf{0} \\ \mathbf{0} & \mathbf{D}_2 \end{bmatrix}, \tag{5.132}$$

where $\mathbf{A}_1, \mathbf{D}_1 \in \mathbb{R}^{n_1\times n_1}$, $\mathbf{A}_2, \mathbf{D}_2 \in \mathbb{R}^{n_2\times n_2}$, $\mathbf{B} \in \mathbb{R}^{n_1\times n_2}$, and $\mathbf{C} \in \mathbb{R}^{n_2\times n_1}$, with $n_1 + n_2 = n$. The matrix $\mathbf{A}$ is D-stable (i.e., the matrix $\mathbf{AD}$ is stable for all positive diagonal $\mathbf{D}$) if and only if the system $S(\mathcal{P}(\mathbf{D}_1), \mathcal{C}(\mathbf{D}_2))$ (see Figure 1.2), with $\mathcal{P}(\mathbf{D}_1) := \{\mathbf{A}_1\mathbf{D}_1, \mathbf{B}, \mathbf{CD}_1, \mathbf{A}_2\}$ and $\mathcal{C}(\mathbf{D}_2)) := \{\mathbf{0}, \mathbf{D}_2, \mathbf{I}, \mathbf{0}\}$, is internally stable for all $\mathbf{D} = \mathrm{diag}\ (\mathbf{D}_1, \mathbf{D}_2)$.

Proof. The state equations for the feedback interconnection $S(\mathcal{P}(\mathbf{D}_1), \mathcal{C}(\mathbf{D}_2))$ are as follows:

$$\dot{\mathbf{x}} = \mathbf{A}_1\mathbf{D}_1\mathbf{x} + \mathbf{Bu}, \tag{5.133}$$

$$\mathbf{y} = \mathbf{CD}_1\mathbf{x} + \mathbf{A}_2\mathbf{u}, \tag{5.134}$$

$$\dot{\mathbf{u}} = \mathbf{D}_2\mathbf{y}, \tag{5.135}$$

where $\mathbf{x} \in \mathbb{R}^{n_1}$ and $\mathbf{u}, \mathbf{y} \in \mathbb{R}^{n_2}$. Introducing the state vector $\mathbf{z} = (\mathbf{x}, \mathbf{u})$ of the closed-loop system, the above equations can be rewritten as

$$\dot{\mathbf{z}} = \begin{bmatrix} \mathbf{A}_1\mathbf{D}_1 & \mathbf{B} \\ \mathbf{D}_2\mathbf{CD}_1 & \mathbf{D}_2\mathbf{A}_2 \end{bmatrix} \mathbf{z} =: \overline{\mathbf{A}}\mathbf{z}. \tag{5.136}$$

Internal stability of the closed-loop system $S(\mathcal{P}(\mathbf{D}_1), \mathcal{C}(\mathbf{D}_2))$ is guaranteed by the stability of the matrix $\overline{\mathbf{A}}$ in (5.136). Observe that the matrix $\overline{\mathbf{A}}$ can be factorized as follows:

$$\overline{\mathbf{A}} := \begin{bmatrix} \mathbf{A}_1\mathbf{D}_1 & \mathbf{B} \\ \mathbf{D}_2\mathbf{CD}_1 & \mathbf{D}_2\mathbf{A}_2 \end{bmatrix} = \begin{bmatrix} \mathbf{I}_{n_1} & \mathbf{0} \\ \mathbf{0} & \mathbf{D}_2 \end{bmatrix} \begin{bmatrix} \mathbf{A}_1 & \mathbf{B} \\ \mathbf{C} & \mathbf{A}_2 \end{bmatrix} \begin{bmatrix} \mathbf{D}_1 & \mathbf{0} \\ \mathbf{0} & \mathbf{I}_{n_2} \end{bmatrix}. \tag{5.137}$$

To complete the proof, note that $\overline{\mathbf{A}}$ is similar to $\mathbf{AD}$, using the similarity transformation $\mathbf{T} = \mathrm{diag}\ (\mathbf{I}_{n_1}, \mathbf{D}_2)$:

$$\begin{aligned} \mathbf{T}^{-1}\overline{\mathbf{A}}\mathbf{T} &= \begin{bmatrix} \mathbf{I}_{n_1} & \mathbf{0} \\ \mathbf{0} & \mathbf{D}_2^{-1} \end{bmatrix} \begin{bmatrix} \mathbf{I}_{n_1} & \mathbf{0} \\ \mathbf{0} & \mathbf{D}_2 \end{bmatrix} \begin{bmatrix} \mathbf{A}_1 & \mathbf{B} \\ \mathbf{C} & \mathbf{A}_2 \end{bmatrix} \begin{bmatrix} \mathbf{D}_1 & \mathbf{0} \\ \mathbf{0} & \mathbf{I}_{n_2} \end{bmatrix} \begin{bmatrix} \mathbf{I}_{n_1} & \mathbf{0} \\ \mathbf{0} & \mathbf{D}_2 \end{bmatrix} \\ &= \begin{bmatrix} \mathbf{A}_1 & \mathbf{B} \\ \mathbf{C} & \mathbf{A}_2 \end{bmatrix} \begin{bmatrix} \mathbf{D}_1 & \mathbf{0} \\ \mathbf{0} & \mathbf{I}_{n_2} \end{bmatrix} \begin{bmatrix} \mathbf{I}_{n_1} & \mathbf{0} \\ \mathbf{0} & \mathbf{D}_2 \end{bmatrix} \\ &= \begin{bmatrix} \mathbf{A}_1 & \mathbf{B} \\ \mathbf{C} & \mathbf{A}_2 \end{bmatrix} \begin{bmatrix} \mathbf{D}_1 & \mathbf{0} \\ \mathbf{0} & \mathbf{D}_2 \end{bmatrix} = \mathbf{AD}, \end{aligned}$$

thus completing the proof. □

In order to use Theorem 5.19, which is stated for single-input, single-output systems, the idea is to use Lemma 5.21, with $n_1 = n - 1$ and $n_2 = 1$, where n is the dimension of the matrix $\mathbf{A}$.

Given $\mathbf{A} = (a_{ij}) \in \mathbb{R}^{n\times n}$ and $\mathbf{D} = \text{diag}\,(d_1, d_2, \ldots, d_n) \in \mathcal{D}^+$, the D-stability problem may then be redefined as that of the stability of the family of standard feedback systems, $S(\mathcal{P}_i, \mathcal{C}_i)$, $i = 1, \ldots, n$, where

$$\mathcal{P}_i = \{\mathbf{A}_i\mathbf{D}_i, \mathbf{b}_i, -\mathbf{c}_i^T\mathbf{D}_i, -a_{ii}\} \text{ for } \mathbf{D}_i = \text{diag}\,(d_1, \ldots, d_{i-1}, d_{i+1}, \ldots, d_n) \tag{5.138}$$

and $\mathbf{A}_i$, $\mathbf{b}_i$, and $\mathbf{c}_i$ are defined from the matrix $\mathbf{A}$ as follows:

$$\mathbf{A}_i = \begin{bmatrix} a_{11} & \cdots & a_{1,i-1} & a_{1,i+1} & \cdots & a_{1n} \\ \vdots & & \vdots & \vdots & & \vdots \\ a_{i-1,1} & \cdots & a_{i-1,i-1} & a_{i-1,i+1} & \cdots & a_{i-1,n} \\ a_{i+1,1} & \cdots & a_{i+1,i-1} & a_{i+1,i+1} & \cdots & a_{i+1,n} \\ \vdots & & \vdots & \vdots & & \vdots \\ a_{n1} & \cdots & a_{n,i-1} & a_{n,i+1} & \cdots & a_{nn} \end{bmatrix}, \quad \mathbf{b}_i = \begin{bmatrix} a_{1,i} \\ \vdots \\ a_{i-1,i} \\ a_{i+1,i} \\ \vdots \\ a_{ni} \end{bmatrix}, \tag{5.139}$$

$$\mathbf{c}_i^T = [a_{i1} \;\; \ldots \;\; a_{i,i-1} \;\; a_{i,i+1} \;\; \ldots \;\; a_{in}],$$

and finally $\mathcal{C}_i$ is defined as

$$\mathcal{C}_i := \{0, d_i, 1, 0\}, \tag{5.140}$$

where d_i is the ith diagonal element of the matrix $\mathbf{D}$.

Lemma 5.22. *Consider the standard feedback system $S(\mathcal{P}_i, \mathcal{C}_i)$, defined in* (5.138)–(5.140). *For each i, the characteristic polynomial of the closed-loop system $S(\mathcal{P}_i, \mathcal{C}_i)$ is* $\det(s\mathbf{I} - \mathbf{AD})$, *where* $\mathbf{D}$ *belongs to the set* $\mathcal{D}^+$.

Proof. Observe that a suitable permutation similarity $\mathbf{P}_i$ will put the matrices $\mathbf{P}_i\mathbf{A}\mathbf{P}_i$, $\mathbf{P}_i\mathbf{D}\mathbf{P}_i$ in the form of (5.132), with $\mathbf{A}_1 = \mathbf{A}_i$, $\mathbf{A}_2 = a_{ii}$, $\mathbf{B} = \mathbf{b}_i$, $\mathbf{C} = \mathbf{c}_i^T$, $\mathbf{D}_1 = \mathbf{D}_i$, and $\mathbf{D}_2 = d_{ii}$. Since $\mathbf{P}_i\mathbf{A}\mathbf{P}_i\mathbf{P}_i\mathbf{D}\mathbf{P}_i = \mathbf{P}_i\mathbf{A}\mathbf{D}\mathbf{P}_i$, which is similar to $\mathbf{AD}$, application of Lemma 5.21 completes the proof. □

The following sufficient condition for D-stability now follows easily.

Theorem 5.23. *If, for any i, the transfer function $H(s)$ of the system $\mathcal{P}_i$ defined in* (5.138)–(5.139) *is strictly positive real for every diagonal matrix* $\mathbf{D}_i$ *in* $\mathcal{D}^+$, *then the matrix* $\mathbf{A} \in \mathbb{R}^{n\times n}$ *is D-stable.*

Proof. The proof follows directly from Theorem 5.19 and Lemma 5.22. □

In order to arrive at conditions for D-stability in terms of the entries of the matrix $\mathbf{A}$, it is necessary to examine the form of the real and imaginary parts of the transfer function $H(s)|_{s=j\omega}$ of the system $\mathcal{P}_i$, for some i. To do this, new variables are defined in terms of the entries of the positive diagonal matrix $\mathbf{D}_i \in \mathcal{D}^+$ and the frequency ω, in the following manner:

$$\begin{aligned}\boldsymbol{\omega} &= (\omega_1, \ldots, \omega_{i-1}, \omega_{i+1}, \ldots, \omega_n), \qquad \text{where } \omega_j := (\omega d_j^{-1}),\\ \Omega &= \operatorname{diag}\,(\omega_1, \ldots, \omega_{i-1}, \omega_{i+1}, \ldots, \omega_n).\end{aligned} \tag{5.141}$$

In terms of the new variables just defined, the real and imaginary parts of the rational function are easily calculated. Note that, strictly speaking, the variables $\boldsymbol{\omega}$ and Ω should also have the subscript i, but, in order to lighten the notation, it will be dropped in what follows.

Proposition 5.24. *The real and imaginary parts, denoted $f(\cdot)$ and $g(\cdot)$, respectively, of the rational function $H(s)|_{s=j\omega}$, $\omega > 0$, for all $\mathbf{D}_i$ in $\mathcal{D}^+$, are functions from $\mathbb{R}_+^{n-1}$ to $\mathbb{R}$ defined by*

$$f(\boldsymbol{\omega}) := \mathbf{c}_i^T \mathbf{L}(\boldsymbol{\omega})^{-1}\mathbf{b}_i - a_{ii}, \tag{5.142}$$

$$g(\boldsymbol{\omega}) := \mathbf{c}_i^T \mathbf{M}(\boldsymbol{\omega})^{-1}\mathbf{b}_i, \tag{5.143}$$

where $\mathbf{L}(\boldsymbol{\omega}) = \Omega\mathbf{A}_i^{-1}\Omega + \mathbf{A}_i$, $\mathbf{M}(\boldsymbol{\omega}) = \mathbf{A}_i\Omega^{-1}\mathbf{A}_i + \Omega$, where $\boldsymbol{\omega}$, Ω are defined in (5.141), *and $\mathbf{A}_i$, $\mathbf{b}_i$, and $\mathbf{c}_i$ are defined in* (5.139).

Proof. The transfer function of the system $\mathcal{P}_i$ is

$$H(s) = -\mathbf{c}_i^T \mathbf{D}_i\,(s\mathbf{I} - \mathbf{A}_i\mathbf{D}_i)^{-1}\,\mathbf{b}_i - a_{ii}.$$

Since $\mathbf{D}_i = \operatorname{diag}\,(d_1, \ldots, d_{i-1}, d_{i+1}, \ldots, d_n)$, $d_j > 0$, then

$$H(s) = -\mathbf{c}_i^T \left(s\mathbf{D}_i^{-1} - \mathbf{A}_i\right)^{-1}\mathbf{b}_i - a_{ii}.$$

In order to examine $H(j\omega)$ for $\omega > 0$, substitute s by $j\omega$ to get

$$H(j\omega) = -\mathbf{c}_i^T \left(j\omega\mathbf{D}_i^{-1} - \mathbf{A}_i\right)^{-1}\mathbf{b}_i - a_{ii},$$

which can be written as

$$H(j\boldsymbol{\omega}) = -\mathbf{c}_i^T\,(j\Omega - \mathbf{A}_i)^{-1}\,\mathbf{b}_i - a_{ii}, \qquad \omega_i \in (0, \infty). \tag{5.144}$$

The real part of (5.144) is given by

$$\begin{aligned}f(\boldsymbol{\omega}) &:= \frac{1}{2}\left[-\mathbf{c}_i^T(j\Omega - \mathbf{A}_i)^{-1}\mathbf{b} - (-\mathbf{c}_i^T)(-j\Omega - \mathbf{A}_i)^{-1}\mathbf{b}_i\right] - a_{ii}\\ &= \frac{1}{2}(-\mathbf{c}_i^T)\left[(j\Omega - \mathbf{A}_i)^{-1} - (j\Omega + \mathbf{A}_i)^{-1}\right]\mathbf{b}_i - a_{ii}\\ &= \frac{1}{2}(-\mathbf{c}_i^T)(j\Omega - \mathbf{A}_i)^{-1}\left[(j\Omega + \mathbf{A}_i) - (j\Omega - \mathbf{A}_i)\right](j\Omega + \mathbf{A}_i)^{-1}\mathbf{b}_i - a_{ii}\\ &= \mathbf{c}_i^T\left[(-j\Omega + \mathbf{A}_i)^{-1}\mathbf{A}_i(j\Omega + \mathbf{A}_i)^{-1}\right]\mathbf{b}_i - a_{ii}.\end{aligned}$$

Define $\mathbf{L}(\boldsymbol{\omega})^{-1}$ as $(-j\Omega + \mathbf{A}_i)^{-1}\mathbf{A}_i(j\Omega + \mathbf{A}_i)^{-1}$. This simplifies to $\mathbf{L}(\boldsymbol{\omega}) = \Omega\mathbf{A}_i^{-1}\Omega + \mathbf{A}_i$ and leads to (5.142). The imaginary part of $H(j\boldsymbol{\omega})$, (5.143), is obtained similarly. $\square$

Remark 5.25. To check if $f(\boldsymbol{\omega}) > 0$, it is, in fact, enough to check the numerator of $f(\boldsymbol{\omega})$; this is so because the denominator is always positive, since $f(\boldsymbol{\omega}) = \mathbf{c}_i^T \mathbf{L}(\boldsymbol{\omega})^{-1}\mathbf{b}_i - a_{ii}$ and

$$\begin{aligned}\mathbf{L}(\boldsymbol{\omega})^{-1} &= (-j\Omega + \mathbf{A}_i)^{-1}\mathbf{A}_i(j\Omega + \mathbf{A}_i)^{-1} \\ &= \frac{\operatorname{adj}(-j\Omega + \mathbf{A}_i)\ \mathbf{A}_i\ \operatorname{adj}(j\Omega + \mathbf{A}_i)}{\det(-j\Omega + \mathbf{A}_i)\det(j\Omega + \mathbf{A}_i)};\end{aligned}$$

thus the denominator of $f(\boldsymbol{\omega})$ is $\det(-j\Omega + \mathbf{A}_i)\det(j\Omega + \mathbf{A}_i) = |\det(j\Omega + \mathbf{A}_i)|^2 > 0$.

5.4.2 D-stability conditions for matrices of orders 2 and 3

From Theorem 5.23, it is clear that if the function of $n-1$ variables, $f(\boldsymbol{\omega})$, defined in (5.142), is positive for all $\boldsymbol{\omega} = (\omega_1, \omega_2, \ldots, \omega_{n-1})$, $\omega_i > 0$, then $H(s)$ is strictly positive real for all $\mathbf{D}$ in $\mathcal{D}^+$ and consequently the matrix $\mathbf{A} \in \mathbb{R}^{n\times n}$ is D-stable. It will be shown, however, that if $\mathbf{A}$ is of order 2 or 3, then this condition is also necessary; i.e., if $\mathbf{A}$ is D-stable, then $f(\boldsymbol{\omega}) > 0$. It is conjectured that this may also be valid for matrices of order n. It should be pointed out that no characterizations of D-stability are yet known for matrices of order greater than 3.

D-stability for matrices of order 2

Let

$$\mathbf{A} = \begin{bmatrix} a_{11} & a_{22} \\ a_{21} & a_{22} \end{bmatrix}, \qquad \mathbf{D} = \begin{bmatrix} d_1 & 0 \\ 0 & d_2 \end{bmatrix}.$$

It is known that in this case $-\mathbf{A} \in \mathcal{P}_0^+$ is a necessary and sufficient condition for D-stability (Theorem 5.17). It can be shown that $-\mathbf{A} \in \mathcal{P}_0^+$ implies $f(\boldsymbol{\omega}) > 0$. Since $-\mathbf{A} \in \mathcal{P}_0^+$, $\det(\mathbf{A}) > 0$, and, without loss of generality, it may be assumed that $a_{11} < 0$ and $a_{22} \leq 0$. Thus, applying Theorem 5.23 with $i = 2$, we get

$$\mathbf{A}_2 = a_{11} < 0, \quad b_2 = a_{12}, \quad c_2 = a_{21}, \quad \text{and} \quad \boldsymbol{\omega} = \Omega = \omega_1 > 0. \tag{5.145}$$

Then

$$\begin{aligned} L(\boldsymbol{\omega}) &= \Omega\mathbf{A}_2^{-1}\Omega + \mathbf{A}_2 \\ &= \frac{\omega_1^2}{a_{11}} + a_{11} = \frac{\omega_1^2 + a_{11}^2}{a_{11}}, \quad \text{i.e.,} \\ L(\boldsymbol{\omega})^{-1} &= \frac{a_{11}}{\omega_1^2 + a_{11}^2}. \end{aligned}$$

Since $f(\boldsymbol{\omega}) = c_2 L\Omega^{-1}b_2 - a_{22}$, it follows that

$$f(\boldsymbol{\omega}) = \frac{a_{11}a_{12}a_{21}}{\omega_1^2 + a_{11}^2} - a_{22} = \frac{a_{11}a_{12}a_{21} - a_{11}^2a_{22} - \omega_1^2a_{22}}{\omega_1^2 + a_{11}^2} = \frac{-a_{11}\det(\mathbf{A}) - a_{22}\omega_1^2}{\omega_1^2 + a_{11}^2}. \tag{5.146}$$

Finally, since $\det(\mathbf{A}) > 0$, it can be seen from (5.145) that (5.146) is always positive.

D-stability for matrices of order 3

Let

$$\mathbf{A} = \begin{bmatrix} a_{11} & a_{12} & a_{13} \\ a_{21} & a_{22} & a_{23} \\ a_{31} & a_{32} & a_{33} \end{bmatrix}, \qquad \mathbf{D} = \begin{bmatrix} d_1 & 0 & 0 \\ 0 & d_2 & 0 \\ 0 & 0 & d_3 \end{bmatrix}.$$

In this case, the condition $-A \in \mathcal{P}_0^+$ is only a necessary condition for D-stability, as the following example shows.

The matrix

$$\mathbf{A} = \begin{bmatrix} -2 & -1 & 0 \\ -1 & -2 & -4 \\ -70 & -1 & -8 \end{bmatrix}$$

is stable and all principal minors of $-\mathbf{A}$ are positive. Thus it belongs to the class $\mathcal{P}_0^+$, however the matrix

$$\mathbf{AD} = \begin{bmatrix} -2 & -1 & 0 \\ -1 & -2 & -4 \\ -70 & -1 & -8 \end{bmatrix} \begin{bmatrix} 1 & 0 & 0 \\ 0 & 1 & 0 \\ 0 & 0 & 0.5 \end{bmatrix}$$

is not stable (i.e., has eigenvalues in the right half plane).

A characterization of D-stability will now be given. The technique will be to analyze the sign of the functions $f(\omega)$ and $g(\omega)$ defined in (5.142) and (5.143). In particular, it can be shown that if the conditions of Theorem 5.26 below are satisfied, then $f(\omega) > 0$ and hence $\mathbf{A}$ is D-stable. On the other hand, if any of the conditions of Theorem 5.26 is not satisfied, then there exists an ω^* such that $f(\omega^*) \le 0$ and $g(\omega^*) < 0$, so that, by Proposition 5.20, matrix $\mathbf{A}$ is not D-stable.

Theorem 5.26. *Let* $\mathbf{A} \in \mathbb{R}^{3\times 3}$ *and* $-\mathbf{A} \in \mathcal{P}_0^+$. *Then* $\mathbf{A}$ *is D-stable if and only if either*

$$\sqrt{\frac{a_{11}m_{11}}{\det(A)}} + \sqrt{\frac{a_{22}m_{22}}{\det(\mathbf{A})}} + \sqrt{\frac{a_{33}m_{33}}{\det(\mathbf{A})}} - 1 > 0 \tag{5.147}$$

or

$$\sqrt{\frac{a_{11}m_{11}}{\det(\mathbf{A})}} + \sqrt{\frac{a_{22}m_{22}}{\det(\mathbf{A})}} + \sqrt{\frac{a_{33}m_{33}}{\det(\mathbf{A})}} - 1 = 0, \tag{5.148}$$

and there exists $j \in \{1, 2, 3\}$ *such that* $a_{jj}m_{jj} = 0$ *with* $a_{jj} \neq 0$ *or* $m_{jj} \neq 0$.

Proof. Only a sketch of the proof will be given. Recall that

$$\mathbf{A} = \begin{bmatrix} a_{11} & a_{12} & a_{13} \\ a_{21} & a_{22} & a_{23} \\ a_{31} & a_{32} & a_{33} \end{bmatrix}, \qquad \mathbf{D} = \begin{bmatrix} d_1 & 0 & 0 \\ 0 & d_2 & 0 \\ 0 & 0 & d_3 \end{bmatrix},$$

and let

$$\mathbf{A}_3 = \begin{bmatrix} a_{11} & a_{12} \\ a_{21} & a_{22} \end{bmatrix}, \quad \mathbf{b}_3 = \begin{bmatrix} a_{13} \\ a_{23} \end{bmatrix}, \quad \mathbf{c}_3 = \begin{bmatrix} a_{31} & a_{32} \end{bmatrix}.$$

Then

$$\boldsymbol{\omega} = \begin{bmatrix} \omega_1 & \omega_2 \end{bmatrix}, \quad \Omega = \begin{bmatrix} \omega_1 & 0 \\ 0 & \omega_2 \end{bmatrix}; \quad \omega_1, \omega_2 > 0.$$

Define $m_{jj} = \det(\mathbf{A}_j)$ for $j \in \{1, 2, 3\}$ (note that $\mathbf{A}_j$ is obtained from $\mathbf{A}$ by deleting the jth row and column). Since $-\mathbf{A} \in \mathcal{P}_0^+$,

$$a_{ii} \le 0, \quad m_{ii} \ge 0, \quad \det(\mathbf{A}) < 0, \quad a_{ii}, m_{jj} \ne 0 \quad \text{for some pair } (i, j).$$

After some algebra, one arrives at the following function $f(\boldsymbol{\omega})$ (actually the *numerator* of $f(\boldsymbol{\omega})$), which should be positive for all $\boldsymbol{\omega}$:

$$\begin{aligned} f(\boldsymbol{\omega}) = & -a_{33}(\omega_1\omega_2)^2 + [a_{33}m_{33} - a_{11}m_{11} - a_{22}m_{22} + \det(\mathbf{A})]\,\omega_1\omega_2 \\ & -a_{11}m_{22}\omega_2^2 - a_{22}m_{11}\omega_1^2 - m_{33}\det(\mathbf{A}). \end{aligned}$$

Defining $x = \omega_1\omega_2$ and $y = \omega_1/\omega_2$ ($x, y > 0$) yields

$$\begin{aligned} f(x, y) = & -a_{33}x^2 + [a_{33}m_{33} - a_{11}m_{11} - a_{22}m_{22} + \det(\mathbf{A}) - a_{11}m_{22}/y - a_{22}m_{11}y]\,x \\ & -m_{33}\det(\mathbf{A}), \end{aligned}$$

i.e.,

$$f(x, y) = ax^2 + b(y)x + c, \tag{5.149}$$

where

$$\begin{aligned} & a = -a_{33}, \quad c = -m_{33}\det(\mathbf{A}), \\ & b(y) = a_{33}m_{33} - a_{11}m_{11} - a_{22}m_{22} + \det(\mathbf{A}) - a_{11}m_{22}/y - a_{22}m_{11}y. \end{aligned}$$

The objective is now to find conditions on a, $b(y)$, and c such that $f(x, y)$ in (5.149) is always positive. Since $f(x, y)$ is a quadratic in x with the coefficient of the degree-1 term, b, dependent on parameter y, analyzing the roots of this equation leads to the desired result. □

This section investigated the problem of D-stability for real matrices by embedding this problem into a suitably defined feedback control system, using the concept of strictly positive real functions and leading to the derivation of a new sufficient condition for the D-stability of $n \times n$ matrices based on the positivity of a function $f(\cdot)$ (defined from the elements of the matrix $\mathbf{A}$) that maps $\mathbb{R}_+^{n-1}$ to $\mathbb{R}$. One possible approach to the problem of checking positivity of a multivariable function such as $f(\cdot)$ would be to use the results of Bose [Bos82] or Šiljak [Šil70]. Since the new sufficient condition has been shown to be necessary also for the D-stability of matrices of orders 2 and 3, this leads to a conjecture that the sufficient condition is also necessary for $n \times n$ matrices.

5.5 Finding Zeros of Two Polynomial Equations in Two Variables via Controllability and Observability

The problem of finding simultaneous solutions of two polynomial equations in two unknowns is important in the area of two-dimensional signal processing [Lim90, Bos82] as well as in various other fields, such as computer graphics, modeling of chemical kinetics, kinematics, and robotics [Mor87, Man94, Gib98].

In this section, the problem of finding simultaneous solutions of two polynomial equations in two unknowns is approached from a control-theoretic viewpoint, based on the paper [BD88]. A polynomial in two variables, x_1 and x_2 can be regarded as a polynomial in one of the variables, x_1 for example, which has coefficients that are polynomials in the other variable, x_2. Taking this point of view, the two single variable polynomials can be considered, respectively, as numerator and denominator of a rational function. The basic concepts of controllability and observability are then used to arrive at a solution method closely related to the method based on the classical resultant method [MS64], but capable of providing additional information, such as the total number of finite solutions, which, for the so-called deficient equations, is below the Bézout number, which is the classical estimate of the number of solutions.

Let $p_1(x_1, x_2)$ and $p_2(x_1, x_2)$ be two polynomials with real coefficients in the variables x_1 and x_2, denoted as $p_i \in \mathbb{R}[x_1, x_2]$, $i = 1, 2$. Let $\mathbf{x} = (x_1, x_2)$, $P(\mathbf{x}) = (p_1(\mathbf{x}), p_2(\mathbf{x}))$. The problem to be solved is that of finding all the isolated zeros of the *polynomial equation* $P(\mathbf{x})$, i.e., to find $\mathbf{x} = (x_1, x_2) \in \mathbb{C}^2$ such that $p_1(x_1, x_2) = p_2(x_1, x_2) = 0$. In the language of algebraic geometry, it is desired to find the (finite) intersections of the varieties determined by the equations $p_1 = 0$ and $p_2 = 0$. The classical result from algebraic geometry is known as *Bézout's theorem* and is stated below in the context of the problem at hand.

Theorem 5.27 [vdW53, Ken77]. *Given coprime polynomials $p_i(x_1, x_2) \in \mathbb{R}[x_1, x_2]$, $i = 1, 2$, with* deg $p_i = d_i$, *the polynomial equation $P(\mathbf{x}) = \mathbf{0}$ has at most $d = d_1 d_2$ zeros, where d is called the* Bézout number *of the system $P(\mathbf{x})$.*

Although Bézout's theorem states that a generic polynomial equation has the Bézout number of solutions, many, if not most, polynomial equations encountered in applications have a smaller (sometimes, much smaller) number of solutions [LSY87].

Definition 5.28. *A polynomial equation that has fewer solutions than its Bézout number is called* deficient.

Some examples of deficient polynomial equations are given below.

Example 5.29. Let $p_1(x_1, x_2) = x_1 + x_2$, $p_2 = x_1 + x_2 - 1$. The Bézout number of this polynomial equation is deg $p_1 \times$ deg $p_2 = 1 \times 1 = 1$. Geometrically it is clear that $p_1 = 0$ and $p_2 = 0$ are the equations of two parallel straight lines that only intersect at infinity.

A less obvious example, taken from [Mar78], is $p_1(x_1, x_2) = x_1^3 + 2x_1^2 x_2 + 2x_2(x_2 - 2)x_1 + x_2^2 - 4$, $p_2(x_1, x_2) = x_1^2 + 2x_1 x_2 + 2x_2^2 - 5x_2 + 2$. The Bézout number is deg $p_1 \times$ deg $p_2 = 3 \times 2 = 6$; however, the equation has only three finite solutions, as will be shown when this example is revisited below.

A final example, in three variables, from [Mor86], is as follows. Let $p_1(x_1, x_2, x_3) = x_1^2 + x_2^2 - a^2$, $p_2 = (x_1 - b)^2 + x_2^2 - c^2$, $p_3 = x_3 - d$. The Bézout number is $2 \times 2 \times 1 = 4$, but the equation has only two finite solutions.

To proceed, two results from control theory are needed.

Theorem 5.30. *Consider a rational (transfer) function $f(s) = n(s)/d(s)$, with $n(s), d(s) \in \mathbb{R}[s]$, $d(s)$ a monic polynomial, and the corresponding triple $\{\mathbf{F}, \mathbf{g}, \mathbf{h}\}$ chosen as the controllable companion form realization of the rational function $f(s)$, which means that the matrix $\mathbf{F} \in \mathbb{R}^{n\times n}$ is in companion form and further that the pair $(\mathbf{F}, \mathbf{g})$ is controllable. Then the pair $(\mathbf{F}, \mathbf{h})$ is observable if and only if the polynomials $n(s)$ and $d(s)$ are coprime.*

Remark. Note that Theorem 5.30 implies that if the controllable companion form realization is unobservable, then $n(s)$ and $d(s)$ must have a nontrivial common divisor; in particular, they must have a greatest common divisor. More can be said about the degree of this greatest common divisor.

Theorem 5.31 [Bar73, p. 3]. *Let the polynomial $d(s) \in \mathbb{R}[s]$ be denoted as*

$$d(s) = s^m + a_{m-1}s^{m-1} + \cdots + a_0,$$

and $\mathbf{A}_d$ be the controllable companion form matrix associated with the polynomial $d(s)$. Also let the polynomial $n(s) \in \mathbb{R}[s]$ be denoted as

$$n(s) = h_{m-1}s^{m-1} + \cdots + h_0,$$

and let $\mathbf{h} := [h_0 \ \ h_1 \cdots h_{m-1}] \in \mathbb{R}^{1\times m}$. Then

$$\deg\,(\gcd(n(s), d(s))) = m - \operatorname{rank} \begin{bmatrix} \mathbf{h} \\ \mathbf{hF} \\ \vdots \\ \mathbf{hF}^{m-1} \end{bmatrix}.$$

Finding the zeros of two polynomials in two variables via controllable realizations

With these preliminaries, the algorithm for finding common zeros is presented below, under the assumption that the polynomials p_1 and p_2 are coprime.

Algorithm 5.5.1 [Finding the zeros of two polynomials in two variables]

Step 1: Make the change of variables $x_1 = s$, $x_2 = s + \alpha$.
Then $p_1(x_1, x_2) = p_1(s, s+\alpha)$ and $p_2(s, s+\alpha)$ become polynomials in $\mathbb{R}[\alpha][s]$ (i.e., polynomials in s whose coefficients are polynomials in α).
The degree in s of one of the polynomials is greater than or equal to the degree of the other; let the first be denoted as $n_1^\alpha(s)$ and the other as $d_1^\alpha(s)$, thus:
$\deg\ n_1^\alpha(s) \geq \deg\ d_1^\alpha(s)$.

Step 2: Divide $n_1^\alpha(s)$ by $d_1^\alpha(s)$ and let the remainder be denoted as $n^\alpha(s)$; then
$\deg\ n^\alpha(s) \leq \deg\ d_1^\alpha(s)$.
Choose $\mu \in \mathbb{R}\backslash\{0\}$ such that $\mu d_1^\alpha(s)$ is monic and define $d^\alpha(s) = \mu d_1^\alpha(s)$.

Define a rational function

$$f^{\alpha}(s) := \frac{n^{\alpha}(s)}{d^{\alpha}(s)} = \frac{\sum_{i=0}^{k} n_i(\alpha)s^i}{s^{\ell} + \sum_{j=0}^{\ell-1} d_j(\alpha)s^j}, \tag{5.150}$$

where $\deg(n^{\alpha}(s)) = k < \ell = \deg(d^{\alpha}(s))$ and $n_i(\alpha), d_j(\alpha) \in \mathbb{R}[\alpha]$. Let

$$\mathbf{F}(\alpha) := \begin{bmatrix} 0 & 1 & 0 & \cdots & 0 \\ 0 & 0 & 1 & & 0 \\ \vdots & & \ddots & & \vdots \\ \vdots & & & \ddots & 1 \\ -d_0(\alpha) & -d_1(\alpha) & & & -d_{\ell-1}(\alpha) \end{bmatrix} \in \mathbb{R}[\alpha]^{\ell\times\ell}, \tag{5.151}$$

$$\mathbf{h}(\alpha) := \begin{bmatrix} n_0(\alpha) & n_1(\alpha) & \cdots & n_k(\alpha) & 0 & \cdots & 0 \end{bmatrix} \in \mathbb{R}^{1\times\ell} \tag{5.152}$$

be the matrices corresponding to a controllable canonical form realization of the rational transfer function $f^{\alpha}(s)$.

Step 3: Form the observability matrix of the pair $(\mathbf{F}(\alpha), \mathbf{h}(\alpha))$ defined in the previous step and calculate its determinant; i.e., calculate the polynomial $q(\alpha)$ as follows:

$$\mathbf{O}(\mathbf{F}(\alpha), \mathbf{h}(\alpha)) := \begin{bmatrix} \mathbf{h}(\alpha) \\ \mathbf{h}(\alpha)\mathbf{F}(\alpha) \\ \vdots \\ \mathbf{h}(\alpha)\mathbf{F}(\alpha)^{\ell-1} \end{bmatrix} \in \mathbb{R}[\alpha]^{\ell\times\ell}, \tag{5.153}$$

$$q(\alpha) := \det \mathbf{O}(\mathbf{F}(\alpha), \mathbf{h}(\alpha)). \tag{5.154}$$

Step 4: Calculate the zeros $\{\alpha_i\}_{i=1}^{m}$ of the polynomial $q(\alpha) \in \mathbb{R}[\alpha]$.

Step 5: For each distinct $\alpha_i \in \mathbb{C}$, define the polynomial $r^{\alpha_i}(s)$ as the greatest common divisor of $n^{\alpha_i}(s)$ and $d^{\alpha_i}(s)$, i.e.,

$$r^{\alpha_i}(s) := \gcd(n^{\alpha_i}(s), d^{\alpha_i}(s)). \tag{5.155}$$

Also, let $g^{\alpha_i} := \deg(r^{\alpha_i}(s))$.

Step 6: Find the roots $\{s_j^{\alpha_i}\}_{j=1}^{g^{\alpha_i}}$ of the polynomial $r^{\alpha_i}(s)$.

Step 7: The finite solutions of the polynomial equation $p_1(x_1, x_2) = 0$, $p_2(x_1, x_2) = 0$ are now written as

$$\begin{aligned} x_1(j) &= s_j^{\alpha_i}, \\ & \qquad\qquad j = 1, \ldots, g^{\alpha_i},\ i \in I, \\ x_2(j) &= s_j^{\alpha_i} + \alpha_i, \end{aligned} \tag{5.156}$$

where $I \subset \{1, 2, \ldots, m\}$ is a set of indices such that for $i, j \in I$, $i \neq j \to \alpha_i \neq \alpha_j$.

The total number of finite solutions is, evidently, $\sum_{i\in I} g^{\alpha_i}$.

Brief justification of Algorithm 5.5.1

A brief outline of the steps that justify Algorithm 5.5.1 is given below. Details can be found in [BD88].

Steps 1 and 2, which can clearly always be carried out, translate the original problem of finding common zeros of a system of polynomial equations into that of finding common zeros between the numerator and denominator of a rational function. The latter problem has been well studied in system and control theory.

More specifically, Steps 1 and 2 are intended to set up a transfer function f^α, parameterized by α, which admits a realization and observability matrix of the smallest order possible, in order to obtain a polynomial $q(\alpha)$, which is the determinant of the observability matrix of a controllable canonical realization of the transfer function $f^\alpha(s)$ (Step 3). Roots (α_i) of this polynomial are values at which the realization is not observable, since the observability matrix loses rank (Step 4). From Theorem 5.30, this means that, for the α_i's, a numerator-denominator (i.e., pole-zero) cancellation occurs; i.e., there exists a nontrivial greatest common divisor, called $r^{\alpha_i}(s)$ in Step 5 of Algorithm 5.5.1. Clearly, this means that the roots (in s) of this greatest common divisor $r^{\alpha_i}(s)$ are the simultaneous solutions of $n^{\alpha_i}(s)$ and $d^{\alpha_i}(s)$, thus explaining Steps 6 and 7.

Some examples at this point will help the reader to follow the steps of Algorithm 5.5.1.

Example 5.32. Let $p_1(x_1, x_2) = x_1^2 + x_2^3 - 9 = 0$, $p_2(x_1, x_2) = x_1 + x_2 - 3 = 0$. Making the substitution $x_1 = s$, $x_2 = s + \alpha$ (Step 1) leads to the rational function $f^\alpha(s) = (2s + \alpha - 3)/(s^3 + (3\alpha + 1)s^2 + 3\alpha^2 s + \alpha^3 - 9)$ (Step 2), with a corresponding controllable canonical form realization, which leads to $q(\alpha) = \det[\mathbf{O}(\mathbf{A}(\alpha), \mathbf{c}(\alpha))] = -(\alpha - 1)(\alpha + 3)(\alpha + 9)$ (Step 3). For each root of $q(\alpha)$, $\alpha = 1, -3, -9$ (Step 4), the observability matrix drops rank by 1. Thus, the number of finite solutions which is, by Theorem 5.31, equal to the sum of the rank drops, is $1 + 1 + 1 = 3$. In this case, in fact, the Bézout number of the equation is also 3, implying that there are no solutions at infinity. Finally, the solutions themselves are easily calculated as follows (gcd denotes greatest common divisor):

Step 5:

$$
\begin{aligned}
\alpha = 1 &\to r^\alpha(s) = \gcd(n^\alpha(s), d^\alpha(s)) = s - 1, \\
\alpha = -3 &\to r^\alpha(s) = \gcd(n^\alpha(s), d^\alpha(s)) = s - 3, \\
\alpha = -9 &\to r^\alpha(s) = \gcd(n^\alpha(s), d^\alpha(s)) = s - 6.
\end{aligned}
$$

Steps 6 *and* 7:

$$
\begin{aligned}
&\alpha = 1, \text{root } r^\alpha(s) : s = 1, \to x_1 = s = 1, x_2 = s + \alpha = 2, \\
&\alpha = -3, \text{root } r^\alpha(s) : s = 3, \to x_1 = s = 3, x_2 = s + \alpha = 3 + (-3) = 0, \\
&\alpha = -9, \text{root } r^\alpha(s) : s = -9, \to x_1 = s = 6, x_2 = s + \alpha = 6 + (-9) = -3.
\end{aligned}
$$

Example 5.33. Let $p_1(x_1, x_2) = x_1^3 + 2x_1^2 x_2 + 2x_2(x_2 - 2)x_1 + x_2^2 - 4$, $p_2(x_1, x_2) = x_1^2 + 2x_1 x_2 + 2x_2^2 - 5x_2 + 2$. Then

$$
f^\alpha(s) = \frac{3\alpha s + \alpha^2 + 10\alpha - 24}{s^2 + (\frac{6}{5}\alpha - 1)s + (\frac{2}{5}\alpha^2 - \alpha + \frac{2}{5})}.
$$

The polynomial $q(\alpha) = \det \mathbf{O} = (\alpha-2)(\alpha-6)^2(\alpha-8)$. For each value of $\alpha = 2, 6, 8$, the observability matrix drops rank by 1, so that there are three solutions corresponding to the greatest common divisors given by s, $s+4$, $s+5$. The finite solutions are thus calculated as $(0, 2)$, $(-4, 2)$, and $(-5, 3)$. In [Mar78, p. 241ff] the resultant method is used to find the solutions resulting in a determinant of a matrix in $\mathbb{R}[x_1]^{4\times 4}$, whereas the algorithm proposed above leads to $q(\alpha)$ (the resultant in α), which is the determinant of the observability matrix in $\mathbb{R}[\alpha]^{2\times 2}$. Besides, Algorithm 5.5.1 can also provide, a priori, the number of solutions of this deficient equation with Bézout number 6, but possessing only three finite solutions. The three missing solutions are located at the hyperplane at infinity in the projective space $\mathbb{CP}^2$ and can be found by homogenization [Mor86]. Finally, note that $\alpha = 6$ is a repeated root of the polynomial $q(\alpha)$, but the corresponding solution $(-4, 2)$ of the original polynomial equation $p_1 = 0$, $p_2 = 0$ has multiplicity 1.

We now give some examples of limiting behavior of Algorithm 5.5.1 as follows: (i) polynomials that are factor-coprime, but not zero-coprime (of interest in two-dimensional system theory); (ii) inconsistent equations (here $q(\alpha) =$ nonzero constant); (iii) infinitely many solutions (here $q(\alpha) \equiv 0$). In this case the polynomials are not factor-coprime, and thus the common factor can be extracted and the algorithm (which assumes factor-coprimeness) can be rerun to extract any other finite solutions.

Example 5.34. Let $p_1(x_1, x_2) = x_2 - x_1$, $p_2 = x_1x_2 - 1$. For this problem, set $x_1 = s$, $x_2 = s + \alpha$ (Step 1), the rational function $f^\alpha(s) = \alpha/(s^2 + \alpha s - 1)$ (Step 2), and the polynomial $q(\alpha) = \det \mathbf{O} = \alpha^2$ (Step 3). For the only root $\alpha = 0$ of $q(\alpha)$ (Step 4), $n^\alpha(s) = 0$, so that the greatest common divisor polynomial $r^{\alpha=0}(s) = d^\alpha(s) = s^2 - 1$ (Step 5), so that the two solutions are immediately found to be $(1, 1)$ and $(-1, -1)$ (Step 6 and 7). The Bézout number is also 2 for this example. The polynomials p_1 and p_2 are factor-comprime (have no common factors), but are not zero-coprime (have common zeros). In contrast to the previous example, a repeated root of $q(\alpha)$ gives rise to two distinct roots of the original system of polynomial equations.

Example 5.35. Let $p_1(x_1, x_2) = x_1^2 + x_2^2 - 2$, $p_1(x_1, x_2) = x_1^2 + x_2^2 - 1$, $x_1 = s$, $x_2 = s + \alpha$ (Step 1). Then $f^\alpha(s) = 1/(s^2 + \alpha s + \frac{1}{2}\alpha^2 - 1)$ (step 2) and $q(\alpha) = 1$. Since the polynomial $q(\alpha)$ is a nonzero constant, it has no roots. The interpretation is that the system of polynomial equations $p_1 = 0$, $p_2 = 0$ has no finite solutions. It is clear that the system of polynomial equations in this example is inconsistent: Geometrically, the loci of $p_1 = 0$ and $p_2 = 0$ correspond to two concentric circles that do not intersect. In fact, it can be shown that the polynomial $q(\alpha)$ reduces to a nonzero constant if and only if the original system of polynomial equations is inconsistent. Note that the system of equations is deficient, since the Bézout number is $2 \times 2 = 4$.

Example 5.36. Let $p_1(x_1, x_2) = x_1^2 + x_2^2 + 2x_1x_2 + x_1 + x_2$, $p_2(x_1, x_2) = x_1^2 + x_2^2 + 2x_1x_2 - 1$; $x_1 = s$, $x_2 = s + \alpha$. Then $d_1^\alpha(s) = 4s^2 + 4\alpha s + \alpha^2 + \alpha$, $n_1^\alpha(s) = d_1^\alpha(s) + 2s + (\alpha + 1)$, implying that

$$f^\alpha(s) = \frac{(2s + \alpha + 1)}{s^2 + \alpha s + \frac{1}{4}(\alpha^2 - 1)}$$

so that

$$\mathbf{O}(\alpha) = \begin{bmatrix} \alpha + 1 & 2 \\ -\frac{1}{2}(\alpha^2 - 1) & -\alpha + 1 \end{bmatrix},$$

which implies $q(\alpha) \equiv 0$. The interpretation of an identically zero polynomial $q(\alpha)$ is that the numerator of f^α, which is $2s + \alpha + 1$, divides its denominator $s^2 + \alpha s + \frac{1}{4}(\alpha^2 - 1)$ for all values of α. In other words, $2s + \alpha + 1$ is a common factor of the polynomials $n_1^\alpha(s)$ and $d_1^\alpha(s)$. Since $2s + \alpha + 1 = s + (s + \alpha) + 1 = x_1 + x_2 + 1$, the conclusion is that $x_1 + x_2 + 1$ is a common factor of p_1 and p_2, which are therefore revealed to not be factor-coprime. This common factor can be extracted and, once this is done, Algorithm 5.5.1, which assumes factor-coprimeness of p_1 and p_2, can be applied again. However, the point of this example is to show that, even when the factor-coprimeness assumption does not hold, Algorithm 5.5.1 indicates this by the manner in which it breaks down (i.e., by the generation of an identically zero polynomial $q(\alpha)$ in Step 3).

This section has shown how the concepts of controllability, observability, and canonical form realizations can be brought to bear on the problem of finding common zeros of two polynomials in two variables. Even though the algorithm that emerges cannot be regarded as a competitor for modern methods used in contemporary symbolic manipulation software, which rely on sophisticated Gröbner basis algorithms [CLO96, CLO05], it does give insight into the structure of finite solutions of a polynomial equation, using standard (and elementary) tools from classical control theory.

5.6 Notes and References

Numerical methods for ODEs

There are several excellent texts on numerical methods for ODEs; some of our personal favorites are [Ise96, DB02, HNW93, HW96]. Shooting methods are covered in the classic reference [AMR95] as well as in [Vic81, LP82, KH83, DB02].

An application of PID control ideas to time-stepping control for PDEs appears in [VCC05].

Decentralized control

Comprehensive references for this topic are [Šil78, Šil91].

Preconditioning of matrices

Two textbooks that contain fundamental material on preconditioning of matrices are [Axe96, Gre97].

D-stability

These conditions for matrices of orders 2 and 3 were obtained earlier by Johnson [Joh74a] (order 2) and Cain [Cai76] (order 3) using matrix theory techniques and later by Yu and Fan

[YF90] (order 3) using optimization techniques. An approach to minimizing the condition number measured in the 2-norm using linear matrix inequalities is given in [BM94a] and may be regarded as a "control-inspired" approach.

Zeros of systems of polynomial equations

Continuation methods for polynomial systems are treated in [Mor87]. The algebraic geometry background as well as Gröbner basis algorithms are insightfully treated in [CLO96, CLO05].

Chapter 6
Epilogue

The mind is but a barren soil; a soil which is soon exhausted, and will produce no crop, or only one, unless it be continually fertilized and enriched with foreign matter.

—Joshua Reynolds (1723–1792)

As mentioned in the preface, existing applications of system theory and dynamical system theory do not use the control theory idea that some inputs can be chosen as controls and used to modify the dynamical behavior of the system under investigation. This book focused on applications of this control theory idea to numerical algorithms and a few related matrix theory problems. The choice of topics was influenced by the authors' recent work and constrained, to a considerable extent, by the authors' interests and knowledge and, to a lesser extent, by space and time limitations. There are numerous other applications of both control theory and system theory and, more generally, of dynamical system theory, to problems in the general area of numerical algorithms, some of which are mentioned briefly below to show the ample scope of the book's theme.

Selected control theory applications to numerical problems

Optimal control and Bézier curves

In a series of papers (see, e.g., [ZTM97, SEM00]) Martin and coworkers relate the problem of fitting smooth curves through preassigned points, also known as spline interpolation, to an optimal control problem for a linear system. The L_2 norm of a control signal is minimized, while the scalar output of a suitably defined control system is driven close to given, prespecified interpolation points. A more specific example appeared recently in [EM04], where the class of Bézier curves that are useful in a number of applications, notably computer graphics and computer-aided design, was investigated. This is a class of approximating curves defined using the so-called control points, which define their shape, although the curves are not necessarily required to pass through these points. It is shown in [EM04] that Bézier curves can be related to certain Hermite interpolation problems and the latter, in turn, are shown to be linear optimal control problems. This shows several things. First, Bézier curves are revealed to be the solution to a linear optimal control problem. Second, reinforcing the message of this book, these facts open the door to using linear control theory in the construction of interpolating polynomials, in addition to relating Bézier curves to dynamic smoothing splines. Third, these facts offer a computational view of curves that differs from the standard de Casteljau algorithm heavily used in computer graphics and computer-aided design. As pointed out in [EM04], further research is required before it can be claimed that, due to the optimality property, Bézier curves offer better performance than other spline methods. In this sense too, there are parallels to the material presented in this book.

Optimal control theory is used in [AB01] to provide a unified framework for stating and solving a variety of problems in computer-aided design. It furnishes a new approach for handling, analyzing, and building curves and surfaces and leads to the definition of new classes of curves and surfaces, as well as the analysis of known problems from a new viewpoint. When the optimal control method is applied to the classical problems of knot selection of cubic splines and parameter correction, it yields new algorithms.

Optimal least squares fitting

Kozlov and Samsonov [KS03] use optimal control theory to get a least squares fit of a parameter-dependent state space model to observed data. The parameter vector is to be adjusted to observed values of the state vector at given time instants. Introducing a quadratic interpolation error function and interpreting the parameter vector as a control, this problem is formulated as an optimal control problem and solved using a modified gradient method. Due to the simplicity of this modified gradient algorithm, it is likely to have significant advantages over random search methods such as simulated annealing when large data sets are involved. This is essentially because, for random search methods, finding each new approximation to the parameter vector requires as many evaluations of the objective function as the number of parameters involved, whereas in the gradient method, the objective function is evaluated only once for each new approximation to the parameter vector.

Real-time optimization without derivative information

Korovin and Utkin [KU74] used the idea of sliding modes, discussed in Chapter 4, to define a class of continuous-time dynamical systems that solve convex programming problems without using explicit derivative information. This dynamical system can be viewed as a control system with a variable structure and relay controller. It was further developed in [TŻ98], and similar ideas have recently been proposed under the name of *extremum-seeking control* (see [AK03] and references therein).

Selected applications of system theory ideas to numerical algorithms

Numerical eigenvalue methods with eigenvalue shifts as control inputs

In general, numerical eigenvalue methods such as the QR algorithms, or inverse power iterations, are examples of nonlinear discrete dynamical systems defined on Lie groups or homogeneous spaces. In numerical linear algebra, convergence of such algorithms is improved using suitable shift strategies. Batterson and others [BS89, Bat95, PS95], in the spirit of [Shu86], carried out a study of matrix eigenvalue algorithms as nonlinear discrete dynamical systems. Helmke and coworkers (see [HF00, HW01] and references therein) took up this theme by viewing the eigenvalue shifts as feedback control variables, but only carrying out an analysis of the controllability properties of the inverse power method [HF00] and the real shifted inverse power iteration [HW01]. The idea is to start a wider investigation that will eventually lead to better understanding and consequently better numerical algorithms. As Helmke and Wirth [HW01] write:

> *So far the analysis and design of shift strategies in numerical eigenvalue algorithms has been more a kind of an art rather than being guided by systematic design principles. The situation here is quite similar to that of control theory in the 1950s before the introduction of state space methods. The advance made during the past two decades in nonlinear control theory indicates that the time may now be ripe for a more systematic investigation of control theoretic aspects of numerical linear algebra.*

A-stability of Runge–Kutta methods characterized by the positive real lemma

When applying numerical methods to stiff systems of ODEs, it is important to examine the stability of these methods. The concept of A-stability, introduced by Dahlquist [Dah63], is generally considered a minimal property to be imposed on any integration method. This concept deals with the behavior of a method applied to a linear autonomous differential equation. The A-stability of a Runge–Kutta method is determined by an analytic property of the so-called stability function, which describes a rational approximation to the exponential function. It is natural to ask for algebraic conditions on the coefficients of the method that characterize A-stability. Although this was regarded as a difficult problem, Scherer and Wendler [SW94] recently gave a complete characterization of A-stability, using a system-theoretic tool known as the positive real lemma, also called the Kalman–Yakubovich–Popov lemma [AV73]. The positive real lemma can be written in terms of linear matrix inequalities amd recent progress in interior point semidefinite programming methods [BGFB94] implies that this characterization is also effectively computable. Finally, in [KAY05], these ideas have also been applied to the estimation of the stability region of explicit Runge–Kutta methods which are not necessarily A-stable.

Probability and estimation theory applications of system theory

Hanzon and Ober [HO02] consider the class of all discrete probability densities on the set $\{0, 1, 2, \ldots\}$ that can be represented as the impulse response (i.e., convolution kernel) of a finite-dimensional discrete-time state space system. They show that all standard probability theory operations such as calculation of moments, convolutions, scaling, translation, products, etc., can be carried out using system representations, making connections between fundamental objects in the two theories (e.g., generating functions of probability densities and transfer functions of system theory). As they point out, these connections bring the well-developed theory of linear systems to bear on the calculus of discrete probabilities. In a similar vein, Ober [Obe02] shows that the Fisher information matrix, the inverse of the Cramer–Rao lower bound, can be calculated from a system-theoretic point of view and, in a certain special case, from the solution of a Liapunov equation, so that the well-developed numerical linear algebra techniques of Liapunov equation solution can be used in the calculation of the Fisher information matrix.

A solution of the machine shop problem using polynomial matrices

The theory of matrices with rational function entries, which can therefore be written as matrix fractions in polynomial matrices, has been developed in the circuit and system theory literature in the last four decades. An extraordinary application of matrix fractions, discovered by Bart and coworkers [BKZ98], concerns the two machine flow shop problem (2MFSP). Suppose that there are k jobs that have to be processed by two machines and that each job consists of two operations, denoted O_j^1 and O_j^2 for the jth job. The first operation O_j^1 must be processed on the first machine, and the second O_j^2 on the second machine. The restrictions that apply are as follows: (i) Each machine can process at most one operation at a time; (ii) processing O_j^2 on the second machine cannot start until processing O_j^1 on the first machine has been completed; (iii) the given, fixed processing times s_j for operation O_j^1 and t_j for O_j^2 are assumed to be nonnegative integers and, furthermore, for each j, either s_j or t_j is positive. Thus an *instance* J of the 2MFSP consists of k pairs (s_j, t_j) specifying the processing times of the operations. Given a schedule that satisfies all the restrictions, the length of time required to carry out all the jobs is called the *makespan* of the schedule. In the standard 2MFSP the objective is to find a feasible schedule with a minimum makespan. A *rational matrix function* is one that can be written as ND^{-1}, where N and D are polynomial matrices. Bart and coworkers showed that one can associate an instance of 2MFSP with each *companion-based matrix function* (a certain type of rational matrix function) and vice-versa. More specifically, they showed that if W is a companion-based matrix function and J is the associated instance of 2MFSP, then W admits a complete factorization if and only if the minimum makespan of J is less than or equal to the McMillan degree of W plus 1.

Wider vistas of control and system theory applications

Doing quantum mechanics with control theory

Bellman, one of the founders of modern system and control theory, showed how classical mechanics can be obtained from Hamilton's principle by dynamic programming [BD64]. Rosenbrock, another of the founders of modern system and control theory, has shown (see [Ros00] and references therein) that, if noise is added in a particular way, then Schrödinger's equation and many other results, some new, from the elementary theory of quantum mechanics can be obtained. Indeed, it is shown that the "noise-modified" version of Hamilton's principle leads to an equation, to which a closed-loop solution can be obtained by dynamic programming; this closed-loop solution is Schrödinger's equation. Several interesting physical and philosophical consequences of this observation are discussed in [Ros00], which we highly recommend to the interested reader.

The Quillen–Suslin theorem and polynomial matrix theory

Youla observed that "some of the most impressive accomplishments in circuits and systems have been obtained by an in-depth exploitation of the properties of elementary polynomial matrices." In the paper [YP84], the famous Quillen–Suslin theorem [Qui76, Sus76] (which proved Serre's conjecture, "finitely generated projective modules over a polynomial ring are free") was translated into a problem of row-bordering a polynomial matrix (which is a familiar operation for control and circuit theorists), and then solved by an ingenious algorithm for obtaining an invertible matrix. Interested readers should consult [LS92, LB01] for further developments.

Closure

In conclusion, it is hoped that this book as well as this long epilogue will convince the reader that the control approach to numerical algorithms and matrix problems has a bright future. This book has only made a beginning and is the proverbial tip of the iceberg. In the words of Pólya and Szegö, "There is something common in the orientations in a city and in any scientific area: from every given point we must be able to reach any other one," and we hope to have started the reader on a journey that will build new bridges between some of the areas touched on in this book and strengthen old ones.

Bibliography

[AABR02] F. Alvarez, H. Attouch, J. Bolte, and P. Redont. A second-order gradient-like dissipative dynamical system with Hessian-driven damping. Application to optimization and mechanics. *J. Math. Pures Appl.*, 81:747–779, 2002.

[AB01] M. Alhanaty and M. Bercovier. Curve and surface fitting design by optimal control methods. *Computer-Aided Design*, 33(2):167–182, February 2001.

[AF66] M. Athans and P. L. Falb. *Optimal Control.* McGraw-Hill, New York, 1966.

[AGR00] H. Attouch, X. Goudou, and P. Redont. The heavy ball with friction method, I. The continuous dynamical system: global exploration of the global minima of a real-valued function by asymptotic analysis of a dissipative dynamical system. *Comm. Contemp. Math.*, 2(1):1–34, 2000.

[AHU58] K. J. Arrow, L. Hurwicz, and H. Uzawa. *Studies in Linear and Nonlinear Programming.* Stanford University Press, Stanford, CA, 1958.

[AK03] K. B. Ariyur and M. Krstić. *Real-time optimization by extremum seeking control.* John Wiley, Hoboken, NJ, 2003.

[Alb71] Ya. I. Alber. Continuous processes of the Newton type. *Differential Equations*, 7(11):1461–1471, November 1971.

[AMNC97] G. L. Amicucci, S. Monaco, and D. Normand-Cyrot. Control Lyapunov stabilization of affine discrete-time systems. In *Proc. of the* 36th *IEEE Conference on Decision and Control*, pages 923–924, December 1997.

[AMR95] U. Ascher, R. M. M. Mattheij, and R. D. Russell. *Numerical Solution of Boundary Value Problems for Ordinary Differential Equations.* Classics in Applied Mathematics. SIAM, Philadelphia, 1995.

[Ant00] A. S. Antipin. From optima to equilibria. In *Proceedings of ISA Russian Academy of Sciences*, volume 3, pages 35–64, 2000.

[AV73] B. D. O. Anderson and S. Vongpanitlerd. *Network Analysis and Synthesis: A Modern Systems Theory Approach.* Prentice–Hall. Englewood Cliffs, NJ, 1973.

[ÅW84] K. J. Åström and B. Wittenmark. *Computer-Controlled Systems*. Prentice-Hall, Englewood Cliffs, NJ, 1984.

[Axe96] O. Axelsson. *Iterative Solution Methods*. Cambridge University Press, Cambridge, U.K., 1996.

[Bar73] S. Barnett. Matrices, polynomials and linear time-invariant systems. *IEEE Trans. Automat. Control*, AC-18(1):1–10, 1973.

[Bat95] S. Batterson. Dynamical analysis of numerical systems. *Numerical Linear Algebra Appl.*, 2(3):297–310, 1995.

[Bau63] F. L. Bauer. Optimally scaled matrices. *Numer. Math.*, 5:73–87, 1963.

[BD64] R. E. Bellman and S. E. Dreyfus. *Applied Dynamic Programming*. Princeton University Press, Princeton, NJ, 1964.

[BD76] P. T. Boggs and J. E. Dennis, Jr. A stability analysis for perturbed nonlinear analysis methods. *Math. Comp.*, 30(134):199–215, April 1976.

[BD88] A. Bhaya and R. J. Dias. Finding zeros of polynomial systems in two variables via controllability and observability. *Controle e Automação*, 2(1):71–75, March 1988. (In Portuguese; *Soluções de equações polinomiais através da teoria de controle*.)

[Bel57] R. E. Bellman. *Dynamic Programming*. Princeton University Press, Princeton, NJ, 1957.

[Ber83] D. P. Bertsekas. Distributed asynchronous computation of fixed points. *Math. Programming*, 27:107–120, 1983.

[Ber99] D. P. Bertsekas. *Nonlinear Programming*. Optimization and Computation series. Athena Scientific, Belmont, MA, 2nd edition, 1999.

[BGFB94] S. Boyd, L. El Ghaoui, E. Feron, and V. Balakrishnan. *Linear Matrix Inequalities in System and Control Theory*. SIAM, Philadelphia, 1994.

[BH69] A. E. Bryson and Y. C. Ho. *Applied Optimal Control*. Blaisdell, Waltham, MA, 1969. (Revised printing, Hemisphere, Washington, D.C., 1975.)

[Bha86] A. Bhaya. *Issues in the Robust Control of Large Flexible Structures*. Ph. D. thesis, Dept. of Electrical Engineering and Computer Sciences, University of California, Berkeley, CA, 1986.

[BJ75] D. J. Bell and D. H. Jacobson. *Singular Optimal Control Problems*. Academic Press, New York, 1975.

[BK03] A. Bhaya and E. Kaszkurewicz. Iterative methods as dynamical systems with feedback control. In *Proc. 42nd IEEE Conference on Decision and Control*, pp. 2374–2380, Maui, HI, December 2003.

[BK04a] A. Bhaya and E. Kaszkurewicz. Newton algorithms via control Liapunov functions for polynomial zero finding. In *Proc. 43rd IEEE Conference on Decision and Control*, pp. 1629–1634, Bahamas, December 2004.

[BK04b] A. Bhaya and E. Kaszkurewicz. Steepest descent with momentum for quadratic functions is a version of the conjugate gradient method. *Neural Networks*, 17(1):65–71, 2004.

[BKB96] B. Barán, E. Kaszkurewicz, and A. Bhaya. Parallel asynchronous team algorithms: convergence and performance analysis. *IEEE Trans. Parallel and Distributed Systems*, 7(7):677–688, 1996.

[BKK96] A. Bhaya, E. Kaszkurewicz, and V. S. Kozyakin. Existence and stability of equilibria in continuous-variable discrete-time neural networks. *IEEE Trans. Neural Networks*, 7(3):620–628, May 1996.

[BKZ98] H. Bart, L. Kroon, and R. Zuidwijk. Quasicomplete factorization and the two machine flow shop problem. *Linear Algebra Appl.*, 278(1-3):195–219, July 1998.

[BM94a] A. Bhaya and F. C. Mota. Equivalence of stability concepts for discrete time-varying systems. *Internat. J. Robust Nonlinear Control*, 4:725–740, Nov.-Dec. 1994.

[BM94b] R. D. Braatz and M. Morari. Minimizing the Euclidean condition number. *SIAM J. Control Optim.*, 32(6):1763–1768, November 1994.

[Bog71] P. T. Boggs. The solution of nonlinear systems of equations by *A*-stable integration techniques. *SIAM J. Numer. Anal.*, 8(4):767–785, December 1971.

[Bog76] P. T. Boggs. The convergence of the Ben-Israel iteration for nonlinear least squares problems. *Math. Computation*, 30(135):512–522, July 1976.

[Bos82] N. K. Bose. *Applied Multidimensional Systems Theory*. Van Nostrand Reinhold Company, New York, 1982.

[Bos03] N. K. Bose. *Multidimensional Systems Theory and Applications*. Kluwer Academic Publishers, Dordrecht, The Netherlands, second edition, 2003.

[BR05] A. Bacciotti and L. Rosier. *Liapunov Functions and Stability in Control Theory*. Communications and Control Engineering. Springer-Verlag, Berlin, 2nd edition, 2005.

[Bra72] F. H. Branin. Widely convergent method for finding multiple solutions of simultaneous nonlinear equations. *IBM J. Res. Develop.*, 16(5):504–522, September 1972.

[Bre01] C. Brezinski. Dynamical systems and sequence transformations. *J. Phys. A.*, 34(48):10659–10669, December 2001. Special Issue dedicated to "Symmetries and Integrability of Difference Equations (SIDE IV)."

[Bro89] R. W. Brockett. Least squares matching problems. *Linear Algebra Appl.*, 122/123/124:710–777, 1989.

[Bro91] R. W. Brockett. Dynamical systems that sort lists, diagonalize matrices and solve linear programming problems. *Linear Algebra Appl.*, 146:76–91, 1991.

[Bru31] O. Brune. Synthesis of a finite two-terminal network whose driving point impedance is a prescribed function of frequency. *J. Math. Phys.*, 10:191–236, 1931.

[BRZ94] C. Brezinski and M. Redivo-Zaglia. Hybrid procedures for solving linear systems. *Numer. Math.*, 67(1):1–19, 1994.

[BS83] P. Bloomfield and W. L. Steiger. *Least Absolute Deviations: Theory, Applications and Algorithms*. Birkhäuser, Boston, 1983.

[BS89] S. Batterson and J. Smillie. The dynamics of Rayleigh quotient iteration. *SIAM J. Numer. Anal.*, 26(3):624–636, 1989.

[BT89] D. P. Bertsekas and J. N. Tsitsiklis. *Parallel and Distributed Computation—Numerical Methods*. Prentice Hall, Englewood Cliffs, NJ, 1989.

[BT00] D. P. Bertsekas and J. N. Tsitsiklis. Gradient convergence in gradient methods with errors. *SIAM J. Optim.*, 10(3):627–642, 2000.

[BV04] S. Boyd and L. Vandenberghe. *Convex Optimization*. Cambridge University Press, Cambridge, U.K., 2004.

[CA02] A. Cichocki and S. I. Amari. *Adaptive Blind Signal and Image Processing: Learning Algorithms and Applications*. John Wiley, Baffins Lane, Chichester, U.K., 2002.

[Cai76] B. E. Cain. Real, 3×3, D-stable matrices. *J. Res. Nat. Bur. Standards*, Sect. B 80:75–77, 1976.

[Cam01] S. L. Campbell. Numerical analysis and systems theory. *Int. J. Appl. Math. Comput. Sci.*, 11(5):1025–1033, 2001.

[CCP70] M. D. Cannon, C. D. Cullum, Jr., and E. Polak. *Theory of Optimal Control and Mathematical Programming*. McGraw-Hill, New York, 1970.

[CD91] F. M. Callier and C. A. Desoer. *Linear System Theory*. Springer Verlag, New York, 1991.

[CdF73] K. S. Chao and R. J. P. de Figueiredo. Optimally controlled iterative schemes for obtaining the solution of a non-linear equation. *Internat. J. Control*, 18(2):377–384, 1973.

[Che99] C.-T. Chen. *Linear System Theory and Design*. Oxford University Press, New York, 3rd edition, 1999.

[Chu88] M. T. Chu. On the continuous realization of iterative processes. *SIAM Rev.*, 30(3):375–387, 1988.

[Chu92] M. T. Chu. Matrix differential equations: A continuous realization process for linear algebra problems. *Nonlinear Anal. Theory Meth. Appl.*, 18(12):1125–1146, 1992.

[CHZ99] E. K. P. Chong, S. Hui, and S. H. Żak. An analysis of a class of neural networks for solving linear programming problems. *IEEE Trans. Automat. Control*, 44(11):1995–2006, November 1999.

[Cla83] F. H. Clarke. *Optimization and Nonsmooth Analysis*. John Wiley, New York, 1983.

[CLO96] D. A. Cox, J. B. Little, and D. O'Shea. *Ideals, Varieties and Algorithms: An Introduction to Computational Algebraic Geometry and Commutative Algebra*. Springer-Verlag, New York, second edition, 1996.

[CLO05] D. A. Cox, J. B. Little, and D. O'Shea. *Using Algebraic Geometry*. Springer-Verlag, New York, second edition, 2005.

[CLSW98] F. H. Clarke, Yu. S. Ledyaev, R. J. Stern, and P. R. Wolenski. *Nonsmooth Analysis and Control Theory*, volume 178 of Graduate Texts in Mathematics. Springer-Verlag, New York, 1998.

[CM00] B. D. Calvert and C. A. Marinov. Another k-winners-take-all analog neural network. *IEEE Trans. Neural Networks*, 11(4):829–838, July 2000.

[CST00] N. Cristianini and J. Shawe-Taylor. *An introduction to support vector machines and other kernel-based learning methods*. Cambridge University Press, Cambridge, U.K., 2000.

[CU92] A. Cichocki and R. Unbehauen. Neural networks for solving systems of linear equations and related problems. *IEEE Trans. Circuits Systems I. Fund. Theory Appl.*, 39(2):124–138, February 1992.

[CU93] A. Cichocki and R. Unbehauen. *Neural Networks for Optimization and Signal Processing*. John Wiley, Chichester, U.K., 1993.

[CV95] C. Cortes and V. Vapnik. Support-vector networks. *Machine Learning*, 20(3):273–297, 1995.

[CW00] P. S. Chang and A. N. Willson, Jr. Analysis of conjugate gradient algorithms for adaptive filtering. *IEEE Trans. Signal Processing*, 48(2):409–418, 2000.

[Dah63] G. Dahlquist. A special stability problem for linear multistep methods. *BIT*, 3:27–43, 1963.

[Dat95] B. N. Datta. *Numerical Linear Algebra and Applications*. Brooks/Cole, Pacific Grove, CA, 1995.

[Dav53a] D. Davidenko. On a new method of numerical solution of systems of nonlinear equations. *Dokl. Akad. Nauk. SSSR*, 88:601–604, 1953 (in Russian).

[Dav53b] D. Davidenko. On approximate solution of systems of nonlinear equations. *Ukraine Mat. Z.*, 5:196–206, 1953 (in Russian).

[DB02] P. Deuflhard and F. Bornemann. *Scientific Computing with Ordinary Differential Equations*, volume 42 of Texts in Applied Mathematics. Springer-Verlag, New York, 2002.

[DB06] O. Diene and A. Bhaya. Adaptive filtering algorithms designed using control Liapunov functions. IEEE Signal Processing Letters, vol. 13, no. 4, April 2006, to appear.

[Del88] D. F. Delchamps. *State Space and Input-Output Linear Systems*. Springer-Verlag, New York, 1988.

[Dem97] J. W. Demmel. *Applied Numerical Linear Algebra*. SIAM, Philadelphia, 1997.

[Des70] C. A. Desoer. *Notes for a second course on linear systems*. Van Nostrand Reinhold, New York, 1970.

[Die93] I. Diener. Newton leaves and the continuous Newton method. In E. Deutsch, B. Brosowski, and J. Guddat, editors, *Parametric Optimization and Related Topics III*, Peter Lang Verlag, Frankfurt, 1993.

[DK80a] D. W. Decker and C. T. Kelley. Newton's method at singular points. I. *SIAM J. Numer. Anal.*, 17(1):66–70, 1980.

[DK80b] D. W. Decker and C. T. Kelley. Newton's method at singular points. II. *SIAM J. Numer. Anal.*, 17(3):465–471, 1980.

[DS75] L. C. W. Dixon and G. P. Szegö, editors. *Towards Global Optimisation*. North-Holland, Amsterdam, 1975.

[DS80] L. C. W. Dixon and G. P. Szegö. *Numerical Optimisation of Dynamic Systems*, see chapter by F. Aluffi, S. Incerti, and F. Zirilli, "Systems of Equations and A-Stable Integration of Second Order O.D.E.'s," pp. 289–307. North-Holland, Amsterdam, 1980.

[DS00] J. Dongarra and F. Sullivan. Guest editors' introduction to the top 10 algorithms. *Computing in Science and Engineering*, 2(1):22–23, 2000.

[dSB81] E. de Souza and S. P. Bhattacharyya. Controllability and the linear equation $Ax = b$. *Linear Algebra Appl.*, 36:97–101, 1981.

[DV85] V. F. Demyanov and L. V. Vasilev. *Nondifferentiable Optimization*. Optimization Software. Springer-Verlag, New York, 1985.

[EKP74] I. V. Emelin, M. A. Krasnosel'skiĭ, and N. P. Panskih. The spurt method of constructing successive approximations. *Soviet Math. Dokl.*, 15(6):1602–1607, 1974. (In Russian: *Dokl. Akad. Nauk.*, Tom 219, no. 3, 1974.)

[Ela96] S. N. Elaydi. *An Introduction to Difference Equations*. Springer-Verlag, New York, 1996.

[EM04] M. B. Egerstedt and C. F. Martin. A note on the connection between Bézier curves and linear optimal control. *IEEE Trans. Automat. Control*, 49(10):1728–1731, October 2004.

[ES98] C. Edwards and S. K. Spurgeon. *Sliding Mode Control: Theory and Applications*, volume 7 of Systems and Control Book Series. Taylor and Francis, London, 1998.

[EZ75] Yu. G. Evtushenko and V. G. Zhadan. Application of the method of Lyapunov functions to the study of the convergence of numerical methods. *USSR Comput. Math. and Math. Phys.*, 15(1):96–108, 1975. *Zh. Vychisl. Mat. Mat. Fiz.*, pp. 101–112 (Russian edition).

[Fil88] A. F. Filippov. *Differential Equations with Discontinuous Righthand Sides*. Mathematics and Its Applications (Soviet Series). Kluwer Academic, Dordrecht, 1988. Translation of *Differentsial'nye uravneniia razryvnoi pravoi chast'iu*.

[FKB02] L. V. Ferreira, E. Kaszkurewicz, and A. Bhaya. Convergence analysis of neural networks that solve linear programming problems. In *Proc. of the International Joint Conference on Neural Networks*, pp. 2476–2481, Honolulu, HI, May 2002.

[FKB03] L. V. Ferreira, E. Kaszkurewicz, and A. Bhaya. Synthesis of a k-winners-take-all neural network using linear programming with bounded variables. In *Proc. of the International Joint Conference on Neural Networks*, pp. 2360–2365, Portland, OR, July 2003.

[FKB04] L. V. Ferreira, E. Kaszkurewicz, and A. Bhaya. Support vector classifiers via gradient systems with discontinuous righthand sides. In *Procedings of the International Joint Conference on Neural Networks 2004, Budapest, Hungary*, pp. 1–6, July 2004. Paper number 1547.pdf on CD-ROM.

[FKB05] L. V. Ferreira, E. Kaszkurewicz, and A. Bhaya. Solving systems of linear equations via gradient systems with discontinuous righthand sides: Application to LS-SVM. *IEEE Trans. Neural Networks*, 16(2):501–505, 2005.

[FKB06a] L. V. Ferreira, E. Kaszkurewicz, and A. Bhaya. A linear programming based k-winners-take-all network via a discontinuous gradient system. Submitted, 2006. Technical report available at http://www.nacad.ufrj.br/~amit.

[FKB06b] L. V. Ferreira, E. Kaszkurewicz, and A. Bhaya. Support vector classifiers via gradient systems with discontinuous righthand sides. Submitted, 2006. Technical report available at http://www.nacad.ufrj.br/~amit.

[FP98] A. L. Fradkov and A. Yu. Pogromsky. *Introduction to Control of Oscillations and Chaos*, volume A 35 of World Scientific Series on Nonlinear Science. World Scientific, Singapore, 1998.

[FS55] G. E. Forsythe and E. G. Straus. On best conditioned matrices. *Proc. Amer. Math. Soc.*, 6:340–345, 1955.

[FW75] B. A. Francis and W. M. Wonham. The internal model principle for linear multivariable regulators. *Appl. Math. Optim.*, 2:170–194, 1975.

[Gav58] M. K. Gavurin. Nonlinear functional equations and continuous analogs of iterative methods. *Izv. Vyssh. Uchebn. Zaved. (Matematika)*, 5(6):18–31, 1958 (in Russian).

[Gea71] C. W. Gear. *Numerical Initial Value Problems in Ordinary Differential Equations*. Prentice-Hall, Englewood Cliffs, NJ, 1971.

[GGS01] G. C. Goodwin, S. F. Graebe, and M. E. Salgado. *Control System Design*. Prentice-Hall, Upper Saddle River, NJ, 2001.

[GHŻ98] M. P. Glazos, S. Hui, and S. H. Żak. Sliding modes in solving convex programming problems. *SIAM J. Control Optim.*, 36(2):680–697, 1998.

[Gib98] C. G. Gibson. *Elementary Geometry of Algebraic Curves: An Undergraduate Introduction*. Cambridge University Press, Cambridge, U.K., 1998.

[GL89] G. H. Golub and C. F. Van Loan. *Matrix Computations*. The John Hopkins University Press, Baltimore, 1989.

[GLS88] K. Gustafsson, M. Lundh, and G. Söderlind. A PI stepsize control for the numerical solution of ordinary differential equations. *BIT*, 28:270–287, 1988.

[Goh66] B. S. Goh. Necessary conditions for singular extremals involving multiple control variables. *SIAM J. Control*, 4:716–731, 1966.

[Goh97] B. S. Goh. Algorithms for unconstrained optimization via control theory. *J. Optim. Theory Appl.*, 92(3):581–604, March 1997.

[Gom75] J. Gomulka. Remarks on Branin's method for solving nonlinear equations. In L. C. W. Dixon and G. P. Szegö, editors, *Towards Global Optimisation*, North-Holland, Amsterdam, 1975.

[Gra03] W. J. Grantham. Trajectory following by gradient transformation differential equations. In 42nd *IEEE Conference on Decision and Control*, pp. 5496–5501, Maui, HI, December 2003.

[Gre97] A. Greenbaum. *Iterative Methods for Solving Linear Systems*. SIAM, Philadelphia, 1997.

[Grü02] L. Grüne. *Asymptotic Behavior of Dynamical and Control Systems under Perturbation and Discretization*, volume 1783 of Lecture Notes in Mathematics. Springer-Verlag, Berlin, 2002.

[GS97] K. Gustafsson and G. Söderlind. Control strategies for the iterative solution of nonlinear equations in ODE solvers. *SIAM J. Sci. Comput.*, 18:23–40, 1997.

[Gus91] K. Gustafsson. Control theoretic techniques for stepsize selection in explicit Runge–Kutta methods. *ACM Trans. Math. Software*, 17:533–554, 1991.

[Gus94] K. Gustafsson. Control theoretic techniques for stepsize selection in implicit Runge–Kutta methods. *ACM Trans. Math. Software*, 20:496–517, 1994.

[GV74] G. H. Golub and J. M. Varah. On a characterization of the best l_2-scaling of a matrix. *SIAM J. Numer. Anal.*, 11(3):472–479, 1974.

[GYA89] J. C. Geromel, A. Yamakami, and V. A. Armentano. Structural constrained controllers for discrete-time linear systems. *J. Optim. Theory Appl.*, 61(1):73–94, 1989.

[Hah67] W. Hahn. *Stability of Motion*. Springer-Verlag, New York, 1967.

[Hay94] S. Haykin. *Neural Networks: A Comprehensive Foundation*. Macmillan, New York, 1994.

[HCB92] L. Hsu, M. Chan, and A. Bhaya. Automated synthesis of decentralized tuning regulators for systems with measurable D.C. gain. *Automatica*, 28(1):185–191, 1992.

[Her92] D. Hershkowitz. Recent directions in matrix stability. *Linear Algebra Appl.*, 171:161–186, 1992.

[HF00] U. Helmke and P. A. Fuhrmann. Controllability of matrix eigenvalue algorithms: the inverse power method. *Systems Control Lett.*, 41:57–66, 2000.

[HH90] D. Higham and G. Hall. Embedded Runge–Kutta formulae with stable equilibrium states. *J. Comput. Appl. Math.*, 29(1):25–33, 1990.

[HJ88] R. A. Horn and C. R. Johnson. *Matrix Analysis*. Cambridge University Press, Cambridge, U.K., 1988.

[HJ91] R. A. Horn and C. R. Johnson. *Topics in Matrix Analysis*. Cambridge University Press, Cambridge, U.K., 1991.

[HKB00] L. Hsu, E. Kaszkurewicz, and A. Bhaya. Matrix-theoretic conditions for the realizability of sliding manifolds. *Systems Control Lett.*, 40(3):145–152, 2000.

[HM94] U. Helmke and J. B. Moore. *Optimization and Dynamical Systems*. Springer-Verlag, London, 1994.

[HN04] R. Hauser and J. Nedić. On the relationship between the convergence rates of discrete and continuous iterative processes. Numerical Analysis Group Research Report NA-04/10, Oxford University Computing Laboratory, Oxford, 2004. *SIAM J. Optim.*, submitted.

[HN05] R. Hauser and J. Nedić. The continuous Newton–Raphson method can look ahead. *SIAM J. Optim.*, 15(3):915–925, 2005.

[HNW87] E. Hairer, S. P. Nørsett, and G. Wanner. *Solving Ordinary Differential Equations I*, volume 8 of Springer series in Computational Mathematics. Springer-Verlag, Berlin, 1987.

[HNW93] E. Hairer, S. P. Nørsett, and G. Wanner. *Solving Ordinary Differential Equations I: Nonstiff Problems*. Springer-Verlag, Berlin, 2nd revised edition, 1993.

[HO02] B. Hanzon and R. J. Ober. State space calculations for discrete probability densities. *Linear Algebra Appl.*, 350:67–87, 2002.

[HS74] M. W. Hirsch and S. Smale. *Differential Equations, Dynamical Systems and Linear Algebra*. Academic Press, New York, 1974.

[HS79] M. W. Hirsch and S. Smale. On algorithms for solving $f(x) = 0$. *Comm. Pure Appl. Math.*, XXXII:281–312, 1979.

[HS95] M. Hagiwara and A. Sato. Analysis of momentum term in back-propagation. *IEICE Trans. Inform. Systems*, E78-D(8):1080–1086, 1995.

[HSŻG92] K. J. Hunt, D. Sbarbaro, R. Żbikowski, and P. J. Gawthrop. Neural networks for control systems—a survey. *Automatica*, 28(6):1083–1112, 1992.

[HT86] J. J. Hopfield and D. W. Tank. Computing with neural circuits: A model. *Science*, 233:625–633, 1986.

[Hur67] J. Hurt. Some stability theorems for ordinary difference equations. *SIAM J. Numer. Anal.*, 4(4):582–596, 1967.

[HW96] E. Hairer and G. Wanner. *Solving Ordinary Differential Equations II: Stiff and Differential-Algebraic Problems*. Springer-Verlag, Berlin, 2nd revised edition, 1996.

[HW01] U. Helmke and F. Wirth. On controllability of the real shifted inverse power iteration. *Systems Control Lett.*, 43:9–23, 2001.

[IM98] I. C. F. Ipsen and C. D. Meyer. The idea behind Krylov methods. *Amer. Math. Monthly*, 105(10):889–899, December 1998.

[IPZ79] S. Incerti, V. Parisi, and F. Zirilli. A new method for solving nonlinear simultaneous equations. *SIAM J. Numer. Anal.*, 16(5):779–789, 1979.

[Ise96] A. Iserles. *A First Course in the Numerical Analysis of Differential Equations*. Cambridge University Press, Cambridge, U.K., 1996.

[Itk76] U. Itkis. *Control Systems of Variable Structure*. John Wiley, New York, 1976.

[JM70] D. H. Jacobson and D. Q. Mayne. *Differential Dynamic Programming*, volume 24 of Modern Analytic and Computational Methods in Science and Mathematics. American Elsevier, New York, 1970.

[Joh74a] C. R. Johnson. Second, third, and fourth order D-stability. *J. Res. Nat. Bur. Standards*, 78B(1):11–13, 1974.

[Joh74b] C. R. Johnson. Sufficient conditions for D-stability. *J. Econom. Theory*, 9:53–62, 1974.

[KA82] L. V. Kantorovich and G. Akilov. *Functional Analysis*. Pergamon Press, Oxford, 1982.

[Kai80] T. Kailath. *Linear Systems*. Prentice Hall, Englewood Cliffs, NJ, 1980.

[KAY05] K. Kashima, S. Ashida, and Y. Yamamoto. System theory for numerical analysis. In *Preprints of the 16th IFAC World Congress*, pp. 1–6, Prague, Czech Republic, July 2005, IFAC. Paper 01980.pdf on Preprints CD-ROM.

[KB60] R. E. Kalman and J. E. Bertram. Control systems analysis and design via the "second method" of Lyapunov: Parts I and II. *Trans. ASME J. Basic Engineering*, 82:371–400, 1960.

[KB94] E. Kaszkurewicz and A. Bhaya. On a class of globally stable neural circuits. *IEEE Trans. Circuits Systems I Fund. Theory Appl.*, 41(2):171–174, 1994.

[KB00] E. Kaszkurewicz and A. Bhaya. *Matrix Diagonal Stability in Systems and Computation*. Birkhäuser, Boston, 2000.

[KBR95] E. Kaszkurewicz, A. Bhaya, and P. R. V. Ramos. A control-theoretic view of diagonal preconditioners. *Internat. J. Systems Science*, 26(9):1659–1672, September 1995.

[KBŠ90] E. Kaszkurewicz, A. Bhaya, and D. D. Šiljak. On the convergence of parallel asynchronous block-iterative computations. *Linear Algebra Appl.*, 131:139–160, 1990.

[Kel95] C. T. Kelley. *Iterative Methods for Linear and Nonlinear Equations*, volume 16 of Frontiers in Applied Mathematics. SIAM, Philadelphia, 1995.

[Ken77] K. Kendig. *Elementary Algebraic Geometry*. Springer-Verlag, New York, 1977.

[KH83] M. Kubiček and V. Hlavaček. *Numerical Solution of Nonlinear Boundary Value Problem with Applications*. Prentice-Hall, Englewood Cliffs, NJ, 1983.

[KH95] A. Katok and B. Hasselblatt. *Introduction to the Modern Theory of Dynamical Systems*, volume 54 of Encyclopedia of Mathematics and Its Applications. Cambridge University Press, New York, 1995.

[Kha02] H. K. Khalil. *Nonlinear Systems*. Prentice-Hall, Englewood Cliffs, NJ, 3rd edition, 2002.

[KK94] S. Kaski and T. Kohonen. Winner-take-all networks for physiological models of competitive learning. *Neural Networks*, 7(6):973–984, 1994.

[KLS89] M. A. Krasnosel'skii, Je. A. Lifshits, and A. V. Sobolev. *Positive Linear Systems: The Method of Positive Operators*. Heldermann Verlag, Berlin, 1989.

[KP99] S. V. Kamarthi and S. Pittner. Accelerating neural network training using weight extrapolations. *Neural Networks*, 12:1285–1299, 1999.

[KR76] N. N. Karpinskaya and M. V. Rybashov. On continuous algorithms of solution of convex programming problems. *Automat. Remote Control*, 37(1, Part 2):143–147, 1976.

[Kry98] A. V. Kryazhimskii. Convex optimization via feedbacks. *SIAM J. Control Optim.*, 37(1):278–302, 1998.

[KŠ64] P. Kokotović and D. Šiljak. Automatic analog solution of algebraic equations and plotting of root loci by generalized Mitrović method. *IEEE Trans. Appl. Ind.*, 83:324–328, 1964.

[KS72] H. Kwakernaak and R. Sivan. *Linear Optimal Control Systems*. Wiley Interscience, New York, 1972.

[KS03] K. N. Kozlov and A. M. Samsonov. New data processing technique based on the optimal control theory. *Technical Physics*, 48(11):1364–1371, 2003 (translated from *Zhurnal Tekhnicheskoĭ Fiziki*, 73 (11):6–14, 2003).

[KU74] S. K. Korovin and V. I. Utkin. Using sliding modes in static optimization. *Automatica*, 10:525–532, 1974.

[KZ93] J. E. Kurek and M. B. Zaremba. Iterative learning control synthesis based on 2D system theory. *IEEE Trans. Automat. Control*, 38(1):121–125, 1993.

[LB01] Z. Lin and N. K. Bose. A generalization of Serre's conjecture and some related issues. *Linear Algebra Appl.*, 338:125–138, 2001.

[Lei92] J. R. Leigh. *Control Theory: A Guided Tour*, volume 45 of IEE Control Series. Peter Peregrinus/IEE, London, 1992.

[Lia49] A. M. Liapunov. *Problème général de la stabilité du mouvement*. Princeton University Press, Princeton, NJ, 1949. French translation of 1892 Russian original. Also published as Liapunov Centenary Issue of *Intern. J. Control*, 55 (3), 1992.

[Lim90] J. S. Lim. *Two Dimensional Signal and Image Proccessing*. Prentice Hall, Englewood Cliffs, NJ, 1990.

[LP82] L. Lapidus and G. Pinder. *Numerical Solution of Partial Differential Equations in Science and Engineering*. John Wiley, New York, 1982.

[LQQ04] L. Z. Liao, H. D. Qi, and L. Q. Qi. Neurodynamical optimization. *J. Global Optim.*, 28(2):175–195, 2004.

[LS92] A. Logar and B. Sturmfels. Algorithms for the Quillen–Suslin theorem. *J. Algebra*, 145:231–239, 1992.

[LSY87] T.-Y. Li, T. Sauer, and J. A. Yorke. Numerical solution of a class of deficient polynomial systems. *SIAM J. Numer. Anal.*, 24(2):435–451, 1987.

[Lue69] D. G. Luenberger. *Optimization by Vector Space Methods*. John Wiley, New York, 1969.

[Lue84] D. G. Luenberger. *Linear and Nonlinear Programming*. Addison-Wesley, Reading, MA, 1984.

[LV92] M. Lemmon and B. V. K. Vijayakumar. Competitive learning with generalized winner-take-all activation. *IEEE Trans. Neural Networks*, 3(2):167–175, 1992.

[MÅ06] R. Murray and K. J. Åström. *Analysis and Design of Feedback Systems*. Forthcoming, 2006. See the Website http://www.cds.caltech.edu/~murray/books /am04/index.html.

[Man94] D. Manocha. Solving systems of polynomial equations. *IEEE Comput. Graphics Applications*, 14(2):46–55, 1994.

[Mar78] M. Marcus. *Introduction to Modern Algebra*. Marcel Dekker, New York, 1978.

[Mar94] J. M. Martínez. Algorithms for solving nonlinear systems of equations. In E. Spedicato, editor, *Continuous Optimization. The State of Art*, pp. 81–108. Kluwer Academic Publishers, Dordrecht, the Netherlands, 1994. Available as a technical report at http://www.ime.unicamp.br/~martinez/nato.pdf.

[McN93] J. McNamee. A bibliography on roots of polynomials. *J. Comput. Appl. Math.*, 47:391–394, 1993. For an updated bibliography, see http://www1.elsevier.com/homepage/sac/cam/mcnamee/ .

[MEAM89] E. Majani, R. Erlanson, and Y. Abu-Mostafa. On the k-winners-take-all network. In D. S. Touretzky, editor, *Advances in Neural Information Processing Systems*, volume 1, pp. 634–642. Morgan Kauffmann, San Mateo, CA, 1989.

[Moo67] J. B. Moore. A convergent algorithm for solving polynomial equations. *J. Assoc. Comput. Mach.*, 14(2):311–315, 1967.

[Mor86] A. P. Morgan. A transformation to avoid solutions at infinity for polynomial systems. *Appl. Math. Comput.*, 18:77–86, 1986.

[Mor87] A. P. Morgan. *Solving Polynomial Systems Using Continuation for Engineering and Scientific Problems*. Prentice-Hall, Englewood Cliffs, NJ, 1987.

[MQR98] R. I. McLachlan, G. R. W. Quispel, and N. Robidoux. Unified approach to Hamiltonian systems, Poisson systems, gradient systems, and systems with Lyapunov functions or first integrals. *Phys. Rev. Lett.*, 81(12):2399–2403, September 1998.

[MS64] A. Mostowski and M. Stark. *Introduction to Higher Algebra.* A Pergamon Press book. Macmillan Co., New York, 1964.

[MS73] C. McCarthy and G. Strang. Optimal conditioning of matrices. *SIAM J. Numer. Anal.*, 10(2):370–388, 1973.

[Mur78] D. M. Murray. *Differential Dynamic Programming for the Efficient Solution of Optimal Control Problems.* Ph.D. thesis, University of Arizona, 1978.

[MW92] P. Mehra and B. W. Wah. *Artificial neural networks: Concepts and theory.* IEEE Computer Society Press, Los Alamitos, CA, 1992.

[MY81] D. M. Murray and S. J. Yakowitz. The application of optimal control methodology to nonlinear programming problems. *Math. Programming*, 21(3):331–347, 1981.

[MZ89] M. Morari and E. Zafiriou. *Robust Process Control.* Prentice-Hall, Englewood Cliffs, NJ, 1989.

[Nai03] D. S. Naidu. *Optimal Control Systems.* CRC Press, Boca Raton, FL, 2003.

[Neu05] J. W. Neuberger. *Sobolev gradients and differential equations.* Available from author's Website, Dept. of Mathematics, Univ. of North Texas, http://www.math.unt.edu/˜jwn/run.pdf (2005).

[Neu97] J. W. Neuberger. *Sobolev Gradients and Differential Equations*, volume 1670 of Lecture Notes in Mathematics. Springer-Verlag, New York, 1997.

[Neu99] J. W. Neuberger. Continuous Newton's method for polynomials. *Math. Intelligencer*, 21(3):18–23, 1999.

[NS96] S. G. Nash and A. Sofer. *Linear and Nonlinear Programming.* McGraw-Hill, New York, 1996.

[NW99] J. Nocedal and S. J. Wright. *Numerical Optimization.* Springer Series in Operations Research. Springer-Verlag, New York, 1999.

[Obe02] R. J. Ober. The Fisher information matrix for linear systems. *Systems Control Lett.*, 47:221–226, 2002.

[OP05] R. C. L. F. Oliveira and P. L. D. Peres. A simple and less conservative test for $\mathbb{D}$-stability. *SIAM J. Matrix Anal. Appl.*, 26(2):415–425, 2005.

[OR70] J. M. Ortega and W. C. Rheinboldt. *Iterative solutions of nonlinear equations in several variables.* Academic Press, New York, 1970.

[Ore74] S. S. Oren. Self-scaling variable metric (SSVM) algorithms. *Management Sci.*, 20:264–280, 1974.

[Ort73] J. M. Ortega. Stability of difference equations and convergence of iterative processes. *SIAM J. Numer. Anal.*, 10(2):268–282, 1973.

[Oza86] T. M. Ozan. *Applied Mathematical Programming for Production and Engineering Management.* Prentice-Hall, Englewood Cliffs, NJ, 1986.

[Par92] B. N. Parlett. Reduction to tridiagonal form and minimal realizations. *SIAM J. Matrix Anal. Appl.*, 13(2):567–593, 1992.

[PB06] F. A. Pazos and A. Bhaya. Design of dynamic controller based iterative methods for nonlinear equations. Submitted, 2006. Technical report available at http://www.nacad.ufrj.br/˜amit.

[PBGM62] L. S. Pontryagin, V. Boltyanskii, R. Gamkrelidze, and E. Mishchenko. *The Mathematical Theory of Optimal Processes.* Interscience, New York, 1962.

[Per69] S. K. Persidskii. Problem of absolute stability. *Automat. Remote Control*, 12:1889–1895, 1969.

[Pol63] B. T. Polyak. Gradient methods for minimisation of functionals. *USSR Comput. Math. and Math. Phys.*, 3(4):864–878, 1963. *Zh. Vychisl. Mat. Mat. Fiz.*, pp. 643–653 (Russian edition).

[Pol64] B. T. Polyak. Some methods of speeding up the convergence of iterative methods. *USSR Comput. Math. and Math. Phys.*, 4(5):1–17, 1964. *Zh. Vychisl. Mat. Mat. Fiz.*, pp. 791–803 (Russian edition).

[Pol71] E. Polak. *Computational Methods in Optimization: A Unified Approach.* Academic, New York, 1971.

[Pol76] B. T. Polyak. Convergence and convergence rate of iterative stochastic algorithms. 1. general case. *Automat. Remote Control*, 37(12):1858–1868, 1976.

[Pol87] B. T. Polyak. *Introduction to Optimization.* Optimization Software, New York, 1987.

[Pol02] B. T. Polyak. History of mathematical programming in the USSR: Analyzing the phenomenon. *Math. Programming*, ser. B, 91:401–416, 2002.

[PS94] V. V. Phansalkar and P. S. Sastry. Analysis of the back-propagation algorithm with momentum. *IEEE Trans. Neural Networks*, 5(3):505–506, 1994.

[PS95] R. D. Pantazis and D. B. Szyld. Regions of convergence of the Rayleigh quotient iteration method. *Numer. Linear Algebra Appl.*, 2(3):251–269, 1995.

[PWZ00] L. Pronzato, H. P. Wynn, and A. A. Zhigljavsky. *Dynamical Search: Applications of Dynamical Systems in Search and Optimization.* Chapman and Hall/CRC, Boca Raton, FL, 2000.

[Pyn56] I. B. Pyne. Linear programming on an electronic analogue computer. *Trans. AIEE*, 75:139–143, May 1956.

[Qia99] N. Qian. On the momentum term in gradient descent learning algorithms. *Neural Networks*, 12:145–151, 1999.

[Qua03] W. Quapp. A valley following method. *Optimization*, 52(3):317–331, 2003.

[Qui76] D. Quillen. Projective modules over polynomial rings. *Invent. Math.*, 36:167–171, 1976.

[Qui80] J. P. Quinn. Stabilization of bilinear systems by quadratic feedback controls. *J. Math. Anal. Appl.*, 75(1):66–80, 1980.

[RB83] E. P. Ryan and N. J. Buckingham. On asymptotically stabilizing feedback control of bilinear systems. *IEEE Trans. Automat. Control*, AC-28(8):863–864, 1983.

[Ros60] H. H. Rosenbrock. An automatic method for finding the greatest or least value of a function. *Comput. J.*, 3:175–184, 1960.

[Ros00] H. H. Rosenbrock. Doing quantum mechanics with control theory. *IEEE Trans. Automat. Control*, 45(1):73–77, 2000.

[Ryb65a] M. V. Rybashhov. Gradient method of solving linear and quadratic programming problems on electronic analog computers. *Automat. Remote Control*, 26(12):2079–2089, 1965.

[Ryb65b] M. V. Rybashov. The gradient method of solving convex programming problems on electronic analog computers. *Automat. Remote Control*, 26(11):1886–1898, 1965.

[Ryb69a] M. V. Rybashov. Circuits for analog computer solution of systems of linear inequalities, inequalities, linear programming problems, and matrix games (survey). *Automat. Remote Control*, 30(3):141–164, 1969.

[Ryb69b] M. V. Rybashov. Method of differential equations in the problem of finding the extremum of a function using analog computers. *Automat. Remote Control*, 30(5):181–194, 1969.

[Ryb74] M. V. Rybashov. Stability of gradient systems. *Automat. Remote Control*, 35(9):1386–1393, 1974.

[RZ99] R. Riaza and P. J. Zufiria. Weak singularities and the continuous Newton method. *J. Math. Anal. Appl.*, 236(2-3):438–462, 1999.

[Saa96] Y. Saad. *Iterative Methods for Sparse Linear Systems*. The PWS series in Computer Science. PWS Publishing Co., Boston, 1996.

[SEM00] S. Sun, M. B. Egerstedt, and C. F. Martin. Control theoretic smoothing splines. *IEEE Trans. Automat. Control*, 45(12):2271–2279, 2000.

[SGB+02] J. A. Suykens, T. V. Gestel, J. D. Brabanter, B. D. Moor, and J. Vandewalle. *Least Squares Support Vector Machines*. World Scientific, Singapore, 2002.

[SH98] A. M. Stuart and A. R. Humphries. *Dynamical Systems and Numerical Analysis*, volume 2 of Cambridge Monographs on Applied and Computational Mathematics. Cambridge University Press, Cambridge, U.K., 1998.

[Sha82] A. Shapiro. Optimally scaled matrices, necessary and sufficient conditions. *Numer. Math.*, 39:239–245, 1982.

[She94] J. R. Shewchuk. *An introduction to the conjugate gradient method without the agonizing pain.* Technical Report CMU-CS-94-125, Carnegie Mellon University, Pittsburgh, PA, 1994. Available online at http://www.cs.cmu.edu/~quake/papers.html.

[Shu86] M. Shub. Some remarks on dynamical systems and numerical analysis. In *Dynamical systems and Partial Differential Equations*, pp. 69–92, Proc. Seventh ELAM, Equinoccio, Universidad Simon Bolivar, Caracas, Venezuela, 1986.

[Šil69] D. D. Šiljak. *Nonlinear Systems: The Parameter Analysis and Design.* John Wiley, New York, 1969.

[Šil70] D. D. Šiljak. Nonnegative polynomials: A criterion. *IEEE Proceedings*, 58(9):1370–1371, 1970.

[Šil78] D. D. Šiljak. *Large Scale Dynamic Systems: Stability and Structure.* Elsevier North-Holland, Amsterdam, The Netherlands, 1978.

[Sil80] H. M. Silveira. Stability in the determination of algorithms to find roots of nonlinear systems of equations. In *Proc. of the 3rd Brazilian Conference on Automatic Control*, pp. 87–91, Rio de Janeiro, 1980 (in Portuguese). (Original title: *Estabilidade em determinação de algoritmos para pesquisa de raízes de sistemas de equações não-lineares.*)

[Šil91] D. D. Šiljak. *Decentralized Control of Complex Systems.* Academic Press, San Diego, 1991.

[SK01] C. Schaerer and E. Kaszkurewicz. The shooting method for the solution of ordinary differential equations: A control-theoretical perspective. *Internat. J. Systems Sci.*, 32(8):1047–1053, 2001.

[SKM04] C. Schaerer, E. Kaszkurewicz, and N. Mangiavacchi. A multilevel Schwarz shooting method for the solution of the Poisson equation in two dimensional incompressible flow simulations. *Appl. Math. Comput.*, 153(3):803–831, 2004.

[SL91] J.-J. E. Slotine and W. Li. *Applied Nonlinear Control.* Prentice-Hall, Englewood Cliffs, NJ, 1991.

[SM03] E. Süli and D. F. Mayers. *An Introduction to Numerical Analysis.* Cambridge University Press, Cambridge, U.K., 2003.

[Sma76] S. Smale. A convergent process of price adjustment and global Newton methods. *J. Math. Econom.*, 3:107–120, 1976.

[SO91] T. Sugie and T. Ono. An iterative learning control law for dynamical systems. *Automatica*, 27(4):729–732, 1991.

[Söd98] G. Söderlind. The automatic control of numerical integration. *Centrum voor Wiskunde en Informatica (CWI) Quarterly*, 11(1):55–74, 1998. Available online at http://www.cwi.nl/cwi/publications_bibl/QUARTERLY/.

[Söd02] G. Söderlind. Automatic control and adaptive time-stepping. *Numer. Algorithms*, 31:281–310, 2002.

[Söd03] G. Söderlind. Digital filters in adaptive time-stepping. *ACM Trans. Math. Software*, 29(1):1–26, 2003.

[Son89] E. D. Sontag. A universal construction of Artstein's theorem on nonlinear stabilization. *Systems Control Lett.*, 13:117–123, 1989.

[Son98] E. D. Sontag. *Mathematical Control Theory: Deterministic Finite Dimensional Systems*, volume 6 of Texts in Applied Mathematics. Springer-Verlag, New York, 2nd edition, 1998.

[Son03] E. D. Sontag. Adaptation and regulation with signal detection implies internal model. *Systems Control Lett.*, 50:119–126, 2003.

[SP94] D. Shevitz and B. Paden. Lyapunov stability of nonsmooth systems. *IEEE Trans. Automat. Control*, 39(9): 1910–1914, September 1994.

[SS02] B. Schölkopf and A. Smola. *Learning with Kernels*. MIT Press, Cambridge, MA, 2002.

[SS04] A. J. Smola and B. Schölkopf. A tutorial on support vector regression. *Statist. Comput.*, 14(3):199–222, August 2004.

[SSWB00] B. Schölkopf, A. J. Smola, R. C. Williamson, and P. L. Bartlett. New support vector algorithms. *Neural Computation*, 12(5):1207–1245, May 2000.

[Sto95] J. A. Stolan. An improved Šiljak's algorithm for solving polynomial equations converges quadratically to multiple zeros. *J. Comput. Appl. Math.*, 64(3):247–268, 1995.

[Sus76] A. A. Suslin. Projective modules over a polynomial ring are free. *Soviet Math. Dokl.*, 17:1160–1164, 1976.

[SvdV00] Y. Saad and H. A. van der Vorst. Iterative solution of linear systems in the 20th century. *J. Comput. Appl. Math.*, 123(1-2):1–33, 2000.

[SW94] R. Scherer and W. Wendler. Complete algebraic characterization of A-stable Runge–Kutta methods. *SIAM J. Numer. Anal.*, 31(2):540–551, 1994.

[SW95] W. Schönauer and R. Weiss. An engineering approach to generalized conjugate gradient methods and beyond. *Appl. Numer. Math.*, 19(3):175–206, 1995.

[Tan80] K. Tanabe. A geometric method in nonlinear programming. *J. Optim. Theory Appl.*, 30(2):181–210, 1980.

[Ter99] W. J. Terrell. Some fundamental control theory I: Controllability, observability, and duality. *Amer. Math. Monthly*, 106(8):705–719, 1999.

[TG86] M. Takeda and J. W. Goodman. Neural networks for computation: Number representations and programming complexity. *Appl. Optics*, 25(18):3033–3046, 1986.

[TH02] M. Torii and M. T. Hagan. Stability of steepest descent with momentum for quadratic functions. *IEEE Trans. Neural Networks*, 13(3):752–756, 2002.

[Tre92] L. N. Trefethen. The definition of numerical analysis. *SIAM News*, November 1992.

[Tru55] J. G. Truxal. *Automatic Feedback Control System Synthesis*. McGraw-Hill, New York, 1955.

[Tsy71] Ya. Z. Tsypkin. *Adaptation and Learning in Automatic Systems*, volume 73 of Mathematics in Science and Engineering. Academic Press, New York, 1971. First published in Russian under the title *Adaptatsia i obuchenie v avtomaticheskikh sistemakh*, Nauka, Moscow, 1968.

[Tsy73] Ya. Z. Tsypkin. *Foundations of the Theory of Learning Systems*, volume 80 of Mathematics in Science and Engineering. Academic Press, New York, 1973. First published in Russian under the title *Osnovy teorii obuchayuschchikhsya sistem*, Nauka, Moscow, 1970.

[TŻ98] M. C. M. Teixeira and S. H. Żak. Analog neural nonderivative optimizers. *IEEE Trans. Neural Networks*, 9(4):629–638, 1998.

[UN95] K. Urahama and T. Nagao. K-winners-take-all circuit with o(n) complexity. *IEEE Trans. Neural Networks*, 6(3):776–778, 1995.

[Utk78] V. I. Utkin. *Sliding Modes and Their Applications in Variable Structure Systems*. Mir, Moscow, 1978.

[Utk92] V. I. Utkin. *Sliding Modes in Control and Optimization*. Springer-Verlag, Berlin, 1992.

[UTK96] M. Utumi, R. Takaki, and T. Kawai. Optimal time step control for the numerical solution of ordinary differential equations. *SIAM J. Numer. Anal.*, 33(4):1644–1653, 1996.

[Val60] M. E. Van Valkenburg. *Introduction to Modern Network Synthesis*. John Wiley, New York, 1960.

[Vap95] V. Vapnik. *The Nature of Statistical Learning Theory*. Springer-Verlag, New York, 1995.

[Vap98] V. Vapnik. *Statistical Learning Theory*. John Wiley, New York, 1998.

[Var00] R. S. Varga. *Matrix Iterative Analysis*, volume 27 of Springer Series in Computational Mathematics. Springer-Verlag, Berlin, 2nd edition, 2000.

[VCC05] A. M. P. Valli, G. F. Carey, and A. L. G. A. Coutinho. Control strategies for timestep selection in finite element simulation of incompressible flows and coupled reaction-convection-diffusion processes. *Internat. J. Numer. Methods Fluids*, 47:201–231, 2005.

[vdS69] A. van der Sluis. Condition numbers and equilibration of matrices. *Numer. Math.*, 14:14–23, 1969.

[vdV00] H. A. van der Vorst. Krylov subspace iteration. *Computing in Science and Engineering*, 2(1):32–37, 2000.

[vdW53] B. L. van der Waerden. *Modern Algebra*. Ungar, New York, 1953.

[VG97] T. L. Vincent and W. J. Grantham. *Nonlinear and Optimal Control Systems*. John Wiley, New York, 1997.

[Vic81] R. Vichnevetsky. *Computer Methods for Partial Diferential Equations*. Prentice-Hall, Englewood Cliffs, NJ, 1981.

[Vid93] M. Vidyasagar. *Nonlinear Systems Analysis*. Prentice-Hall, Englewood Cliffs, NJ, 2nd edition, 1993.

[Vid95] M. Vidyasagar. Minimum-seeking properties of analog neural networks with multilinear objective functions. *IEEE Trans. Automat. Control*, 40:1359–1375, 1995.

[VR77] V. I. Venets and M. V. Rybashov. The method of Lyapunov functions in the study of continuous algorithms of mathematical programming. *USSR Comput. Math. and Math. Phys.*, 17(3):64–73, 1977. (Russian edition: *Zh. Vychisl. Mat. Mat. Fiz.*, pp. 622–633.)

[War84] R. K. Ward. An on-line adaptation for discrete l_1 linear estimation. *IEEE Trans. Automat. Control*, AC-29(1):67–71, 1984.

[WCXC00] Z. Wang, J. Y. Cheung, Y. S. Xia, and J. D. Chen. Neural implementation of unconstrained minimum l_1-norm optimization—least absolute deviation model and its application to time delay estimation. *IEEE Trans. Circuits and Systems II: Analog Digital Signal Processing*, 47(11):1214–1226, 2000.

[WMA$^+$91] W. J. Wolfe, D. Mathis, C. Anderson, J. Rothman, M. Gottler, G. Brady, R. Walker, G. Duane, and G. Alaghband. K-winner networks. *IEEE Trans. Neural Networks*, 2:310–315, 1991.

[Yak86] S. J. Yakowitz. The stagewise Kuhn–Tucker condition and differential dynamic programming. *IEEE Trans. Automat. Control*, AC-31(1):25–30, 1986.

[YC97] X.-H. Yu and G.-A. Chen. Efficient backpropagation learning using optimal learning rate and momentum. *Neural Networks*, 10(3):517–527, 1997.

[YCC95] X.-H. Yu, G.-A. Chen, and S.-X. Cheng. Dynamic learning rate optimization of the backpropagation algorithm. *IEEE Trans. Neural Networks*, 6(3):669–677, 1995.

[YF90] C.-C. Yu and M. K. H. Fan. Decentralized integral controllability and D-stability. *Chemical Engineering Science*, 45(11):3299–3309, 1990.

[YGC98] J. C. Yen, J. I. Guo, and H.-C. Chen. A new k-winners-take all neural network and its array architecture. *IEEE Trans. Neural Networks*, 9:901–912, 1998.

[YHSD00] T.-M. Yi, Y. Huang, M. I. Simon, and J. Doyle. Robust perfect adaptation in bacterial chemotaxis through integral feedback control. *Proc. Nat. Acad. Sci. (USA)*, 97:4649–4653, 2000.

[You80] L. C. Young. *Lectures on the Calculus of Variations and Optimal Control Theory*. Chelsea, New York, 1980.

[You89] D. M. Young. A historical overview of iterative methods. *Comput. Phys. Comm.*, 53(1-3):1–17, 1989.

[You71] D. M. Young. *Iterative Solutions of Large Linear Systems*. Academic Press, London, 1971.

[YP84] D. C. Youla and P. F. Pickel. The Quillen–Suslin theorem and the structure of normal-dimensional elementary polynomial matrices. *IEEE Trans. Circuits and Systems*, 31(6):513–518, 1984.

[Żak03] S. H. Żak. *Systems and Control*. Oxford University Press, New York, 2003.

[Zan67] W. Zangwill. Nonlinear programming via penalty functions. *Management Sci.*, 13:344–358, 1967.

[ZG02] P. J. Zufiria and R. S. Guttalu. On the role of singularities in Branin's method from dynamic and continuation perspectives. *Appl. Math. Comput.*, 130(2-3):593–618, 2002.

[ZTM97] Z.-M. Zhang, J. Tomlinson, and C. F. Martin. Splines and linear control theory. *Acta Appl. Math.*, 49(1):1–34, 1997.

Index